Lecture Notes in Computer Science 14727

Founding Editors

Gerhard Goos
Juris Hartmanis

The series Lecture Notes in Computer Science (LNCS), including its subseries Lecture Notes in Artificial Intelligence (LNAI) and Lecture Notes in Bioinformatics (LNBI), has established itself as a medium for the publication of new developments in computer science and information technology research, teaching, and education.

LNCS enjoys close cooperation with the computer science R & D community, the series counts many renowned academics among its volume editors and paper authors, and collaborates with prestigious societies. Its mission is to serve this international community by providing an invaluable service, mainly focused on the publication of conference and workshop proceedings and postproceedings. LNCS commenced publication in 1973.

Robert A. Sottilare · Jessica Schwarz
Editors

Adaptive Instructional Systems

6th International Conference, AIS 2024
Held as Part of the 26th HCI International Conference, HCII 2024
Washington, DC, USA, June 29 – July 4, 2024
Proceedings

Editors
Robert A. Sottilare
Soar Technology, Inc.
Orlando, FL, USA

Jessica Schwarz
Fraunhofer FKIE
Wachtberg, Germany

ISSN 0302-9743   ISSN 1611-3349 (electronic)
Lecture Notes in Computer Science
ISBN 978-3-031-60608-3   ISBN 978-3-031-60609-0 (eBook)
https://doi.org/10.1007/978-3-031-60609-0

Foreword

This year we celebrate 40 years since the establishment of the HCI International (HCII) Conference, which has been a hub for presenting groundbreaking research and novel ideas and collaboration for people from all over the world.

The HCII conference was founded in 1984 by Prof. Gavriel Salvendy (Purdue University, USA, Tsinghua University, P.R. China, and University of Central Florida, USA) and the first event of the series, "1st USA-Japan Conference on Human-Computer Interaction", was held in Honolulu, Hawaii, USA, 18–20 August. Since then, HCI International is held jointly with several Thematic Areas and Affiliated Conferences, with each one under the auspices of a distinguished international Program Board and under one management and one registration. Twenty-six HCI International Conferences have been organized so far (every two years until 2013, and annually thereafter).

Over the years, this conference has served as a platform for scholars, researchers, industry experts and students to exchange ideas, connect, and address challenges in the ever-evolving HCI field. Throughout these 40 years, the conference has evolved itself, adapting to new technologies and emerging trends, while staying committed to its core mission of advancing knowledge and driving change.

As we celebrate this milestone anniversary, we reflect on the contributions of its founding members and appreciate the commitment of its current and past Affiliated Conference Program Board Chairs and members. We are also thankful to all past conference attendees who have shaped this community into what it is today.

The 26th International Conference on Human-Computer Interaction, HCI International 2024 (HCII 2024), was held as a 'hybrid' event at the Washington Hilton Hotel, Washington, DC, USA, during 29 June – 4 July 2024. It incorporated the 21 thematic areas and affiliated conferences listed below.

A total of 5108 individuals from academia, research institutes, industry, and government agencies from 85 countries submitted contributions, and 1271 papers and 309 posters were included in the volumes of the proceedings that were published just before the start of the conference, these are listed below. The contributions thoroughly cover the entire field of human-computer interaction, addressing major advances in knowledge and effective use of computers in a variety of application areas. These papers provide academics, researchers, engineers, scientists, practitioners and students with state-of-the-art information on the most recent advances in HCI.

The HCI International (HCII) conference also offers the option of presenting 'Late Breaking Work', and this applies both for papers and posters, with corresponding volumes of proceedings that will be published after the conference. Full papers will be included in the 'HCII 2024 - Late Breaking Papers' volumes of the proceedings to be published in the Springer LNCS series, while 'Poster Extended Abstracts' will be included as short research papers in the 'HCII 2024 - Late Breaking Posters' volumes to be published in the Springer CCIS series.

I would like to thank the Program Board Chairs and the members of the Program Boards of all thematic areas and affiliated conferences for their contribution towards the high scientific quality and overall success of the HCI International 2024 conference. Their manifold support in terms of paper reviewing (single-blind review process, with a minimum of two reviews per submission), session organization and their willingness to act as goodwill ambassadors for the conference is most highly appreciated.

This conference would not have been possible without the continuous and unwavering support and advice of Gavriel Salvendy, founder, General Chair Emeritus, and Scientific Advisor. For his outstanding efforts, I would like to express my sincere appreciation to Abbas Moallem, Communications Chair and Editor of HCI International News.

July 2024   Constantine Stephanidis

HCI International 2024 Thematic Areas
and Affiliated Conferences

- HCI: Human-Computer Interaction Thematic Area
- HIMI: Human Interface and the Management of Information Thematic Area
- EPCE: 21st International Conference on Engineering Psychology and Cognitive Ergonomics
- AC: 18th International Conference on Augmented Cognition
- UAHCI: 18th International Conference on Universal Access in Human-Computer Interaction
- CCD: 16th International Conference on Cross-Cultural Design
- SCSM: 16th International Conference on Social Computing and Social Media
- VAMR: 16th International Conference on Virtual, Augmented and Mixed Reality
- DHM: 15th International Conference on Digital Human Modeling & Applications in Health, Safety, Ergonomics & Risk Management
- DUXU: 13th International Conference on Design, User Experience and Usability
- C&C: 12th International Conference on Culture and Computing
- DAPI: 12th International Conference on Distributed, Ambient and Pervasive Interactions
- HCIBGO: 11th International Conference on HCI in Business, Government and Organizations
- LCT: 11th International Conference on Learning and Collaboration Technologies
- ITAP: 10th International Conference on Human Aspects of IT for the Aged Population
- AIS: 6th International Conference on Adaptive Instructional Systems
- HCI-CPT: 6th International Conference on HCI for Cybersecurity, Privacy and Trust
- HCI-Games: 6th International Conference on HCI in Games
- MobiTAS: 6th International Conference on HCI in Mobility, Transport and Automotive Systems
- AI-HCI: 5th International Conference on Artificial Intelligence in HCI
- MOBILE: 5th International Conference on Human-Centered Design, Operation and Evaluation of Mobile Communications

List of Conference Proceedings Volumes Appearing Before the Conference

1. LNCS 14684, Human-Computer Interaction: Part I, edited by Masaaki Kurosu and Ayako Hashizume
2. LNCS 14685, Human-Computer Interaction: Part II, edited by Masaaki Kurosu and Ayako Hashizume
3. LNCS 14686, Human-Computer Interaction: Part III, edited by Masaaki Kurosu and Ayako Hashizume
4. LNCS 14687, Human-Computer Interaction: Part IV, edited by Masaaki Kurosu and Ayako Hashizume
5. LNCS 14688, Human-Computer Interaction: Part V, edited by Masaaki Kurosu and Ayako Hashizume
6. LNCS 14689, Human Interface and the Management of Information: Part I, edited by Hirohiko Mori and Yumi Asahi
7. LNCS 14690, Human Interface and the Management of Information: Part II, edited by Hirohiko Mori and Yumi Asahi
8. LNCS 14691, Human Interface and the Management of Information: Part III, edited by Hirohiko Mori and Yumi Asahi
9. LNAI 14692, Engineering Psychology and Cognitive Ergonomics: Part I, edited by Don Harris and Wen-Chin Li
10. LNAI 14693, Engineering Psychology and Cognitive Ergonomics: Part II, edited by Don Harris and Wen-Chin Li
11. LNAI 14694, Augmented Cognition, Part I, edited by Dylan D. Schmorrow and Cali M. Fidopiastis
12. LNAI 14695, Augmented Cognition, Part II, edited by Dylan D. Schmorrow and Cali M. Fidopiastis
13. LNCS 14696, Universal Access in Human-Computer Interaction: Part I, edited by Margherita Antona and Constantine Stephanidis
14. LNCS 14697, Universal Access in Human-Computer Interaction: Part II, edited by Margherita Antona and Constantine Stephanidis
15. LNCS 14698, Universal Access in Human-Computer Interaction: Part III, edited by Margherita Antona and Constantine Stephanidis
16. LNCS 14699, Cross-Cultural Design: Part I, edited by Pei-Luen Patrick Rau
17. LNCS 14700, Cross-Cultural Design: Part II, edited by Pei-Luen Patrick Rau
18. LNCS 14701, Cross-Cultural Design: Part III, edited by Pei-Luen Patrick Rau
19. LNCS 14702, Cross-Cultural Design: Part IV, edited by Pei-Luen Patrick Rau
20. LNCS 14703, Social Computing and Social Media: Part I, edited by Adela Coman and Simona Vasilache
21. LNCS 14704, Social Computing and Social Media: Part II, edited by Adela Coman and Simona Vasilache
22. LNCS 14705, Social Computing and Social Media: Part III, edited by Adela Coman and Simona Vasilache

47. LNCS 14730, HCI in Games: Part I, edited by Xiaowen Fang
48. LNCS 14731, HCI in Games: Part II, edited by Xiaowen Fang
49. LNCS 14732, HCI in Mobility, Transport and Automotive Systems: Part I, edited by Heidi Krömker
50. LNCS 14733, HCI in Mobility, Transport and Automotive Systems: Part II, edited by Heidi Krömker
51. LNAI 14734, Artificial Intelligence in HCI: Part I, edited by Helmut Degen and Stavroula Ntoa
52. LNAI 14735, Artificial Intelligence in HCI: Part II, edited by Helmut Degen and Stavroula Ntoa
53. LNAI 14736, Artificial Intelligence in HCI: Part III, edited by Helmut Degen and Stavroula Ntoa
54. LNCS 14737, Design, Operation and Evaluation of Mobile Communications: Part I, edited by June Wei and George Margetis
55. LNCS 14738, Design, Operation and Evaluation of Mobile Communications: Part II, edited by June Wei and George Margetis
56. CCIS 2114, HCI International 2024 Posters - Part I, edited by Constantine Stephanidis, Margherita Antona, Stavroula Ntoa and Gavriel Salvendy
57. CCIS 2115, HCI International 2024 Posters - Part II, edited by Constantine Stephanidis, Margherita Antona, Stavroula Ntoa and Gavriel Salvendy
58. CCIS 2116, HCI International 2024 Posters - Part III, edited by Constantine Stephanidis, Margherita Antona, Stavroula Ntoa and Gavriel Salvendy
59. CCIS 2117, HCI International 2024 Posters - Part IV, edited by Constantine Stephanidis, Margherita Antona, Stavroula Ntoa and Gavriel Salvendy
60. CCIS 2118, HCI International 2024 Posters - Part V, edited by Constantine Stephanidis, Margherita Antona, Stavroula Ntoa and Gavriel Salvendy
61. CCIS 2119, HCI International 2024 Posters - Part VI, edited by Constantine Stephanidis, Margherita Antona, Stavroula Ntoa and Gavriel Salvendy
62. CCIS 2120, HCI International 2024 Posters - Part VII, edited by Constantine Stephanidis, Margherita Antona, Stavroula Ntoa and Gavriel Salvendy

https://2024.hci.international/proceedings

Preface

The goal of the Adaptive Instructional Systems (AIS) Conference, affiliated to the HCI International Conference, is to understand the theory and enhance the state of practice for a set of technologies (tools and methods) called adaptive instructional systems (AIS). AIS are defined as artificially intelligent, computer-based systems that guide learning experiences by tailoring instruction and recommendations based on the goals, needs, preferences, and interests of each individual learner or team in the context of domain learning objectives. The interaction between individual learners or teams of learners and AIS technologies is a central theme of this conference. AIS observe user behaviors to assess progress toward learning objectives and then act on learners and their learning environments (e.g., problem sets or scenario-based simulations) with the goal of optimizing learning, performance, retention, and transfer of learning to work environments.

The 6th International Conference on Adaptive Instructional Systems (AIS 2024) encouraged papers from academics, researchers, industry, and professionals, on a broad range of theoretical and applied issues related to AIS and their applications. The focus of this conference on instructional tailoring of learning experiences highlights the importance of accurately modeling learners to accelerate their learning, boost the effectiveness of AIS-based experiences, and precisely reflect their long-term competence in a variety of domains of instruction.

The content for AIS 2024 centered on design processes and aspects, individual learner differences, and applications of AISs. More specifically, several works focused on Human-Centered Design, examining facets such as personality traits, improved self-awareness, human performance, learner engagement, trust and acceptance of AISs, as well as the establishment of communities of practice. In addition, a number of papers focused on design strategies and guidelines, discussing topics such as cognitive simulations, gamification, and design based on data intelligence and learner analytics. In the area of individual differences in adaptive learning, contributions elaborated on competency-based training, learner control, knowledge states of learners, stress and coping with task difficulty, as well as identifying individual differences as a predictor of usage of AISs. Finally, a selected number of papers focused on applications of AISs demonstrating issues of high practical value across different domains, such as math courses, architecture and design, simulation-based training, and content improvement services.

One volume of the HCII 2024 proceedings is dedicated to this year's edition of the AIS Conference and focuses on topics related to Designing and Developing Adaptive Instructional Systems, Adaptive Learning Experiences, and AI in Adaptive Learning.

The papers in this volume were accepted for publication after a minimum of two single-blind reviews from the members of the AIS Program Board or, in some cases, from members of the Program Boards of other affiliated conferences. We would like to thank all of them for their invaluable contribution, support and efforts.

July 2024   Robert A. Sottilare
                                                                    Jessica Schwarz

6th International Conference on Adaptive Instructional Systems (AIS 2024)

Program Board Chairs: **Robert A. Sottilare**, *Soar Technology, Inc., USA*, and **Jessica Schwarz**, *Fraunhofer FKIE, Germany*

- Roger Azevedo, *University of Central Florida, USA*
- Benjamin Bell, *Eduworks Corporation, USA*
- Elizabeth Biddle, *The Boeing Company, USA*
- Bruno Emond, *National Research Council Canada, Canada*
- Jim Goodell, *QiP / IEEE LTSC, USA*
- Alexander Streicher, *Fraunhofer IOSB, Germany*
- Wendi L. Van Buskirk, *US Naval Air Warfare – Training Systems Division, USA*
- Joost Van Oijen, *Royal Netherlands Aerospace Centre, Netherlands*
- Elizabeth Whitaker, *Georgia Tech Research Institute, USA*
- Thomas E.F. Witte, *Fraunhofer FKIE, Germany*
- Ryan Wohleber, *Soar Technology, USA*

The full list with the Program Board Chairs and the members of the Program Boards of all thematic areas and affiliated conferences of HCII 2024 is available online at:

http://www.hci.international/board-members-2024.php

HCI International 2025 Conference

The 27th International Conference on Human-Computer Interaction, HCI International 2025, will be held jointly with the affiliated conferences at the Swedish Exhibition & Congress Centre and Gothia Towers Hotel, Gothenburg, Sweden, June 22–27, 2025. It will cover a broad spectrum of themes related to Human-Computer Interaction, including theoretical issues, methods, tools, processes, and case studies in HCI design, as well as novel interaction techniques, interfaces, and applications. The proceedings will be published by Springer. More information will become available on the conference website: https://2025.hci.international/.

General Chair
Prof. Constantine Stephanidis
University of Crete and ICS-FORTH
Heraklion, Crete, Greece
Email: general_chair@2025.hci.international

https://2025.hci.international/

Contents

AI in Adaptive Learning

Designing and Developing Adaptive Instructional Systems

Beyond Standalone Systems: Creating an Ecosystem of Adaptive Training Services

Brice Colby(✉) ⓘ, Eric Tucker, and Tim Siggins

Soar Technology, Inc., Ann Arbor, MI 48105, USA
brice.colby@soartech.com

Abstract. This paper explores the development and initial evaluation of SoarTech Adaptive Training Services (STATS), a modular ecosystem designed to address the limitations of traditional intelligent tutoring systems (ITSs). By adopting Gibbon's [1] layers theory, STATS introduces a structured, service-based approach to ITSs, enhancing adaptability and efficiency. The paper reviews the historical evolution of ITSs, highlighting the shift from rigid architectures to modular, flexible designs. Through a case study in a basic electricity and electronics course, STATS demonstrates its practical application and potential for personalized learning. Initial feedback from 18 participants indicates positive responses towards the system's effectiveness and efficiency, though usability challenges, particularly in interface navigation, were identified. The study underscores the importance of modular design in overcoming the historical challenges of ITS development, such as domain dependence and siloed research. Future directions for STATS include deeper exploration of its layers, exploration of ethical considerations in algorithmic decision-making, and the exploration of interoperability as STATS leverages services from other frameworks like GIFT [2]. This research contributes to the ongoing discourse on the necessity of modular, adaptable ITSs in meeting diverse educational needs and advancing the field of intelligent tutoring.

Keywords: Intelligent Tutoring · Adaptive Training · Service-oriented Architecture

1 Introduction

The quest for individualized instruction has been a driving force in intelligent tutoring research, aimed at creating systems that can match the efficacy of human tutors. However, the adoption of adaptive training technologies is constrained by inflexible architectures and a lack of shared knowledge in the field. Vassileva [3] and Nkambou, Bourdeau, and Psyche [4] have highlighted the challenges of domain dependence and siloed research, which limit the broader application of intelligent tutoring. Nye [5] argues for a shift towards a modular, services-based ecosystem to overcome these limitations, enabling a more tailored approach to intelligent tutoring.

This paper introduces SoarTech Adaptive Training Services (STATS), a solution developed in response to these challenges and in alignment with Nye's call for a services

R. A. Sottilare and J. Schwarz (Eds.): HCII 2024, LNCS 14727, pp. 3–14, 2024.
https://doi.org/10.1007/978-3-031-60609-0_1

approach. STATS is grounded in Colby's [6] framework for the design and development of individual services, expanding on the traditional architectures of intelligent tutoring systems (ITSs).

The paper begins by outlining the historical context that has led to the current push for a services-based approach, reviewing relevant research and initiatives such as Cognitive Tutor's CT+A [7] and GIFT [2]. The focus of this paper is to provide a conceptual (rather than technical) overview of STATS' architecture with a high level explanation of its services. A case study is presented where STATS is applied to a basic electricity and electronics course, illustrating its practical application. The paper concludes with a forward-looking discussion on the future trajectory of STATS, including evaluation strategies, the integration of Large Language Models (LLMs) for content creation, and addressing interoperability challenges with other services, among other issues.

1.1 Historical Context

To better understand the need for intelligent tutoring services, it would be beneficial to understand the evolution of intelligent tutoring throughout its generations. Nkambou et al. [4] introduced 3 distinct generations: 1970–1990, 1990–2010, 2010–Present.

First Generation (1970–1990). The first generation of ITSs was characterized by the transition from basic computer-assisted instruction (CAI) to more sophisticated systems incorporating elements of artificial intelligence (AI). Early CAI systems, largely linear in their approach, evolved to include branching programs that considered student responses, leading to more personalized learning paths [7]. The introduction of generative CAI, capable of creating and solving its own problems, marked a significant advancement [8, 9]. However, these systems still lacked the adaptive intelligence necessary to emulate the effectiveness of one-on-one human tutoring, as evidenced by Bloom's 2-sigma effect [10]. The integration of artificial intelligence (AI) into CAI, initiated by Carbonell's SCHOLAR, laid the foundation for the first ITS, focusing on replicating the personalized guidance of human tutors [11].

Second Generation (1990–2010). This generation was marked by a push for establishing a scientific foundation for the broader field of artificial intelligence in education (AIED), advocated by Self [12]. This time period saw the creation of specialized conferences and journals dedicated to ITSs, reflecting a growing academic interest and community engagement in the field. Indeed, during this period, Colby [13] found 65 unique systems being developed across 31 domains. Despite the proliferation of ITSs, a lack of standardized architecture led to diverse and incompatible systems. The commonality found among them, as Wenger [14] noted, was the four-component architecture: domain, tutor model, student model, and the interface. Yet, the diversity in architecture led to fragmented development and siloed research, underscoring the need for a more unified approach in ITS design.

Third Generation (2010–Present). As these ITSs matured, they began to pioneer novel approaches for adaptive instruction, inching closer to achieving the gold standard (i.e., Bloom's 2-sigma effect). A meta-analysis conducted by VanLehn [15] showed two noteworthy findings. First, the 2-sigma effect size was overstated. VanLehn found that human tutors were actually closer to 0.79 in terms of learning gains when compared to no tutoring. Second, VanLehn found that some ITSs are much closer to approximating human tutors than previously thought for specific domains. In his review, ITSs had an effect size of 0.76 when compared to no tutoring – a significant milestone for the AIED community.

Building on these advancements, Nye [16] proposed a paradigm shift for the field of ITSs, advocating for a transition towards a modular, services-based ecosystem. This shift was driven by the recognition that ITSs could benefit from a more flexible and interoperable framework, allowing for the integration of various educational technologies and methodologies. Nye suggested that ITSs should evolve from being comprehensive, standalone systems to becoming part of a broader ecosystem of reusable infrastructure and platforms [16]. This approach would enable ITSs to draw from a diverse range of services, much like how a basic blog site might utilize authentication services from multiple sources like Google or Facebook.

However, the transition to a services-based ecosystem is still in its early stages, and it remains to be seen how fully ITSs and AIED will embrace this new paradigm. Despite this, the direction proposed by Nye is seen as an existential necessity for the future of ITSs, emphasizing the need for systems that are not only effective, but also flexible and capable of integrating with an ever-evolving technological landscape.

Third Generation: Steps Towards Modularity. Let's consider two approaches to introducing modularity in ITSs. The first approach involves augmenting existing systems for enhanced modularity, while the second entails developing a new architecture with a primary focus on modular design. Two examples illustrate these approaches:

1. CTAT to CT+A (augmentation approach): Holstein et al. [6] noted that new and effective student modeling techniques rarely make the transition from research to practical application. To address this issue, they took an existing authoring tool, CTAT, and extended it to include a modularized student model. While this effort allows authors to add new variables to the student model, it underscores the complexities of adapting systems not initially designed for modularity. Despite progress, CT+A's limitations, including the need for significant re-architecting, underscore the challenges of this augmentation approach

2. GIFT (new architecture approach): GIFT [2] was developed with a focus on overcoming the architectural limitations of previous ITSs, which were not inherently designed for modularity. GIFT's innovative design, featuring independent yet interconnected modules (e.g., domain, pedagogical, and learner modules), addresses the tutoring process comprehensively. GIFT supports domain independence and adapts to diverse contexts, showcasing its alignment with the evolving educational technology landscape. Crucially, GIFT also addresses the issue of siloed research by facilitating the integration of best practices across all modules, exemplifying the shift towards more collaborative and interoperable intelligent tutoring services.

GIFT represents a significant evolution in ITSs, marking a transition from standalone systems to an integrated, adaptable framework. It sets a precedent for future ITS developments, emphasizing the need for systems that are not just effective, but also flexible and capable of integrating within a broader, ever-evolving ecosystem of educational technologies. STATS both builds upon and diverges from GIFT, primarily in its ecosystem architecture. While GIFT is grounded in extensive ITS research, STATS adopts an external perspective, rooted in instructional design principles.

2 STATS Overview

STATS leverages Gibbons' [1] layers theory, which adopts an architectural approach to instructional design. According to this theory, instructional products are composed of components, similar to a building's structure, which age at varying rates. It advocates for a modular approach, enabling updates or replacements with minimal impact on the overall system. This modular design allows ITS developers or end-users to enhance or modify parts without disrupting the overall service. STATS is structured around seven functional layers (see Fig. 1): content, data management, strategy, control, message, representation, and media-logic, each serving a distinct function essential to every instructional artifact. Each layer also has sublayers that are representative of specific services that would need to be created to provide intelligent tutoring.

Next, we briefly present each layer, their sublayers, and examples of research that can be used as the basis for creating new services.

2.1 The Content Layer

The content layer encompasses much of the domain model within a traditional ITS, divided into two sublayers: knowledge representation and learning objects. Knowledge representation involves capturing the domain knowledge in formats such as ontologies, production rules, and constraints [13]. This sublayer deals primarily with the overall structure of the domain knowledge. Learning objects represent the basic units of learning, including facts, concepts, procedures, and principles [17], structured in a way that they can be independent units of instruction. This dual structure provides a foundation for conceptualizing the basic learning unit of an intelligent tutor, coupled with the structure needed to organize and communicate content effectively. Research in this area has focused on developing tools and methodologies for automating the creation of these components, and reducing the time it takes to create content, which has traditionally been the main bottleneck of ITS development [5].

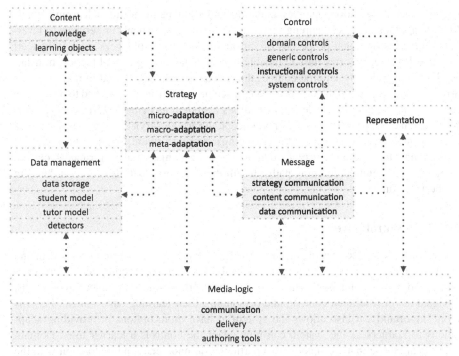

Fig. 1. The layers, sublayers, and their relation to each other. Adapted from [1].

2.2 The Data Management Layer

The data management layer is further subdivided into 4 sublayers: data storage, student and tutor models, and detectors. The purpose of a data storage service is straightforward: the data collected by the training environment needs to be stored. The student model matches ITSs' traditional student model in that it is a cognitive model representing what a student knows or can do. The tutor model (not to be confused with the tutor or pedagogical model of the traditional ITS architecture) parallels the student model. This model analyzes the effectiveness of instructional strategies and content delivery for each student, taking cues from their interactions and performance in the training environment. This allows STATS to determine the success of teaching methods and know when to pivot, ensuring each student's educational path is optimally tailored to their needs. Finally, STATS uses detectors to interpret data from the student and tutor models, identifying key learner states such as engagement, effort, and understanding or measuring tutor performance metrics (e.g., the effectiveness of remediation given this student profile).

2.3 The Strategy Layer

The strategy layer is the intelligence of an intelligent tutor, facilitating adaptation across three sublayers: micro-adaptation, macro-adaptation, and meta-adaptation. Micro-adaptation refers to problem-based adaptation to address immediate learning needs. Macro-adaptation involves planning and sequencing to align learning goals. Meta-adaptation allows swapping between different environments or services to leverage unique instructional properties of various tools. These sublayers work in concert to refine learning theories into actionable instructional strategies and tactics, translating learning goals into tailored courses that are executed and adapted in real-time. An arbiter service makes adaptation decisions based on input from data passed to it from the detectors. For example, if a detector identifies a student as being historically high performing but currently struggling with a concept, the arbiter service will follow a nuanced decision-making process. Rather than applying a straightforward, one-size-fits all rule, it will evaluate the student's current situation, considering historical performance, current struggles, and potential learning paths. It will then suggest a path forward that leads to the best outcomes.

2.4   The Control Layer

The control layer describes the mechanisms by which a learner communicates with and navigates the system. By categorizing controls into domain-specific, generic, instructional, and system controls, this layer ensures that learners can communicate their needs, navigate the instructional content, and personalize their learning environment. Perhaps more importantly, it also allows the system to have more options in providing adaptations to learners as the system can now adapt the controls a learner uses. Domain controls are those that are specific to a particular domain, enabling interaction with the content. Generic controls are ubiquitous tools that can be utilized across various systems and domains. Instructional controls reflect the current instructional goals, strategies, and tactics (e.g., facilitating social learning through group work tools). System controls are for managing system functions, allowing for customization and adaptability to meet learner preferences.

2.5   The Message Layer

Message services answer the question, "What needs to be communicated to the student and how will it be communicated?" Communication comes from three channels: the data, strategy, and content services. The data channel answers these questions by sharing the underlying models (e.g., a skillometer), enabling students to be responsible for their own progress or even influence teaching practices/system responses. The strategy channel concerns itself with how pedagogy is communicated to the learner. For example, how is flag feedback presented to the student (e.g., flag feedback vs. text. The content channel is concerned with how the content is displayed to the student (e.g., text vs. video vs. image). Being able to adapt the messages communicated to the student allows us to vary the level of interventions applied.

2.6   The Representation Layer

The control and message layers are brought together in the representation layer. This is what the learner sees and interacts with. Typically, these three layers are so intertwined that separating them becomes challenging. Researchers typically develop the interface with the strategy baked in, focusing heavily on the representation layer while neglecting the distinct affordances, roles, and designs of the control and message layers [13]. This approach results in a rigid interface that lacks modularity. STATS separates controls and messages from the interface to provide a more customizable learning experience.

2.7 The Media-Logic Layer

The media-logic layer is the glue that brings all the services together within this ecosystem and is the delivery method. By way of analogy, the traditional media-logic in a classroom is a human teacher. A teacher knows different instructional strategies, tracks data, knows how content is related, and interacts with the students. A technological approach needs to consider two things. First, it needs to consider what the delivery platform will be for providing the training. The delivery can range from the web, mobile app, simulations, and more. The constraints and affordances of each delivery platform can have significant impacts on the rest of the layers. Second, determining how the different layers or services communicate with each other is vital in ensuring interoperability. In our use case, STATS is embedded in a virtual training environment (the media-logic) that has its own data storage, control, message, and representation services. STATS fills in the adaptive training gaps by providing student and tutor modeling, detectors, and micro- and macro-adaptations, thus showing how our services can be interoperable in unique training environments.

3 Methods

As was mentioned previously, STATS has been embedded in a virtual training environment that uses high-fidelity 3D simulations for a basic electricity and electronics domain. The initial prototype has been field tested. Feedback was collected from 18 participants through a focus group where they completed a survey aimed at level one of Kirkpatrick's evaluation model [18]. No demographic or personal identifiable information was gathered as part of this process. The survey included nine Likert-scale questions, with three questions each for the following categories: effectiveness, efficiency, and usability. Five open-ended questions were also included to capture qualitative feedback about their experience. The questions that were asked were:

1. **Understanding the material.** The adaptive training helped me understand the material as well as or better than the teacher-led instruction.
2. **Skills mastery.** The adaptive training prepares me for practical application as wee as or better than the teacher-led instruction.
3. **Content relevance.** The adaptive training provided content that felt relevant and tailored to my personal learning needs.
4. **Learning pace.** The adaptive training allowed me to learn at my own pace efficiently.
5. **Enough opportunities.** There were enough opportunities to practice the content for me to feel confident with the material.
6. **Adaptivity.** The adaptivity (e.g., providing remediation) responded well to my learning needs.

7. **System navigation.** The adaptive training interface was easy to navigate and use.
8. **Content presentation.** The content presented in the adaptive training was clear, visually appealing, and easy to comprehend.
9. **Using the system.** I would want to use this training environment in the future.
10. **Like best.** What features did you like best about this training program?
11. **Like least.** What features did you like least about this training program?
12. **Improve.** How would you improve this training program?
13. **Professional development.** Would you like to use this training tool for your professional development? Why?
14. **Other comments.** Do you have any other comments you'd like to share?

The participants spent approximately one hour going through various lessons relating to alternating currents or direct currents in the virtual training environment with STATS providing live adaptations. The hour was split with 10 min spent getting the participants set up and familiar with the environment, 40 min spent interacting with the lessons, and 10 min for the participants to fill out the surveys. These participants had already been exposed to the content in their coursework. Some participants had already done well in the course whereas some were being held back to repeat the content.

At this stage of development, STATS offered two primary modes of adaptations: acceleration and remediation. As students answered questions, the student model would calculate mastery proficiency levels with respect to individual concepts. Detectors would then determine if the student was a high-skill or low-skill level student for that concept. If the student was deemed high-skill, STATS would accelerate the participants through the content, skipping practice opportunities as they had already mastered the content. If the student was struggling to reach mastery, they would receive targeted remediation for the concept they were on. Note, that while the interventions seem straightforward and rules-based, STATS' architecture can handle multiple interventions. For example, if a student state is low-skill, they could receive remediation, or STATS could use a lighter touch to intervene by providing targeted feedback like a hint. The limited use of interventions during this prototype phase was to demonstrate initial adaptive training for an audience generally unfamiliar with concepts like mastery or adaptations.

4   Results

Overall, the results from the participants were positive. The participants found the adaptive training environment to be effective, efficient, and usable. The following figure (see Fig. 2) shows the responses for the Likert-scale questions (questions 1–9 as listed above):

For the effectiveness measures, 89% of participants agreed that the adaptive training helped them better understand the content, 78% felt that the adaptive training prepared them for practical application, and 89% agreed that the adaptive training presented tailored content for their needs.

For the efficiency measures, 94% of participants agreed that the adaptive training helped them learn at their own pace, 89% agreed that the adaptive training gave them enough practice to feel confident with the content, and 89% agreed that the adaptive training responded to their needs.

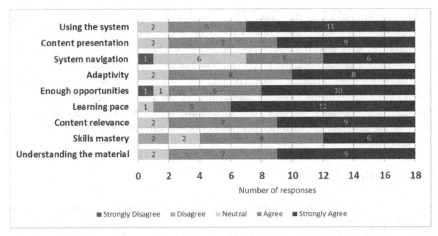

Fig. 2. The spread of student responses for the Likert-scale questions asked. The name of the question corresponds with the list above.

For the usability measures, 61% agreed that the interface was easy to navigate and use, 89% agreed that the content was clear, well-designed and simple, and 89% agreed that they would want to use this system in the future.

The one concerning area was the interface being easy to navigate and use, with only 61% in agreement with that question. The qualitative responses provided some good insight into where some potential issues are with the interface. For example, navigating between lessons was difficult, interacting with the virtual equipment (e.g., the multimeter or the calculator) created some bugs, and overall unfamiliarity with operating in the virtual environment made for a steeper learning curve than expected.

Overall, the data indicates a strong endorsement of the adaptive training environment's effectiveness and efficiency, with a majority of participants recognizing its benefits in terms of personalized learning and pace adjustment. However, the usability of the interface presents an area for improvement, particularly in navigation and interaction with virtual tools. These insights suggest that while the adaptive training system meets educational objectives, enhancing user experience through interface optimization could further increase its effectiveness and user satisfaction. Future iterations of the system will focus on addressing these usability concerns, streamline navigation, and refining virtual interactions to ensure a more intuitive and seamless learning experience. This feedback will be invaluable in guiding the next development phase, aiming to create a more user-friendly and accessible adaptive training environment.

5 Limitations

This study's methodology presents several limitations that should be considered when interpreting the results. First, the sample size of 18 participants, though adequate for a preliminary investigation, may not provide the statistical power necessary to generalize findings across a broader population. Additionally, the participants' prior experience with the content could bias their interactions with the adaptive training environment, potentially skewing results. The study's focus on a specific domain (basic electricity and electronics) may also limit the applicability of findings to other educational contexts or subjects. Furthermore, the prototype phase of the adaptive training environment meant that only a limited range of adaptive interventions and content were tested, which my a not fully represent the system's capabilities or the user's experience with a fully developed system. Finally, the lack of demographic information and the varied backgrounds of participants could introduce unaccounted variables that would affect the study's outcomes.

6 Future Directions

The development and initial evaluation of STATS provide a compelling use case for the creation and implementation of a modular ecosystem of intelligent tutoring services. This approach not only enhances the adaptability and efficiency of educational technologies, but also opens up numerous avenues for future research and development. Below are outlined future directions that could significantly advance the field.

Additional work should be done to detail STATS' approach to each layer. For example, work has been started with relation to the content layer to utilize LLMs to assist with the content design workflow. The use of LLMs could streamline the creation of educational materials, making the process more efficient and less of a bottleneck. Additionally, exploring the potential of incorporating LLMs as virtual agents within a training environment could provide learners with interactive, responsive, and personalized content delivery. Another example includes our development of the tutor model that was described above. Introducing a dedicated tutor model service to track the effectiveness of interventions could profoundly impact adaptive training performance. Key questions to explore include the ethical considerations in algorithmic decision-making, such as whether the system should prioritize optimal outcomes for students or encourage exploration of alternative learning paths. This exploration into the content layer and tutor model is just a starting point; addressing STATS' approach to all the layers and potential services systematically will be crucial for the comprehensive development and refinement of this ecosystem of intelligent tutoring services.

This study provides valuable insights into the implementation and initial evaluation of STATS within a virtual training environment. As the project continues to mature and develop, additional studies will need to be conducted to evaluate the higher levels of Kirkpatrick's evaluation model [18], namely to determine if students actually learn in the adaptive training environment, if the learning transfers into behavioral changes, and if the adaptive training affects greater learning outcomes like pass rates or reduced time to learn.

A key challenge for a modular ecosystem like STATS is ensuring interoperability between different services and layers. Future directions should include alignment with recognized interoperability standards. A tangible approach would involve leveraging services from modular frameworks like GIFT [2], demonstrating the potential for cross-project collaboration and application. This strategy not only underscore the versatility of STATS, but also sets a precedent for the broader adoption and adaptability of modular educational technologies.

7   Conclusion

The exploration of STATS within a virtual training environment underscores the significant potential of a modular ecosystem to provide individualized, efficient, and effective instruction. The positive participant feedback highlights our approach's strengths in content delivery and adaptability, aligning with the evolving needs of modern learners. However, the identified usability challenges present critical areas for improvement, particularly in enhancing interface navigation and interaction. The study's findings reinforce the importance of transitioning from traditional ITS architectures to more flexible, modular designs, as advocated by recent research. Future developments of STATS will focus on addressing these usability concerns, addressing our approaches to the broader STATS ecosystem, and exploring interoperability with other frameworks like GIFT. The journey towards fully realizing the potential of modular ecosystems is ongoing, but each iteration brings us closer to the ideal of truly adaptive, learner-centered education.

Acknowledgments. We would like to acknowledge Nick Englert and Kevin Jones for their contributions to and support of this project. This study was funded by NAWCTSD (contract number N6134022C0024).

Disclosure of Interests. The authors have no competing interests to declare that are relevant to the content of this article.

References

1. Gibbons, A.: An Architectural Approach to Instructional Design. Routledge, New York (2014)
2. Brawner, K., Goodwin, G., Sottilare, R.: Agent-based practices for an intelligent tutoring system architecture. In: Schmorrow, D.D.D., Fidopiastis, C.M.M. (eds.) AC 2016. LNCS (LNAI), vol. 9744, pp. 3–12. Springer, Cham (2016). https://doi.org/10.1007/978-3-319-399 52-2_1
3. Vassileva, J.: An architecture and methodology for creating a domain-independent, plan-based intelligent tutoring system. Educ. Train. Technol. Int. **27**(4), 386–397 (1990)
4. Nkambou, R., Boudreau, J., Psyche, V.: Building intelligent tutoring systems: an overview. In: Nkambou, R., Mizoguchi, R., Bourdeau, J. (eds.) Advances in Intelligent Tutoring Systems. Studies in Computational Intelligence, vol. 308, pp. 361–375. Springer, Berlin (2010). https://doi.org/10.1007/978-3-642-14363-2_18
5. Colby, B.: From systems to services: changing the way we conceptualize intelligent tutoring systems. Dissertation (2020)
6. Holstein, K., Yu, Z., Sewall, J., Popescu, O., McLaren, B., Aleven, V.: Opening up an intelligent tutoring system development environment for extensible student modeling. In: Penstein Rosé, C., et al. (eds.) AIED 2018. LNCS, vol. 10947, pp. 169–183. Springer, Cham (2018). https://doi.org/10.1007/978-3-319-93843-1_13
7. Nwana, H.: Intelligent tutoring systems: an overview. Artif. Intell. Rev. **4**(4), 251–277 (1990)
8. Uhr, L.: Teaching machine programs that generate problems as a function of interaction with students. In: Proceedings of the 1969 24th National Conference, pp. 125–134. ACM, New York (1969)
9. Woods, P., Hartley, J.: Some learning models for arithmetic tasks and their use in computer-based learning. Br. J. Educ. Psychol. **41**(1), 38–48 (1971)

10. Bloom, B.: The 2-sigma problem: the search for methods of group instruction as effective as one-to-one tutoring. Educ. Res. **13**(6), 4–16 (1984)
11. Carbonell, J.: AI in CAI: an artificial-intelligence approach to computer-assisted instruction. IEEE Trans. Man-Mach. Syst. **11**(4), 190–202 (1970)
12. Self, J.: Theoretical foundations for intelligent tutoring systems. J. Artif. Intell. Educ. **1**(4), 3–14 (1990)
13. Colby, B.: A comprehensive literature review of intelligent tutoring systems from 1995–2015. Master's thesis (2017)
14. Wenger, E.: Artificial Intelligence and Tutoring Systems: Computational and Cognitive Approaches to the Communication of Knowledge. M. Kaufmann, Los Altos (1987)
15. VanLehn, K.: The relative effectiveness of human tutoring, intelligent tutoring systems, and other tutoring systems. Educ. Psychol. **46**(4), 197–221 (2011)
16. Nye, B.: AIED is splitting up (into services) and the next generation will be all right. In: AIED Workshops, vol. 1432, no. 4, pp. 62–71 (2015)
17. Merrill, M.: Component display theory. In: Reigeluth, C. (ed.) Instructional-Design Theories and Models: An Overview of Their Current Status, pp. 292–333. Erlbaum, Hillsdale (1983)
18. Kirkpatrick Partners. https://www.kirkpatrickpartners.com/the-kirkpatrick-model. Accessed 17 Feb 2024

Trend-Aware Scenario Authoring: Adapting Training Toward Patterns from Real Operations

Mark G. Core[1]($\boxtimes$) (ID), Benjamin D. Nye[1] (ID), and Brent D. Fegley[2] (ID)

[1] Institute for Creative Technologies, University of Southern California, 12015 Waterfront Drive, Playa Vista, CA 90094, USA
{core,nye}@ict.usc.edu

[2] Aptima, Inc., 12249 Science Drive, Suite 110, Orlando, FL 32826, USA
bfegley@aptima.com

Abstract. An important prerequisite to trend-aware authoring is that scenarios be authorable and inspectable by instructors but also machine-readable such that authoring tools can assist with integrating real-world patterns into training. In this research, we use a semi-structured approach to authoring flight training scenarios in which textual descriptions of related scenario elements (i.e., happening at roughly the same time) are grouped together and assigned training objectives and phases of flight. This same representation can be used to represent real-world emergencies allowing their integration into scenarios for more realistic training. Such a representation is sufficient to support a recommender that ranks possible insertion points for real-world emergencies using constraints (i.e., the phase of flight of the emergency must match the phase of flight of the insertion point) and a ranking score. Our ranking score is currently based on matching training objectives associated with the emergency with training objectives in the scenario (i.e., training the same skills but using a more realistic example). The recommender is integrated into the scenario editor such that instructors can see the ranked injection points and modify the scenario by selecting one of these points.

Keywords: Authoring · Scenario-based training · Recommender

1 Introduction

Scenario-based training offers a powerful tool for trainees to practice skills in simulated or narrative analogs of a real situation, which should increase the transfer of these skills to real-life tasks. However, developing training scenarios and ensuring that they train the intended skills remains challenging and labor intensive. Stacy and Freeman [7] suggest Training Objective Packages (TOPs) as a way to address the challenge of creating opportunities for training objectives during live and simulated exercises. A TOP encodes the conditions necessary for

© The Author(s), under exclusive license to Springer Nature Switzerland AG 2024
R. A. Sottilare and J. Schwarz (Eds.): HCII 2024, LNCS 14727, pp. 15–24, 2024.
https://doi.org/10.1007/978-3-031-60609-0_2

the trainee to meet the training objective and how to measure and assess trainee performance.

Formal representations such as TOPs can enable better record-keeping for scenarios (e.g., knowing which competencies they train) and promote reusability. Different TOPs can be introduced into the same base scenario to train different skills. However, reusability does not ensure relevance: training scenarios are often static (infrequently updated) until replaced at significant cost. A scenario that fails to reflect changes in training needs or operational problems erodes the advantages of scenario-based training to transfer to real-life situations. In this paper, we will explore a potential solution to faster scenario updates, by integrating data from real-world trends, events, and emergencies into a scenario-authoring approach inspired by TOPs. Although this research remains exploratory, our design-based investigation suggests trend-based recommendations can be aligned to real-life training plans and in some cases can offer a quick, drop-in replacement for less relevant training activities.

2   Background

Training Objective Packages (TOPs) are one type of a broader category of representation designed to determine when a training objective can be trained and how performance should be measured. As noted, a TOP specifies: a) necessary conditions for the trainee to meet the training objective and b) how to assess performance on the training objective. Conditions are encoded as "behaviour envelopes" specifying boundaries on conditions, such as spatial or temporal coordinates [7].

A related approach [4] builds upon the Total Learning Architecture [1], a U.S. Department of Defense (DoD) standard for learning ecosystems, by introducing experience events (xEvents). Like TOPS, xEvents encode the conditions under which training objectives may be achieved. TOPS and xEvents can be seen as attempts to improve upon Master Scenario Event Lists (MSELs) in which authors use spreadsheet or word processor documents to outline a plausible sequence of events for a training exercise. It is difficult to trace the origins of MSELs as they appear without citation in training across a variety of organizations (e.g., military, police and other emergency responders) in which teams must work together to accomplish goals in environments which are often unpredictable (e.g., [6] describes best practices for authoring MSELs for US Homeland Security exercises). MSELs have the advantage of being human-readable and thus understandable by instructors, role-players, simulation controllers and trainees. There is generally a clear connection between events (e.g., clearing a room) and training objectives (e.g., practice room clearing) whereas machine-readable scripts for virtual and constructive entities are black boxes in which training objectives may not be represented or included in decision making. The goal of TOPs and xEvents is to retain the ability for instructors to author and inspect this data while using a machine-readable format to facilitate integration with simulators as well as authoring tools.

Authoring tools are particularly important because they can assist instructors in modifying scenarios and in particular introducing new elements based on lessons learned from the field. Scenarios based on real problems can boost trainee engagement and generally training offers the strongest advantage when it aligns closely to the problems and conditions for applying the skills in real situations (e.g., [5] discusses the importance of realistic radio simulations mirroring problems that arise on the battlefield). There needs to be a constant updating of scenarios based on current conditions and emergencies from the field because real-life needs change over time. For example, pilots might train to land an aircraft in the desert for years, but suddenly missions are needed in arctic conditions or where rain storms are frequent. These high-level changes are reflected in specific issues and signals: a different warning code or light might appear to signal that a landing is unsafe and must be aborted.

Unfortunately, there is often a substantial lag between trends changing in the field and instructors tuning practice to better reflect real-life. Experts must first recognize trends and lessons learned from the field, and bundle them into reports. Then course developers and instructors must review these reports and find "teachable moments" such as a real-life malfunction that can be used to practice emergency procedures. Due to the many steps in the process, such updates are infrequent and often limited to major changes (e.g., new equipment models). This could be improved through the use of authoring tools to update training based on real-life "teachable moments".

3   Domain: Flight Training Based on Incident Trends

As part of a broader research effort called TOPMAST (Training Operational Performance via Measure Automation and Scenario-generation Technology), we investigated approaches to speeding up the process for updating training scenarios based on observations from the field, with an emphasis on aviation training. Aircraft are particularly well-suited to a trends-based approach, because they have a well-defined taxonomy of issues (e.g., fault codes), they have multiple types of trends (e.g., aircraft versions, flight routes, subsystem updates), and require operators to quickly diagnose and react to emergencies to maintain safety.

In this work, we studied Navy flight instructors who took a MSEL-style approach to authoring using word processors to generate scenario outlines (e.g., events that should occur and rough time guidelines) and only providing scenario-level training objectives. To address this challenge, we developed a scenario editor for flight training that explicitly represents the structures of these flights starting with individual scenario elements. Figure 1 shows an example with details specific to the aircraft obscured or scrambled in the case of fault codes.

1. **Event:** a normal flight event (e.g., communications, achieving takeoff) or an emergency (a fault code or description of the problem) which is listed under Event Sequence. Faults are highlighted in red and indented.

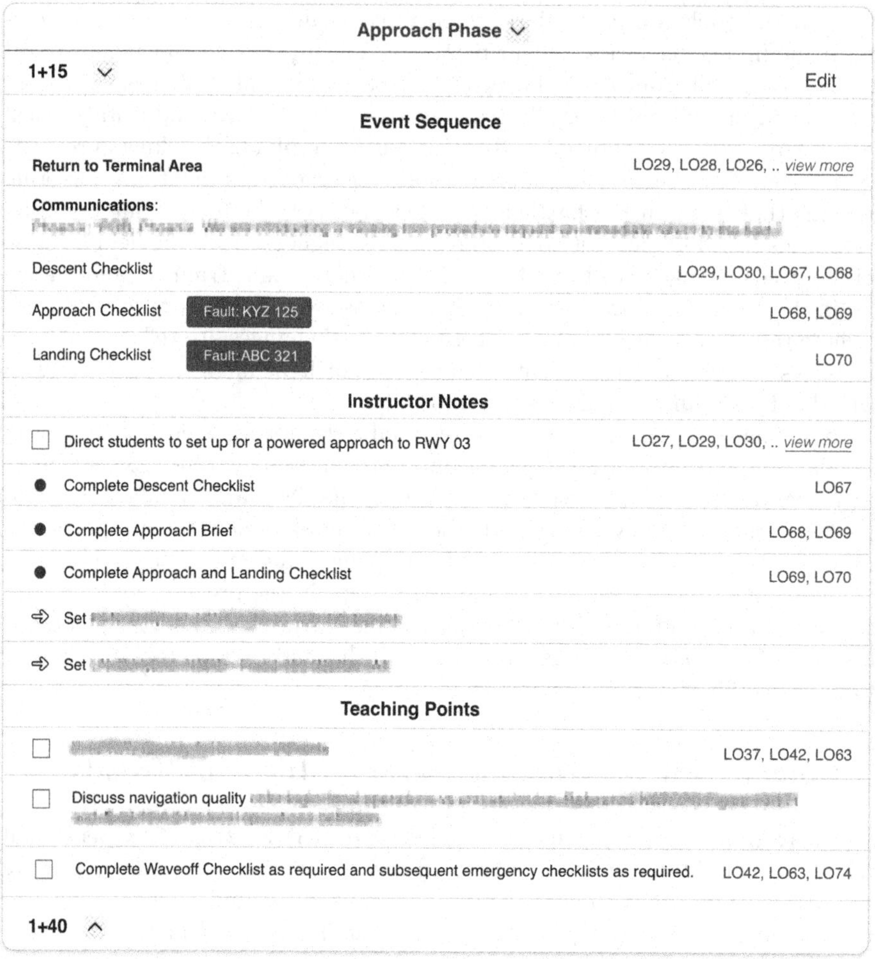

Fig. 1. Sample time point in TOPMAST scenario editor

2. **Expected Crew Action** (e.g., performing a checklist, responding to communications, manipulating controls) which is listed under Instructor Notes using round bullet points.
3. **Simulator Manipulation:** how the instructor triggers or resolves an emergency in the simulator which is listed under Instructor Notes using arrow bullet points.
4. **Instructor Action:** (e.g., giving verbal instructions to trainees) which is listed under Instructor Notes using square bullet points.
5. **Teaching Point:** item to discuss with trainees which is listed under Teaching Points.

Figure 1 also shows how scenario elements are grouped into a row and assigned a time point (e.g., after approximately 1 h and 15 min these scenario elements should occur). These rows are associated with a phase of flight which in this case is the landing approach. The scenarios under study typically assumed the presence of two trainees, with one having principal responsibility for control and safety of flight. Mid-way through the block of instruction, trainees would switch responsibilities, meaning that each scenario had two versions (one for student A, one for B). The student B version allows practice of the same training objectives but includes variations so that it is not a complete repetition. The scenario editor also includes a form in which authors specify initial conditions such as weather and aircraft state.

Following the guidance of our subject matter experts, we used the term "learning objective" in the scenario editor to reflect instructor expectations. In Fig. 1, we see scenario elements annotated with learning objectives (presented as hyperlinks) by the subject matter experts. Each learning objective is assigned a code prefixed by "LO" and authors use a drop-down menu with the full learning-objective names when annotating. In this paper, we will continue to use the term "training objective" to reflect their role as training goals (i.e., "the trainee should be able to do X").

In some cases, the mapping of scenario elements to training objectives is straight-forward and one-to-one (e.g., the crew action of completing the descent checklist is linked to the training objective of completing the descent checklist). In other cases such as a malfunction event, the scenario element can be linked to a set of training objectives (e.g., general troubleshooting, managing navigation issues, executing wave-off procedures). In current practice, the scenarios are well-defined, have clear training objectives, are limited in number, and static (i.e., training objectives for a course are rarely updated based on real-life events or a real-time data feed).

The overall TOPMAST system is designed to be a force multiplier. It is both a scenario library management system and a scenario generator. We envision that TOPMAST will support a set of official (approved) scenarios as well as allowing the creation of variants to keep training relevant and fresh, and mitigate the "gouge" whereby trainees effectively skip the decision-making process having memorized how to respond to scenario events. Version tracking will ensure that variants do not overwrite official scenarios and that the provenance of each scenario is known and preserved.

4   Recommender: Adding Real-World Events to Scenarios

To update training, data sources must be available to track real-world events. Data from the field typically is either an equipment log file or a written hazard/accident report. Equipment logs have the potential to be directly transformed into structured representations (e.g., database tables). However, although aircraft manufacturers may collect and archive such logs they are currently not readily available to training developers. Hazard/accident reports are distributed

Read Board

Fig. 2. Sample Read Board card in the TOPMAST scenario editor

widely and unlike equipment logs indicate causality and lessons to be learned. Each emergency needs to be described using the same format as the scenario (e.g., events, expected crew actions, teaching points, training objectives and phases of flight) which is currently done by subject matter experts. However, advances in natural language processing such as large language models (LLMs) should enable automation of the information extraction process for this domain, due to its relatively well-defined ontology of fault codes, events, and actions.

Once a new real-world emergency is authored in this way, it appears in the scenario editor's read board (Fig. 2). Once an author clicks "Create New Event" to add this emergency to the scenario library, the TOPMAST recommender system identifies and ranks "injection" points for this emergency into a library of training scenarios. Currently we analyze each row of the scenario (i.e., scenario elements occurring at roughly the same time) as a possible injection point, assuming the phase of flight of the emergency matches that of the scenario row. Such injections into a pre-existing scenario ensure the same skills are addressed but in more realistic conditions.

We then generate an injection score, currently the number of overlapping training objectives between the scenario row and the emergency. Such an approach is sufficient for matching real-world emergencies to scenario events that exercise the same training objectives. In future work, we intend to explore including expected crew actions, teaching points and related events in the injection score calculation. For an emergency occurring across two phases of flight, possible insertion points are constrained to these boundary points in the scenario

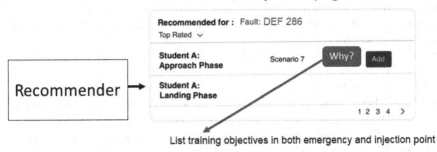

Fig. 3. Annotated mockup of recommender interface

library; injection scores for these boundary points can be calculated separately for each phase of flight and then summed.

Currently our recommender considers rows in isolation as possible locations to inject an emergency (i.e., only supports cause and effect in the same row). Once the author picks an injection point, the recommender appends the emergency elements to the row, which the author may need to modify subsequently. For example, in some cases the real-world emergency can serve as an explanation for an instructor-triggered fault and should immediately precede it (e.g., a fault in the navigation system could explain why the flight computer triggered a wave-off). In other cases, the real-world emergency should serve as a replacement for a fake event (e.g., a navigation sensor error triggers a wave-off instead of a vehicle blocking the runway). We will address this issue in future work by allowing authors to identify causal links (e.g., all crew and instructor actions related specifically to the vehicle on the runway) and enforcing the use of consistent vocabulary for checklists and procedures. In some cases background knowledge will be needed (e.g., fault A often causes fault B) to infer when a real-world event might be linked to a pre-existing scenario element.

The recommender is integrated into the scenario editor such that instructors can see the ranked injection points and modify the scenario by selecting one of these points. The interface is still a work-in-progress. Figure 3 is a mockup of how the interface would present an emergency impacting two phases of flight (i.e., approach and landing). The recommender maintains the set of common training objectives between the emergency and each injection point such that an author can inspect them before clicking the "add" button to inject the emergency.

5   Discussion: Scaling Up Trend-Aware Authoring

Training objective packages (TOPs) [7] are a general approach to authorable, machine-readable scenario representations. TOPs represent both the necessary scenario conditions for a trainee to achieve a training objective as well as the criteria for achieving that training objective. Our flight instructors took an approach similar to a Master Scenario Event List and used our authoring tool to specify a sequence of events annotated with training objectives. The conditions for a particular event and associated training objective are assumed to be the accomplishment of all the previous events (i.e., there is no representation of causal links between events).

In the context of Navy air training, a potential resource for this missing information is the Naval Air Training and Operating Procedures Standardization (NATOPS) manual for the target aircraft. Electronic versions of some NATOPS manuals have been available since 2002 [3] but focus on supporting pilots (e.g., quick access to emergency procedures) rather than representing knowledge for machine use. The rows of a scenario group related elements (e.g., a system fault may be in the same row as the corresponding wave-off) but in some cases, a minor malfunction may not cause an emergency until a phase of flight such as "approach" in which precise control and navigation are critical.

The introduction of flight recorder data would also present a number of opportunities. Stacy et al. [8] discuss a tool to analyze flight recorder data from an emergency and author a corresponding scenario event list. The availability of such data would also allow analysis of trends to suggest real-world malfunctions that are more frequent and identify possible training gaps. A trends-based approach would allow a human expert or a computational model to analyze data from many flights to measure malfunction frequencies and how they co-occur and vary based on conditions. This is especially critical with a newer aircraft as it could lead to changes in procedures and training such that early corrective actions can prevent more serious consequences.

Large language models (LLMs) such as GPT-4 have been used for natural language processing and common sense reasoning tasks [2] and could potentially aid the human author by processing text and making connections (e.g., event A causes event B). Newer LLMs could also help to extract real-life trends from noisier data sources (e.g., flight logs, mission reports). It may even be possible to suggest modified scenarios by having LLMs modify a standard flight to meet new training objectives.

Another opportunity would be the ability to integrate authoring tools such as TOPMAST with the simulator. Currently scenarios include guidance for instructors on how to manipulate the simulator to introduce faults to test trainees and then resolve those faults when the trainees have performed the appropriate procedures. Ideally, TOPMAST could serve as both an authoring tool and simulation controller such that it could handle such details and allow the instructor to focus on observing and guiding the trainee. More complex logic could be introduced to make the scenarios adapt to individual trainees (e.g., giving more difficult, realistic challenges to high performing trainees).

6   Conclusion

This paper describes an effort to introduce machine-readable, instructor-authorable scenarios to flight training currently using scenario outlines authored with word processors. A semi-structured approach was taken in which related scenario elements (i.e., happening at roughly the same time) were given textual descriptions, grouped together and assigned training objectives and phases of flight. This same representation can be used to encode real-world emergencies allowing their integration into scenarios for more realistic training. Such a representation is sufficient to support a recommender that ranks possible insertion points for real-world emergencies using constraints (i.e., the phase of flight of the emergency must match the phase of flight of the insertion point) and a ranking score. Our ranking score is currently based on matching training objectives associated with the emergency with training objectives in the scenario (i.e., training the same skills but using a more realistic example).

The recommender is integrated into the scenario editor such that instructors can see the ranked injection points and modify the scenario by selecting one of these points. Our subject matter experts indicated that this type of tool would be beneficial over the current approach of using a word processor to add real-world emergencies to the scenario outlines. Although aircraft equipment logs are not available at this time, it is an important issue to address in the future as instructors must now review hazard/accident reports and select representative emergencies rather than the recommendation system directly measuring trends and associated conditions (e.g., weather, equipment readings). Another important future consideration is integration with the simulators used in training which must be configured to trigger emergencies matching their real-world counterparts.

Acknowledgments. Research reported in this paper was supported by the U.S. Naval Warfare Center under award number N68335-19-C-0583, as a collaboration between Aptima, Inc. and the USC ICT University Affiliated Research Center (U.S. Army W911NF-14D0005). The content does not necessarily reflect the position or the policy of the Government, and no official endorsement should be inferred.

References

1. ADL Initiative: Total learning architecture functional requirements document (2021). https://adlnet.gov/assets/uploads/2021%20TLA%20Functional%20Requirements%20Document%20w%20SF298.pdf
2. Bubeck, S., et al.: Sparks of artificial general intelligence: early experiments with GPT-4 (2023). https://arxiv.org/abs/2303.12712
3. Deaton, J., Glenn, F., Burke, C.S., Good, M., Dorneich, M.: Evaluation of an interactive electronic NATOPS (IE-NATOPS) and associated graphic interaction concepts. Final report, Office of Naval Research (2002)
4. Hernandez, M., Blake-Plock, S., Owens, K., Goldberg, B., Robson, R., Ray, T.W.F.: Enhancing the total learning architecture for experiential learning. In: Proceedings of the Interservice/Industry Training, Simulation, and Education Conference (I/ITSEC) (2022)
5. Khooshabeh, P., Choromanski, I., Neubauer, C., Krum, D.M., Spicer, R., Campbell, J.: Mixed reality training for tank platoon leader communication skills. In: IEEE Virtual Reality (VR) (2017)

6. Renger, R., Wakelee, J., Bradshaw, J., Hites, L.: Steps in writing an effective master scenario events list. J. Emerg. Manage. **7**(6), 51–60 (2009)
7. Stacy, W., Freeman, J.: Training objective packages: enhancing the effectiveness of experiential training. Theor. Issues Ergon. Sci. **17**(2), 149–168 (2016)
8. Stacy, W., Picciano, P., Sullivan, K., Sidman, J.: From flight logs to scenarios: flying simulated mishaps. In: Proceedings of the Interservice/Industry Training, Simulation, and Education Conference (I/ITSEC) (2010)

Verification and Validation of Adaptive Instructional Systems: A Text Mining Review

Bruno Emond[✉][iD]

National Research Council Canada, Ottawa, Canada
bruno.emond@nrc-cnrc.gc.ca
http://www.nrc-cnrc.gc.ca

Abstract. The current paper aims at qualifying the distribution of academic papers related to verification and validation of Adaptive Instructional Systems (AIS) and Adaptive software (ASOFT). For the purpose of the literature review, the theme of software verification and validation is divided into three sub-themes: 1) software verification, 2) empirical validation, and 3) model and simulation validation. In order to maintain the paper broad objectives, the approach will apply text mining techniques to analyze the literature. The corpus contained 33 546 documents extracted from Scopus (January 2024) in 5 document sets that were analyzed using TF-IDF vectors to measure cosine similarity between documents. The similarity distributions as well as the statistical tests indicate a difference in attention given to verification and validation issues between publications in AIS and ASOFT. The more important difference between the two sets of documents is the role given to software verification. The stronger emphasis on empirical, and model and simulation validation in the AIS literature points to the high dependency of considering human factors in the success of adaptive instructional systems. In regard to software verification in AIS, the few publications addressing this issue in comparison to publications looking at tutoring software engineering techniques indicate a possible research and practical gap to explore. The brief overview of the extended number of publications in software verification of ASOFT did not provide an immediate and clear set of methods and techniques that could be applied in the context of AIS. A further analysis could look in more depth in the ASOFT literature or at the AI development engineering literature.

Keywords: Adaptive Instructional Systems · Verification and Validation · Adaptive Software · Literature Review · Text Mining Literature Review · Software Verification · Empirical Validation · Modelling and Simulation

1 Introduction

Adaptive Instructional Systems (AIS) are tailored to facilitate personalized learning experiences, they include a range of technologies such as intelligent

R. A. Sottilare and J. Schwarz (Eds.): HCII 2024, LNCS 14727, pp. 25–43, 2024.
https://doi.org/10.1007/978-3-031-60609-0_3

tutoring systems, intelligent mentors, recommender systems, and intelligent instructional media [38]. These systems dynamically adapt to individual learners, employing mechanisms that respond to specific goals, needs, preferences, and interests in alignment with educational objectives [40]. For instance, learner models within AIS can predict and respond to learner behaviour, enhancing the system's ability to personalize learning. The primary objective of AIS is to optimize learning outcomes, engagement, and retention through this personalization. A key component of AIS is the use of adaptive instructional agents [4], software entities that actively engage with both the learner and the learning environment. These agents, often using artificial intelligence, are crucial in realizing the allocated learning objectives by modifying instructional strategies based on real-time assessments of learner progress and needs.

Verification and validation (V&V) are critical processes in the development and implementation of every software systems, including adaptive instructional systems. Verification ensures that a system is built correctly according to its specifications and requirements. In the context of AIS, this means ensuring that the system accurately identifies learners' knowledge states, preferences, and learning styles, and adapts its instructional strategies accordingly. Validation, on the other hand, involves ensuring that a system is fulfilling its intended purpose and meeting the needs of its users. For AIS, this means assessing whether the adaptive mechanisms enhance learning outcomes. The relevance of V&V methods for AIS cannot be overstated. As these systems increasingly influence educational outcomes and potentially critical applications, ensuring their reliability, effectiveness, and safety becomes essential for the advancement of personalized learning and the development of more effective and reliable educational technologies.

The current paper aims at qualifying the distribution of academic papers related to verification and validation of adaptive instructional systems. One objective is to provide a high-level view of the literature on the topic, and a second object is to identify possible research gaps. As a means to help identify gaps, the review will include and compare results about *adaptive instructional systems* to results about *adaptive software*. Adaptive software is a more generic class of software which dynamically modifies at runtime its own internal structure and hence its behaviour in response to changes in its operating environment [20]. Adaptive software includes dynamic and context-aware software, autonomic computing, adaptive user interface, as well as software controlling autonomous systems.

These objectives faces two challenges. A first challenge is the large amount of learning technologies covered under the umbrella of adaptive instructional systems (intelligent tutoring systems, intelligent mentors, recommender systems, and intelligent instructional media). A second challenge, related to the breadth of adaptive instructional systems, is the difficulty of formulating a document search strategy to retrieve documents that does not result in either a too small or too large set of relevant documents. A small set would lack meaningful coverage, while a large set would demand a significant effort by human experts to sort, analyze and synthesize documents.

In order to address the coverage and search strategy challenges, the app-roach will apply text mining techniques to analyze the literature [30, 42, 43]. This approach allows for quantitatively reproducible review on a large volume of literature compared to expert systematic reviews. The current approach uses a simple text mining technique to represent every document as a vector of terms with associated numerical "Term Frequency-Inverse Document Frequency" val-ues (TF-IDF) [31]. Based on these vectors, documents are compared to one another using a cosine similarity measure. The detail of this technique is pre-sented in the methodology section.

2   Problem Statement

The current paper aims at qualifying the distribution of academic papers related to verification and validation of adaptive instructional systems, and to compare this distribution to the publications on verification and validation of adaptive software. This review will provide a high level view of the literature, and identify possible research gaps.

Figure 1 shows (A) the number of publications per year from a query on Scopus (https://www.scopus.com/) using the AIS search string of Listing 1.1, and (B) the number of publications using ("'validation" OR "verification") added to the same search string. As the figure indicates, the quantity of publications over the years for the whole field is much larger than the publications related to verification or validation. The years 2014 to 2016 are particularly low in this respect. Understanding the conditions that might have impacted the drop of publications during those years is beyond the scope of the current paper.

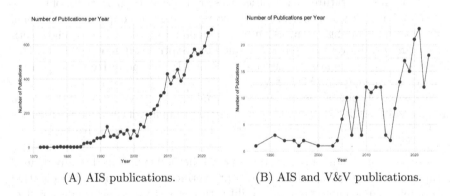

(A) AIS publications.   (B) AIS and V&V publications.

Fig. 1. Number of publications per year related to (A) Adaptive instructional systems, and (B) Verification or validation and Adaptive instructional systems. The data was obtained from Scopus (https://www.scopus.com/) in January 2024 with the AIS search string of Listing 1.1 for figure (A), and with ("verification" OR "validation") added to the AIS search string for figure (B).

For the purpose of the literature review, the theme of software verification and validation is divided into three sub-themes: 1) software verification, 2) empirical validation, and 3) model and simulation validation. Each sub-theme focuses on a distinct aspect of ensuring that software systems perform as expected and meet users needs.

Software verification involves checking the software to ensure that it is correctly implementing the specified requirements. Verification activities typically include reviews, inspections, and static analysis of computer code. The goal is to test that the software product is built correctly according to the design and requirements, often summarized by the question: "Are we building the product right?".

Empirical validation involves testing the software in a real-world or realistic setting to gather empirical data about its performance and intended use. Empirical validation aims to demonstrate that the software works in practice, addressing the question: "Does the software do what the user needs?" Empirical validation consists of conducting experiments, collecting data, or using observational studies to determine whether the empirical results support or refute a claim being tested.

Model and simulation validation involves testing that a computational model or simulation accurately reflects the real-world system that is meant to represent. Models and simulations are essential for adaptive systems as they allow to answer questions like: "How will the software perform under expected and unexpected conditions?". Validation of models and simulations aims to ensure that a model or simulation can be used reliably for its intended purposes.

3   Review Scope

The text mining literature review objectives are to provide an overview of the use of verification and validation methods across a range of adaptive instructional systems such as intelligent tutors, and recommender systems, but also how verification and validation methods are applied to adaptive software in general. One objective of this comparison between V&V in AIS and adaptive software is to identify possible gaps in AIS V&V.

The scope of each of these adaptive technology types is defined by the search strings used to retrieved documents, which will form part of the document corpus set. Listing 1.1 presents the search string for each of the adaptive technology types.

A set of verification and validation (V&V) method types was broadly defined to capture different perspective 1) software verification, 2) empirical validation with human participants such as usability studies and experimental data collection, and 3) modelling and simulation including learner models (learning analytics), and simulated learners. Listing 1.2 presents the search string for each of verification and validation methods explored in the current study. The single keyword "software verification" provided a sufficient number of relevant documents. Too many documents were retrieved when adding other keywords like "software testing" to the search string, and too few documents were retrieved when seeking documents relevant to software verification in the context of software users or learners.

Listing 1.1. Search strings for adaptive technology types (AIS: Adaptive instructional systems; ASOFT: Adaptive software).

```
AIS   ("adaptive instruction*"
      OR "intelligent instruction*"
      OR "intelligent tutor*"
      OR ("intelligent system*" AND "tutor*")
      OR ("recommender"
          AND ("learning"
              AND ("student" OR "learner")))))
ASOFT ("adaptive software"
      OR "adaptive user interface*"
      OR "context-aware software"
      OR "autonomic computing"
      OR ("autonomous system*" AND "software"))
```

Listing 1.2. Search strings for V&V method types (VER: software verification; EMP: data collection of human performance and learning; and ModSim: modeling and simulation).

```
VER     ("software verification")
EMP     ("methodology"
        AND ("participants" OR "subjects"
            AND ("male" OR "female"))
        AND "results"
        AND ("user stud*" OR "experiment"
            OR "data collection"))
ModSim (("interface model*" OR "student model*"
        OR "learner model*" OR "domain model*"
        OR "teacher model*" OR "pedagogical model*"
            ) AND "simulation")
```

4   Methodology

4.1   Research Questions

Given the scope defined for the adaptive technology and verification and validation method types, the text mining literature review aims at answering the following two questions:

Q.1 What is the distribution of academic publications for three sub-themes of verification and validation being 1) software verification, 2) empirical validation, and 3) modelling and simulation in the domain of adaptive instructional systems and adaptive software?

Q.2 Given the observed distribution, are there any trends and research gaps that can be identified?

4.2   Data Collection

The data collection was performed at the end of January 2024 on the Scopus document database (https://www.scopus.com/) using the five search strings presented in Sect. 3 (Listings 1.1 and 1.2). The search strings were applied to the documents title, abstract, and keywords. All documents had to be in English, with no limitation regarding the publication dates. The retrieved document records were exported in the BibTeX format by groups of 2 000 or less records at a time. Prior to processing the BibTeX records to extract terms, duplicated entries (identical DOI or title), and records with no authors were deleted. Records deletion was performed only within and not across each of the five document sets. Table 2 presents the number of document records per set for a total of 33 546 document records in the whole corpus.

4.3   Data Extraction

After duplicated and without authors records were removed from the document record sets, each BibTeX record title, abstract, keywords, and author keywords were concatenated before tokenization. Tokens were generated by splitting the concatenated string into words, trimming words of special characters, removing punctuation, and stop words using English words from stopwords-iso (https://github.com/stopwords-iso/stopwords-iso). The list of tokens were added to a Common Lisp structure representing a document record with other properties such as the total number of tokens, individual token counts, and the number of times a document was cited.

4.4   Text Mining Method

The text mining method in the literature review analysis is based on the technique known as the "Term Frequency-Inverse Document Frequency" (TF-IDF) calculation. For each document the frequency (TF) of each token is calculated and normalized by the length of the document to prevent bias towards longer documents. The equation for calculating term frequencies is given in Eq. 1.

$$TF(t, d) = \frac{\text{Number of times token } t \text{ appears in a document } d}{\text{Total number of tokens in the document}} \tag{1}$$

The Inverse Document Frequency (IDF) calculates the inverse frequency for each token across all document sets. IDF measures how important a term is across the entire collection of document sets by giving higher scores to terms that are rare across documents. The equation for calculating the inverse document frequencies is given in Eq. 2.

$$IDF(t, D) = \log \left(\frac{\text{Total number of documents } D}{\text{Number of documents with term } t \text{ in it}} \right) \tag{2}$$

The TF-IDF value is the product of the TF by the IDF for each token in each document to produce an TF-IDF score. This score represents the importance of a term in a document relative to its importance across all documents. The equation for calculating a TF-IDF value is given in Eq. 3.

$$\text{TF-IDF}(t, d, D) = TF(t, d) \times IDF(t, D) \tag{3}$$

where t represents a token (word),
d represents a document in the corpus, and
D represents the corpus of documents.

Using these equations, a document is converted into a vector of its TF-IDF scores. Each dimension of the vector corresponds to a unique token in the entire corpus (the collection of all document sets), and the value in each dimension is the TF-IDF score of the token in that document. For tokens not present in a document, the corresponding value in the vector is zero. For the current text mining analysis, each document TF-IDF vector was normalized by dividing each dimension of the vector by the vector's magnitude. The equations to normalize a vector and calculate its magnitude are given in Eqs. 4 (magnitude) and 5 (normalization). Normalized vectors have a unit length, which allows the cosine similarity to effectively measure the angle between vectors.

$$\|\mathbf{A}\| = \sqrt{\sum_{i=1}^{n} a_i^2} \tag{4}$$

$$\mathbf{V}_{\text{norm}} = \frac{\mathbf{V}}{\|\mathbf{V}\|} = \left(\frac{v_1}{\|\mathbf{V}\|}, \frac{v_2}{\|\mathbf{V}\|}, \ldots, \frac{v_n}{\|\mathbf{V}\|} \right) \tag{5}$$

Once a document is represented as a TF-IDF vector, it can be compared for its similarity to another document vector using a pairwise cosine similarity calculation. This pairwise comparison can be extended to compare document sets by applying the cosine similarity calculation between each pair of documents (one from each set). Cosine similarity measures the cosine of the angle between two vectors and is a commonly used metric for comparing documents in high-dimensional space. The cosine similarity ranges from -1 (exactly opposite) to 1 (exactly the same), with 0 indicating no similarity. The equation for calculating the cosine similarity is given in Eq. 6.

$$\text{Cosine Similarity}(\mathbf{A}, \mathbf{B}) = \frac{\mathbf{A} \cdot \mathbf{B}}{\|\mathbf{A}\| \times \|\mathbf{B}\|} \tag{6}$$

where $\mathbf{A}$ and $\mathbf{B}$ are TF-IDF vectors,
$\mathbf{A} \cdot \mathbf{B}$ is the dot product of vectors $\mathbf{A}$ and $\mathbf{B}$,
and $\|\mathbf{A}\|$ and $\|\mathbf{B}\|$ are the magnitudes of vectors $\mathbf{A}$ and $\mathbf{B}$ respectively.

4.5   Analysis Procedure

The text mining procedure consists of two separate analyses: 1) pairwise document record sets comparaison using cosine similarity (Eq. 6), and 2) classification of document records with highest similarity scores in categories relevant to the research questions. This second step did not use text mining techniques and is explained later in the current section.

Using text mining methods, in addition to comparing every document to documents from its own set to measure internal similarity, source sets (software verification (VER), empirical validation (EMP) and modelling and simulation (ModSim)), were compared in a pairwise fashion with the two target sets: Adaptive instructional systems (AIS) and Adaptive software (ASOFT). Assuming that every document record is already represented as a normalized TF-IDF vector, the steps in carrying the analysis are:

1. Keeping intact every document TF-IDF vectors, create an aggregated document set vector with the arithmetic mean of every vector dimensions of every documents in the set.
2. Calculate the cosine similarity between the aggregated source vector and every document vector of the target set.
3. Use the Kruskal-Wallis H test to determine if there are statistically significant differences in the distribution of cosine similarity scores across the different target document sets when compared to the source set. The Kruskal-Wallis test is a non-parametric method that does not assume a normal distribution and is suitable for comparing two or more groups. The null hypothesis ($H0$) is that the median similarity scores are the same across all groups. Given the large number of documents, the alpha threshold is set to 0.001.
4. If the Kruskal-Wallis test indicates significant differences, a Dunn's post-hoc test is performed to identify which specific document sets have significantly different similarity scores when compared to the source set. Dunn's test compares the pairwise differences between groups with a correction for multiple comparisons, such as the Bonferroni correction. Dunn's test will provide p-values for each pair of comparisons. Significant p-values indicate that the similarity distribution between those sets and the source set is significantly different.
5. Select documents with high-similarity scores for content and classification analysis, 600 documents (100 for each V&V-Adaptive sets combination) were selected.
6. Classify (content analysis) high-similarity score documents using the categories provided in Table 1. Documents that did not fit into one of these categories were classified as "other". The classification activity does not involve text mining techniques.
7. Analyze the document classification frequencies, briefly describe the content of documents within each category.

Table 1. Classification categories for content analysis of high-similarity adaptive instructional systems (AIS) and adaptive software (ASOFT) target documents in relation to the source document of: 1) verification (VER), 2) validation by empirical studies (EMP), and 3) validation of modelling and simulation (ModSim). Documents were classified as either "directly" or "indirectly" addressing the verification and validation of AIS/ASOFT software. Documents that did not fit into one of these categories were classified as "other".

Source Set	Direct application of V&V *Document classification criteria*	Indirect application to V&V *Document classification criteria*
VER Software verification	Verification of AIS/ASOFT: *The document is about verification/testing of AIS/ASOFT, or a generic verification method*	AIS/ASOFT for verification: *The document is about AIS as a tutor for verification or software engineering; The document is about ASOFT verification of autonomous devices*
EMP Empirical validation	Empirical studies of AIS/A-SOFT: *The document is about data collected for the purpose of AIS/ASOFT validation*	Literature review of AIS/A-SOFT: *The document is about a literature review for the purpose of AIS/ASOFT validation or identification of research directions*
ModSim Models and simulations validation	Validation of AIS/ASOFT models and simulations: *The document is about processes and methodologies for validating AIS/ASOFT models and simulations against the real-world systems they represent*	AIS/ASOFT models and simulations, but no reference to validation: *The document is about AIS/ASOFT models and simulations (architecture, systems, or development) without reference to their validation*

5   Results

5.1   TF-TDF Values

Figure 2 presents (A) the distribution of TF-IDF vector values on any dimension, and (B) within document sets similarity values. For plotting purposes, the logarithm of the TF-IDF values was used to reduce the distance between low and high TF-IDF values.

A Kruskal-Wallis test was used to assess the variability of TF-IDF values across all vector dimensions per document set (Fig. 2(A)). The test revealed significant differences in the distributions of TF-IDF values between the five document sets ($H(4) = 40879.29, p < 2.2e - 16$), leading to the rejection of the null hypothesis that there is no difference in the median TF-IDF values. In the subsequent analysis to further explore the specific pairwise differences among the document sets following a significant Kruskal-Wallis test, Dunn's post-hoc tests were conducted with Bonferroni correction to adjust for multiple comparisons.

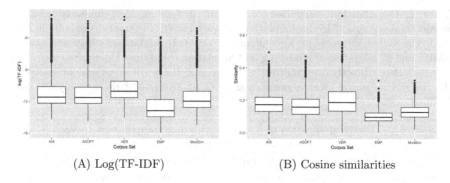

(A) Log(TF-IDF)   (B) Cosine similarities

Fig. 2. (A) Distribution of the logarithm of TF-IDF values across all vector dimensions per document set; and (B) Distribution of internal document similarity values between every document of a given set compared to the mean vector of all documents in a set. Acronyms are: Adaptive instructional systems (AIS), Adaptive software (ASOFT), software verification (VER), empirical validation (EMP) and modelling and simulation (ModSim).

The comparison between the AIS document set and the ASOFT document set yielded a p-value of 0.011. Given our predetermined alpha level of 0.001, this p-value exceeds the threshold required to declare statistical significance.

Table 2 list the highest 25 token TF-IDF values per corpus set. The table also gives for each set; 1) the number of document records, 2) the number of unique tokens (vector dimension), and 3) the total number of tokens. The number of tokens cited in a document was weighted by the number of times the document was cited in order to give more weight to highly cited tokens in the corpus set. The whole corpus contained 114 382 unique tokens and 3 709 395 tokens including repetitions and multiplication by citation counts. Document records with no citations were assigned a count of 1.

5.2 Within Document Sets Similarity

Figure 2(B) presents a box plot of the internal similarity distribution within every document set. A Kruskal-Wallis test was used to assess the variability of within similarity distributions between the group of sets. The test revealed significant differences in the distributions of internal similarity scores between the sets $(H(4) = 9304.51, p < 2.2e - 16)$. Subsequent Dunn's post-hoc tests, with Bonferroni correction for multiple comparisons, indicated that the null hypothesis could be rejected on all pairwise comparisons, all with very low p values.

5.3 Between Document Sets Similarity

Figure 3(A) shows the distribution of similarity values for all six combination of V&V types: verification (VER), empirical validation (EMP), and validation with modelling and simulation (ModSim); by adaptive technology types: adaptive instructional systems (AIS), and adaptive software (ASOFT).

Table 2. Highest 25 token TF-IDF values per corpus set. Token counts (TC) were weighted by the number of times a citation was cited.

Corpus set	Highest 25 TF-ITF tokens/words values
AIS 9 419 documents 31 035 tokens 876 902 tokens count	learning, tutoring, intelligent, students, student, knowledge, systems, recommendation, learners, recommender, educational, elearning, education, learner, adaptive, tutor, teaching, model, training, feedback, online, personalized, cognitive, instruction, data
ASOFT 5 987 documents 28 106 tokens 575 132 tokens count	autonomic, computing, software, autonomous, systems, adaptive, user, management, self-adaptive, architecture, service, cloud, adaptation, applications, runtime, control, services, approach, framework, network, interfaces, distributed, application, requirements, development
VER 2 386 documents 15 440 tokens 214 386 tokens count	verification, software, testing, checking, formal, program, validation, programs, code, tools, model, analysis, specification, safety, logic, development, techniques, tool, approach, systems, hardware, requirements, abstraction, correctness, properties
EMP 9 892 documents 65 374 tokens 1 453 275 tokens count	participants, health, subjects, study, data, care, patients, studies, women, clinical, trials, methodology, methods, age, activity, interviews, effects, healthy, trial, medical, training, blood, analysis, treatment, social
ModSim 5 525 documents 34 311 tokens 589 700 tokens count	model, simulation, interface, modeling, domain, models, numerical, time-domain, simulations, method, frequency, power, modelling, non-linear, time, energy, flow, finite, wave, dynamics, proposed, analysis, phase, surface, control

A Kruskal-Wallis test was used to assess the variability of similarity distributions between each V&V types and adaptive technology types. The test revealed significant differences in the distributions of similarity scores among the sets ($H(5) = 449.92$, $p < 2.2e-16$). Subsequent Dunn's post-hoc tests, with Bonferroni correction for multiple comparisons, indicated that the null hypothesis that there is no difference in the distribution of similarity values was rejected on all pair comparisons, all with very low p values. The box plot of Fig. 3(A) shows that the combination verification methods and of adaptive software has the highest similarity distribution among the other combinations.

5.4   Content Analysis

Content analysis was performed by extracting the 100 most similar documents in the similarity comparison between source and target sets. Source document

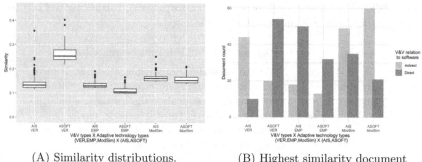

(A) Similarity distributions.

(B) Highest similarity document
count per V&V sub-themes.

Fig. 3. (A) Distribution of similarity values for all six combination of V&V types (VER, EMP, and ModSim) by adaptive technology types (AIS, and ASOFT). (B) Frequency counts following the content analysis of highly-similar documents. Acronyms are: Adaptive instructional systems (AIS), Adaptive software (ASOFT), software verification (VER), empirical validation (EMP) and modelling and simulation (ModSim).

sets were defined as referring to verification and validation methods. The target documents were either the set of document under adaptive instructional systems or adaptive software. Table 1 presents the criteria for classifying a document has having either a direct, or an indirect verification and validation relation to the software object. Document records that did not satisfy one of those two labels were classified as "other".

Figure 3(B) shows a bar chart giving the count of documents being either in a direct or indirect relationship to the verification and validation themes. Direct relation being when the verification and validation methods are being applied to either an adaptive instructional system or an adaptive software. The figure shows a higher number of verification ASOFT documents than AIS documents directly addressing the issue of software verification. This result is congruent with the distribution of high-similarity documents presented in Fig. 3(A). AIS validation by empirical means, or of model and simulation show a higher number of documents addressing theses issues than for ASOFT. Keeping in mind the subjective classification with its limitations, the analysis indicates that the subset of AIS classified literature shows a diminishing order of attention from empirical validation, to model and simulation validation, and finally software verification. The order for ASOFT is software verification, empirical verification, and last, model and simulation validation.

A Chi-squared statistic of $\chi^2 = 148.7$ with 10 degrees of freedom was obtained on the contingency table behind Fig. 3(B). The p-value for this test was <2.2e−16, which is below the significance level of 0.001. Listing all references in each categories is beyond the scope and available space of the current paper. However, the next paragraphs provide a sense of the high-similarity values and most recent documents revealed by the content analysis.

Software Verification and Adaptive Instructional Systems. Among the 100 high-similarity AIS documents with the software verification set, very few were addressing directly the issue of verifying AIS software. This short list includes the application of formal methods in human-computer interaction to adaptive instructional systems [9], formal semantic model for verification of property preservation in model-driven agent-based software development [18], mixture of software engineering techniques for verification, and both qualitative and quantitative techniques for validation [2], and verification of an ITS multi-agent plans based on model checking and Petri nets [8]. The group of indirect references to verification issues are mostly about learning software engineering techniques. The tutoring applications include ensuring source code quality, detecting anomalies, and identifying potential improvements [35], representation of code with attentional neural network for tutoring programming [17], development of procedural knowledge for analyzing and refactoring code [15,27], recommendation for fault prediction [7], tutor for the semantics of logic programs and deductive databases [3], and an intelligent tutoring system for learning security [21].

Software Verification and Adaptive Software. The ASOFT documents had the larger number of high-similarity documents with the software verification set. This result is congruent with the distribution of high-similarity documents presented in Fig. 3 (A). Example of publications directly addressing the issue of software verification includes domain-specific language for tracing expressions for both static formal verification and runtime verification [11], design and verification of self-adaptive software using finite state machines [25], runtime verification of probabilistic systems [13], and automatic construction of tests for evaluating code correctness [16].

Empirical Verification and Adaptive Instructional Systems. Example of publications about the empirical validation of AIS include evaluating the role of artificial intelligence in surgical skills training [10], evaluation of a neural network algorithm to dynamically identifies learning difficulties [29], comparison of personalized content with non-adaptive material [34], and video recommendation system for algebra [26]. The body of AIS literature related to empirical validation also contains a number of literature reviews [1,5,12,23,45].

Empirical Verification and Adaptive Software. Example of publications about the empirical validation of ASOFT include measuring user experience of adaptive user interfaces using EEG [14], assessing walking ability using a robotic gait trainer [28], non-intrusive stress detection in off-the-shelf desktop environments [46], and emotion recognition using sensor data fusion [48].

Modelling and Simulation and Adaptive Instructional Systems. Example of publications about models and simulations validation of AIS include latent variable modelling for tracking learning flow during computer-interactive artificial tutoring [22], differentiated learning environment for children with learning disabilities [44], dynamics of self-regulation, emotion, grit, and student performance in cyber-learning environments [24], and knowledge level tracking with data-driven student models and collaborative filtering [6].

Modelling and Simulation and Adaptive Software. Example of publications about models and simulations validation of ASOFT include: model-based adaptive user interface based on context and user experience [19], evaluation metrics of an adaptive user interface with Bayesian user models [32], usability evaluation of adaptive web interface using user goal achievements [33], and ontology-based approach for user interface adaptation [41].

6   Discussion

TF-TDF Values. The main result from the text mining analysis of the TF-IDF values distribution is that the group of five document sets have different distributions, but that among those sets, the distribution of AIS and ASOFT TF-IDF values did not differ significantly, which suggests that the two sets are topically less diverse than the other document sets. This does not mean that AIS and ASOFT use the same words, but that the variability of words between the sets is not statistically different.

Table 2 by listing the token/words with the highest TF-IDF values gives an indication of the topical content of each document set. Overall the text mining method seems to be able to capture the general content of each set. However, the current TF-IDF method is limited and other methods could lead to a better model of the corpus set content. It is also worth mentioning that the tokenizer algorithm employ in the current study did not unify singular and plural word forms which adds to the number of tokens.

Within Document Sets Similarity. Overall the low median similarity values for all document sets, indicates a high variability of topic/tokens within each set. In spite of the verification document set (VER) having the highest (0.188), and the empirical studies (EMP) having the lowest (0.098) median similarity values, all document sets have low internal similarity. So in spite of the token/words having high TF-TDF values listed in Table 2, there are many other tokens/words and concepts being present in the title, abstract and keywords of the corpus that are not shared among documents.

Between Document Sets Similarity. The results obtained by comparing between document sets similarities hinges the most on the the main objectives of this study. The text mining literature review intended to answer the following two questions:

Q.1 What is the distribution of academic publications for three sub-themes of verification and validation being 1) software verification, 2) empirical validation, and 3) modelling and simulation in the domain of adaptive instructional systems and adaptive software?

Q.2 Given the observed distribution, are there any trends and research gaps that can be identified?

The similarity distributions of Fig. 3(A) and (B), as well as the statistical tests on the distribution indicate a difference in attention given to verification and validation issues between publications in adaptive instructional system (AIS) and adaptive software (ASOFT). Software verification being addressed more in the literature on adaptive software than in the literature on adaptive instructional systems. This result is supported by both the similarity measures and the conceptual analysis.

On the other hand, the emphasis on empirical, and modelling and simulation validation seems stronger in the AIS than in the ASOFT literature. However, the difference in this case does not appear to be as important than in the case with software verification. The stronger emphasis on empirical, and model and simulation validation in the AIS literature certainly points to the high dependency of considering human factors in the success of adaptive instructional systems.

In regard to software verification in adaptive instructional systems, the few AIS publications addressing this issue in comparison to publications looking at tutoring software engineering techniques indicate a possible research and practical gap to explore. The brief overview and superficial conceptual analysis of the extended number of publications in software verification of ASOFT did not provide an immediate and clear set of methods and techniques that could be applied in the context of AIS. A further analysis could look in more details at the literature on runtime verification methods as well as the AI development engineering literature, the verification or AI systems [17,35,47], and on the use of simulation and testing for complex autonomous systems verification [37].

7   Conclusion

Even if software engineering, qualitative and quantitative techniques have been used for the verification intelligent tutoring systems for some time [2,8,9,18], the amount of research addressing the issue of verification in AIS appears under represented in the literature. To address these challenges, V&V methods for adaptive systems may need to incorporate more iterative, continuous, and multifaceted approaches, such as using simulation environments, incorporating user feedback loops, applying runtime verification and validation, and ensuring transparency and explainability for automated adaptive decision-making processes.

As instances of a human-centred artificial intelligence systems, adaptive instructional systems are also facing issues of reliability, safety, and trustworthiness, which can be addressed through: 1) reliable systems based on sound software engineering practices, 2) safety culture through business management strategies, and 3) trustworthy certification by independent oversight [36]. Current technical challenges affecting adaptive instructional systems research and development include data availability, data quality, personalization, interoperability, scalability, and explainability [39]. These challenges are similar to those faced by other types of autonomous systems like self-driving cars, where complexity is inherent in environment perception, data interpretation, decision-making, and knowledge management [37]. This complexity affects both the reliability and trustworthiness of autonomous systems. Sifakis and Harel argue that traditional model-based techniques are defeated by the complexity of the problem, while solutions based on end-to-end machine learning fail to provide the necessary trustworthiness. They propose a hybrid design approach combining model-based and machine learning techniques, using simulation and testing for complex autonomous systems verification and validation [37].

References

1. Acikgul Firat, E., Firat, S.: Web 3.0 in learning environments: a systematic review. Turk. Online J. Dist. Educ. **22**(1), 148–169 (2021). https://doi.org/10.17718/TOJDE.849898
2. Aguilar, R., Muñoz, V., Noda, M., Bruno, A., Moreno, L.: Verification and validation of an intelligent tutorial system. Expert Syst. Appl. **35**(3), 677–685 (2008). https://doi.org/10.1016/j.eswa.2007.07.024
3. Barker, S., Douglas, P.: An intelligent tutoring system for program semantics, vol. 1, pp. 482–487 (2005). https://doi.org/10.1109/itcc.2005.82
4. Bell, B., Sottilare, R.: Adaptation vectors for instructional agents. In: Sottilare, R.A., Schwarz, J. (eds.) HCII 2019. LNCS, vol. 11597, pp. 3–14. Springer, Cham (2019). https://doi.org/10.1007/978-3-030-22341-0_1
5. Bodily, R., Verbert, K.: Review of research on student-facing learning analytics dashboards and educational recommender systems. IEEE Trans. Learn. Technol. **10**(4), 405–418 (2017). https://doi.org/10.1109/TLT.2017.2740172
6. Cully, A., Demiris, Y.: Online knowledge level tracking with data-driven student models and collaborative filtering. IEEE Trans. Knowl. Data Eng. **32**(10), 2000–2013 (2020). https://doi.org/10.1109/TKDE.2019.2912367
7. Das Dôres, S., Alves, L., Ruiz, D., Barros, R.: A meta-learning framework for algorithm recommendation in software fault prediction, vol. 04-08-April-2016, pp. 1486–1491 (2016). https://doi.org/10.1145/2851613.2851788
8. Oliveira de Almeida, H., Dias da Silva, L., Perkusich, A., de Barros Costa, E.: A formal approach for the modelling and verification of multiagent plans based on model checking and petri nets. In: Choren, R., Garcia, A., Lucena, C., Romanovsky, A. (eds.) SELMAS 2004. LNCS, vol. 3390, pp. 162–179. Springer, Heidelberg (2005). https://doi.org/10.1007/978-3-540-31846-0_10
9. Emond, B.: Formal methods in human-computer interaction and adaptive instructional systems. In: Sottilare, R.A., Schwarz, J. (eds.) HCII 2021. LNCS, vol. 12792, pp. 183–198. Springer, Cham (2021). https://doi.org/10.1007/978-3-030-77857-6_12
10. Fazlollahi, A.: AI in surgical curriculum design and unintended outcomes for technical competencies in simulation training. JAMA network open **6**(9), e2334658 (2023). https://doi.org/10.1001/jamanetworkopen.2023.34658

11. Ferrando, A., Dennis, L., Cardoso, R., Fisher, M., Ancona, D., Mascardi, V.: Toward a holistic approach to verification and validation of autonomous cognitive systems. ACM Trans. Softw. Eng. Methodol. **30**(4) (2021). https://doi.org/10.1145/3447246

12. Fontaine, G., Cossette, S., Maheu-Cadotte, M.A., Mailhot, T., Deschênes, M.F., Mathieu-Dupuis, G.: Effectiveness of adaptive e-learning environments on knowledge, competence, and behavior in health professionals and students: protocol for a systematic review and meta-analysis. JMIR Res. Protocols **6**(7) (2017). https://doi.org/10.2196/resprot.8085

13. Forejt, V., Kwiatkowska, M., Parker, D., Qu, H., Ujma, M.: Incremental runtime verification of probabilistic systems. In: Qadeer, S., Tasiran, S. (eds.) RV 2012. LNCS, vol. 7687, pp. 314–319. Springer, Heidelberg (2013). https://doi.org/10.1007/978-3-642-35632-2_30

14. Gaspar-Figueiredo, D., Abrahao, S., Insfran, E., Vanderdonckt, J.: Measuring user experience of adaptive user interfaces using EEG: a replication study, pp. 52–61 (2023). https://doi.org/10.1145/3593434.3593452

15. Haendler, T., Neumann, G., Smirnov, F.: RefacTutor: an interactive tutoring system for software refactoring. In: Lane, H.C., Zvacek, S., Uhomoibhi, J. (eds.) CSEDU 2019. CCIS, vol. 1220, pp. 236–261. Springer, Cham (2020). https://doi.org/10.1007/978-3-030-58459-7_12

16. Heitmeyer, C.: A model-based approach to testing software for critical behavior and properties. In: Petrenko, A., Simão, A., Maldonado, J.C. (eds.) ICTSS 2010. LNCS, vol. 6435, p. 15. Springer, Heidelberg (2010). https://doi.org/10.1007/978-3-642-16573-3_2

17. Hoq, M., Chilla, S., Ranjbar, M., Brusilovsky, P., Akram, B.: SANN: programming code representation using attention neural network with optimized subtree extraction, pp. 783–792 (2023). https://doi.org/10.1145/3583780.3615047

18. Hou, J.: Formal semantic model for agent-based software system (2010). https://doi.org/10.1109/DBTA.2010.5658949

19. Hussain, J., et al.: Model-based adaptive user interface based on context and user experience evaluation. J. Multimodal User Interfaces **12**(1), 1–16 (2018). https://doi.org/10.1007/s12193-018-0258-2

20. Huynh, N.T., Segarra, M.T., Beugnard, A.: A development process based on variability modeling for building adaptive software architectures, pp. 1715–1718 (2016). https://doi.org/10.15439/2016F170

21. Imtiaz, S., Sultana, K., Varde, A.: Mining learner-friendly security patterns from huge published histories of software applications for an intelligent tutoring system in secure coding, pp. 4869–4876 (2021). https://doi.org/10.1109/BigData52589.2021.9671757

22. Kang, H.A., Sales, A., Whittaker, T.: Flow with an intelligent tutor: a latent variable modeling approach to tracking flow during artificial tutoring. Behav. Res. Methods (2023). https://doi.org/10.3758/s13428-022-02041-w

23. Koedinger, K., Aleven, V.: An interview reflection on "intelligent tutoring goes to school in the big city". Int. J. Artif. Intell. Educ. **26**(1), 13–24 (2016). https://doi.org/10.1007/s40593-015-0082-8

24. Kooken, J., Zaini, R., Arroyo, I.: Simulating the dynamics of self-regulation, emotion, grit, and student performance in cyber-learning environments. Metacogn. Learn. **16**(2), 367–405 (2021). https://doi.org/10.1007/s11409-020-09252-6

25. Lee, E., Kim, Y.G., Seo, Y.D., Seol, K., Baik, D.K.: RINGA: design and verification of finite state machine for self-adaptive software at runtime. Inf. Softw. Technol. **93**, 200–222 (2018). https://doi.org/10.1016/j.infsof.2017.09.008

26. Leite, W., et al.: Heterogeneity of treatment effects of a video recommendation system for algebra, pp. 12–23 (2022). https://doi.org/10.1145/3491140.3528275

27. Luburić, N., et al.: Clean code tutoring: makings of a foundation, vol. 1, pp. 137–148 (2022). https://doi.org/10.5220/0010800900003182

28. Maggioni, S., Lünenburger, L., Riener, R., Curt, A., Bolliger, M., Melendez-Calderon, A.: Assessing walking ability using a robotic gait trainer: opportunities and limitations of assist-as-needed control in spinal cord injury. J. NeuroEng. Rehabil. **20**(1) (2023). https://doi.org/10.1186/s12984-023-01226-4

29. Menor, J.: Design, development and effectiveness of an intelligent tutoring system using neural network, vol. 2602 (2023). https://doi.org/10.1063/5.0125246

30. O'Mara-Eves, A., Thomas, J., McNaught, J., Miwa, M., Ananiadou, S.: Using text mining for study identification in systematic reviews: a systematic review of current approaches. Systems Control Found. Appl. **4**(1), 5 (2015). https://doi.org/10.1186/2046-4053-4-5

31. Rájaraman, A., Ullman, J.D.: Data Mining, p. 1–17. Cambridge University Press (2011)

32. Rebai, R., Maalej, M., Mahfoudhi, A., Abid, M.: Bayesian user modeling: evaluation metrics of an adaptive user interface, vol. 10341 (2017). https://doi.org/10.1117/12.2268568

33. Rim, R., Mohamed Amin, M., Adel, M., Mohamed, A.: Evaluation method for an adaptive web interface: GOMS model. In: Madureira, A.M., Abraham, A., Gamboa, D., Novais, P. (eds.) ISDA 2016. AISC, vol. 557, pp. 116–124. Springer, Cham (2017). https://doi.org/10.1007/978-3-319-53480-0_12

34. Sancenon, V., et al.: A new web-based personalized learning system improves student learning outcomes. Int. J. Virtual Pers. Learn. Environ. **12**(1) (2022). https://doi.org/10.4018/IJVPLE.295306

35. Sghaier, O., Sahraoui, H.: A multi-step learning approach to assist code review, pp. 450–460 (2023). https://doi.org/10.1109/SANER56733.2023.00049

36. Shneiderman, B.: Bridging the gap between ethics and practice: guidelines for reliable, safe, and trustworthy human-centered AI systems. ACM Trans. Interact. Intell. Syst. **10**(4), 1–31 (2020). https://doi.org/10.1145/3419764

37. Sifakis, J., Harel, D.: Trustworthy autonomous system development. ACM Trans. Embed. Comput. Syst. **22**(3), 1–24 (2023). https://doi.org/10.1145/3545178

38. Sottilare, R., Knowles, A., Goodell, J.: Representing functional relationships of adaptive instructional systems in a conceptual model. In: Sottilare, R.A., Schwarz, J. (eds.) HCII 2020. LNCS, vol. 12214, pp. 176–186. Springer, Cham (2020). https://doi.org/10.1007/978-3-030-50788-6_13

39. Sottilare, R.A.: Establishing an effective adaptive instructional systems community of practice. In: Sottilare, R.A., Schwarz, J. (eds.) HCII 2023. LNCS, vol. 14044, pp. 76–93. Springer, Cham (2023). https://doi.org/10.1007/978-3-031-34735-1_6

40. Sottilare, R.A., Barr, A., Graesser, A., Hu, X., Robson, R.: Exploring the opportunities and benefits of standards for adaptive instructional systems (AISs). In: Sottilare, R.A., Barr, A., Graesser, A., Hu, X., Robson, R. (eds.) Proceedings of the 1st Adaptive Instructional System (AIS) Standards Workshop. U.S. Army Research Laboratory (2018)

41. Soui, M., Diab, S., Ouni, A., Essayeh, A., Abed, M.: An ontology-based approach for user interface adaptation. In: Shakhovska, N. (ed.) Advances in Intelligent Systems and Computing. AISC, vol. 512, pp. 199–215. Springer, Cham (2017). https://doi.org/10.1007/978-3-319-45991-2_13

42. Sundaram, G., Berleant, D.: Automating systematic literature reviews with natural language processing and text mining: a systematic literature review. In: Yang, X.S., Sherratt, R.S., Dey, N., Joshi, A. (eds.) ICICT 2023. LNNS, vol. 693, pp. 73–92. Springer, Singapore (2023). https://doi.org/10.1007/978-981-99-3243-6_7

43. Thakur, K., Kumar, V.: Application of text mining techniques on scholarly research articles: methods and tools. New Rev. Acad. Librariansh. **28**(3), 279–302 (2022). https://doi.org/10.1080/13614533.2021.1918190

44. Thapliyal, M., Ahuja, N., Shankar, A., Cheng, X., Kumar, M.: A differentiated learning environment in domain model for learning disabled learners. J. Comput. High. Educ. **34**(1), 60–82 (2022). https://doi.org/10.1007/s12528-021-09278-y

45. Wang, H., et al.: Examining the applications of intelligent tutoring systems in real educational contexts: a systematic literature review from the social experiment perspective. Educ. Inf. Technol. **28**(7), 9113–9148 (2023). https://doi.org/10.1007/s10639-022-11555-x

46. Wang, J., Yang, C., Fu, E., Ngai, G., Leong, H.: Is your mouse attracted by your eyes: non-intrusive stress detection in off-the-shelf desktop environments. Eng. Appl. Artif. Intell. **123** (2023). https://doi.org/10.1016/j.engappai.2023.106495

47. Wang, X., Tang, X., Dong, Z., Zhen, L.: Research on rapid development platform of plc control system. High Technol. Lett. **27**(2), 210–217 (2021). https://doi.org/10.3772/j.issn.1006-6748.2021.02.012

48. Younis, E., Zaki, S., Kanjo, E., Houssein, E.: Evaluating ensemble learning methods for multi-modal emotion recognition using sensor data fusion. Sensors **22**(15) (2022). https://doi.org/10.3390/s22155611

Constructing Compelling Persuasive Messages: A Pilot Study Among University Students Assessing Three Persuasive Technology Strategies

Fidelia A. Orji[1]($\boxtimes$), Francisco J. Gutierrez[2], and Julita Vassileva[1]

[1] Department of Computer Science, University of Saskatchewan, Saskatoon, Canada
fidelia.orji@usask.ca, jiv@cs.usask.ca
[2] Department of Computer Science, University of Chile, Santiago, Chile
frgutier@dcc.uchile.cl

Abstract. Persuasive messages and communications are powerful tools for influencing individuals' behaviours in a specific manner. A range of persuasive strategies are employed to construct persuasive messages to make them more effective. In this research, we investigated the persuasiveness of messages constructed based on three persuasive strategies (social comparison, commitment and consistency, and self-monitoring) in encouraging students to engage actively with their learning management system (LMS) in a real university course-based setting. We constructed persuasive messages based on the strategies and then conducted a survey study among university students who use LMS for a course to evaluate the messages in terms of perceived persuasiveness measured using four factors: motivational, effective, appropriate, and convincing. Our results suggest that messages exploiting the three strategies could be used to influence student engagement with online systems and that message content significantly impacts persuasiveness. Also, the results showed that messages employing the self-monitoring strategy were more persuasive, followed by commitment and consistency, and then social comparison. This study's findings shed light on the varying degrees of effectiveness of the three persuasive technology strategies in education. Additionally, the research highlights the importance of tailoring persuasive messages to specific target audiences and behaviours. The implications of these findings for persuasive technology design and research are discussed.

Keywords: Persuasive Strategies · Persuasive Messages · Social Comparison · Commitment and Consistency · Self-Monitoring · Student Engagement · Online Learning · Persuasive Technology

1 Introduction

Online learning has become a prevalent and convenient mode of education in the 21st century, especially in higher education. However, online learning also poses many challenges and barriers for students, such as a lack of motivation, engagement, interaction, feedback, and support. To overcome these challenges and enhance students' online

R. A. Sottilare and J. Schwarz (Eds.): HCII 2024, LNCS 14727, pp. 44–58, 2024.
https://doi.org/10.1007/978-3-031-60609-0_4

learning outcomes, persuasive technology can be a useful and effective approach. Persuasive technology (PT) is the use of interactive systems to influence human attitudes and behaviours [1]. Persuasion, in its various forms, plays a vital role in influencing human attitudes, decisions, and behaviours across diverse domains, from marketing and advertising to public health campaigns and educational initiatives. In the domain of education, persuasive technology can be used to design and deliver persuasive messages that aim to motivate and persuade students to adopt desirable learning behaviours and outcomes. However, not all persuasive messages are equally effective. Different PT strategies may have different effects on different target audiences and behaviours. Therefore, it is important to understand how to design and evaluate persuasive messages based on the characteristics and preferences of the target users and the goals and contexts of the target behaviours.

Persuasive messages intend to change or reinforce the target user's attitude or behaviour toward a certain issue or action. Persuasive messages can use various strategies and techniques to influence the target user, such as PT strategies, argumentation, emotion, humour, and storytelling. They can also be delivered through various media and channels, such as text, audio, video, multimedia, web, mobile, and social media. Several researchers have examined the effects of persuasive messages in various domains. For example, Goh et al. [2] investigated the influence of persuasive messages on students' self-regulated learning (measured using the Motivated Strategies for Learning Questionnaire). The findings from the evaluation of the messages revealed that they improved students' self-regulated learning. Anagnostopoulou et al. [3] applied persuasive messages that exploit three PT strategies (suggestion, social comparison, self-monitoring) and tailored them based on users' persuadability profiles to influence them to change their mobility behaviour and choose more sustainable options. The findings of the study revealed instances of behaviour change.

Persuasive technology has been widely applied to influence people for specific goals. In that respect, successful persuasive messaging interventions revealed their effectiveness at motivating behaviour change. However, there is still a gap in knowledge on how to employ the three persuasive strategies explored in this study in constructing persuasive messages for improving student engagement in education and learning contexts. To advance research in this area, this study answers the following research questions:

RQ1: What types of persuasive messages could be applied to improve student engagement? Do students perceive all the messages as being persuasive?

RQ2: Do the perceived persuasiveness factors, namely motivational, effective, appropriate, and convincing, differ among messages employing various persuasive strategies?

RQ3: Do the overall perceived persuasiveness of the messages differ among the persuasive strategies?

Our findings advance existing knowledge in the area of persuasive messaging intervention by offering the following main contributions. First, we constructed persuasive messages based on three PT strategies: social comparison, commitment and consistency, and self-monitoring. Second, we conducted a survey among university students to evaluate the messages in terms of perceived persuasiveness, using four factors. Third, we compared the perceived persuasiveness of each message to determine whether messages differ in their persuasiveness. Finally, we evaluate the persuasiveness of messages in each strategy and show that they differ significantly in their overall persuasiveness.

The rest of this paper is organized as follows: Sect. 2 provides related work. Section 3 describes the methodology of this study, including message construction, survey design, data collection, and data analysis. Section 4 presents the results of this study, including statistical analysis. Section 5 discusses the findings and their implications for persuasive technology design and research. Section 6 suggests some directions for future work and Sect. 7 presents the conclusion.

2   Related Work

In this section, we present a brief overview of the persuasive strategies and application of persuasive messaging interventions.

2.1 Persuasive Strategies

Researchers have explored the impact of various persuasive strategies in different domains [4, 5]; however, not all the strategies are relevant and applicable to the education domain. In this paper, we focus on three PT strategies widely used and studied in persuasive technology that are relevant and applicable to the education domain: social comparison, commitment and consistency, and self-monitoring. The social comparison and self-monitoring strategies were selected from Oinas-Kukkonen's twenty-eight design principles [6], while commitment and consistency was selected from the eight persuasive strategies developed by Cialdini [7].

Social comparison is a persuasive technology strategy that uses information about the behaviour or performance of others to influence the target user [8]. Social comparison can be classified into two types: upward and downward. Upward social comparison is when the target user compares themselves with others who are better or superior in some aspect. Downward social comparison is when the target user compares themselves with others who are worse or inferior in some aspect. The main mechanism behind social comparison is social influence [9]. Social influence is the process by which the target user's thoughts, feelings, or actions are affected by the presence or actions of others. Social influence can operate through different principles, such as conformity, compliance, obedience, normative pressure, and information pressure. It can also trigger different psychological processes in the target user, such as motivation, competition, cooperation, emulation, and imitation.

Previous research has shown that a persuasive system that implemented social comparison as a visualization using assessment grades of students inspired them to improve their engagement in learning activities [10]. Social comparison can be used to design persuasive messages that aim to change or reinforce the target user's attitude or behaviour toward a certain issue or action.

Commitment and consistency is a persuasive strategy that increases the target user's sense of obligation and responsibility to perform a behaviour [7]. Previous research has revealed the effectiveness of commitment and consistency strategies in various domains. For instance, Guéguen and Jacob [11] conducted experiments demonstrating how small initial commitments can lead to larger, more significant commitments over time, a phenomenon known as the "foot-in-the-door" technique. In the environmental domain, studies by Katzev and Wang [12] have revealed that commitment strategy could provide desired and lasting changes across a wide range of environmental behaviours. A persuasive message that reminds students of their commitment to education and learning and the benefits of fulfilling it could reinforce their initial commitments to study and succeed.

Self-monitoring as a strategy helps the target user track and evaluate their behaviour and performance [6]. Self-monitoring can be based on various types of user data, such as activity, time, location, and mood. It can use different methods and techniques to collect and display user data, such as sensors, logs, messages, dashboards, and graphs. The main mechanism behind self-monitoring is feedback. Feedback is the information that the target user receives about their behaviour or performance concerning a goal or a standard. Feedback can affect the target user's awareness, motivation, and learning. It can also affect the target user's self-regulation, self-efficacy, and self-improvement. The main effect of self-monitoring on persuasive messages is that it can increase the perceived persuasiveness of the messages by enhancing the goal setting and attainment of the target user [13]. Goal setting is the process by which the target user specifies a desired outcome or state. Goal attainment is the process by which the target user achieves or approaches the desired outcome or state. Self-monitoring can increase goal setting by enabling the target user to define and refine their goals based on their data. Self-monitoring can increase goal attainment by enabling the target user to monitor and adjust their progress and performance based on their data. It could improve learners' awareness, motivation, self-regulation, self-efficacy, and self-improvement for online learning. Research in the domain of persuasive technology has demonstrated the potential of self-monitoring to drive behavioural change [14, 15]. Mobile applications and wearable devices, such as fitness trackers, use self-monitoring to encourage physical activity, healthier eating, and other positive behaviours.

While these three persuasive strategies—social comparison, commitment and consistency, and self-monitoring—have been explored in various contexts, their application and effectiveness in the specific context of university students' engagement and behaviours remain relatively unexplored. This research aims to bridge this gap by investigating how these strategies can be employed through persuasive messages to positively engage and influence university students. By building upon and extending the existing body of knowledge, this study provides insights into the persuasiveness of these strategies in the education domain.

2.2 Persuasive Messages

One of the most compelling aspects of persuasive messages is their ability to shape opinions, influence decisions, and evoke emotions. The messages are carefully constructed, using language, visuals, and psychological techniques to appeal to the target audience's needs, desires, and values. They tap into the human mind, fostering a sense of connection and understanding between the sender and receiver, ultimately prompting the desired response. In the advertising domain, persuasive messages are the driving force behind the success of countless brands. They create narratives that engage customers, forming bonds and loyalty. Political campaigns rely on persuasive messages to sway voters, employing strategies that tap into voters' concerns and aspirations. In interpersonal relationships, persuasive messages can resolve conflicts, build trust, and motivate positive change.

The success of persuasive messages centers on several key elements. First and foremost, they must be clear, concise, and tailored to the intended audience. Understanding the audience's values, beliefs, and emotions is fundamental to constructing a relevant message. Moreover, credibility plays a vital role in the effectiveness of persuasive messages. When the sender is perceived as trustworthy and knowledgeable, the audience is more likely to be influenced. Research has shown that persuasive message interventions could be utilized to inspire people for a specific task. For example, Kaptein et al. [16] in a study to reduce snacking tailored persuasive messages to participants based on their susceptibility to persuasive strategies. The study reported a higher decrease in snack consumption in the tailored group than in the randomized messages group. Hirsh et al. [5] investigated the influence of persuasive messages in advertising and suggested that adapting the messages to people's personality traits has the potential to increase the impact of the messages. The effect of persuasive messages in motivating drivers to register for National Health Service Organ Donor was explored by Sallis et al. [17]. Persuasive messages were added to the prompt that drivers receive at the end of road tax payment transactions online. The messages increased registration in the experimental group than in the control.

In summary, persuasive messaging has gained popularity as a novel approach for encouraging behaviour change. The effectiveness of many persuasive interventions depends on the use of appropriate strategies for a specific user group and context. Employing an unsuitable strategy can serve as a significant obstacle to change [16]. Thus, in this paper, we compare the persuasiveness of three strategies in the education context to determine their suitability for students' education and learning.

3   Study Design and Methods

Our study was designed to offer insights about the ability of persuasive messages employing three distinct strategies in motivating students to actively engage with the Learning Management System (LMS) for their educational needs.

To achieve our research objectives, we first constructed sets of persuasive messages for the three persuasive strategies that we selected. Second, we developed a survey to evaluate the persuasiveness of the messages. We adopted the perceived persuasiveness scale for measuring the persuasiveness of messages. The scale is a 5-point Likert scale [18] that measures the persuasiveness of messages based on four factors motivational, effective, appropriate, and convincing. Third, university students taking a course that used LMS for course content were recruited for this study. A total of 48 students participated in this study.

3.1 Persuasive Message Creation

We constructed 18 messages aimed to promote student engagement with learning activities. Six messages were constructed to reflect social comparison, 7 messages employed commitment and consistency, and 5 messages employed self-monitoring. The construction of the messages closely imitates how the strategies were operationalized in persuasive systems and messages in the literature [3, 18, 19] and they relate to the learning context. The message content and format varied according to the assigned strategy. The messages are shown in Table 1.

The messages are aimed to influence the students' learning behaviours, such as 1) logging in to the LMS regularly and checking the course updates, announcements, and feedback; 2) accessing and studying the course materials, resources, and multimedia content; 3) completing the course assignments, quizzes, and exams on time and with quality; 4) participating in course lab sessions, discussion forums, and chats with peers and instructors; and 5) reviewing course progress and grades.

Table 1. Persuasive Messages grouped according to PT Strategies

Category	Messages	Persuasiveness Mean
Social Comparison	70% of your peers in this course performed more learning activities than you this week to improve their programming skills and academic performance	2.80
	Unlock your full potential for this course! 80% of the high-achieving classmates actively engage with learning contents, lab sessions, and assignments to stay ahead in their studies	4.12
	This week many of your peers increased their interaction with course materials and discussion forums to gain a deeper comprehension of the subject matter	2.81
	The top 10% of your peers with excellent grades interact and engage with course materials and discussion forums on this platform regularly. Increase your interaction with course materials and discussion forums to enhance your knowledge and grade	3.69

(continued)

Table 1. (*continued*)

Category	Messages	Persuasiveness Mean
	See how your peers are excelling academically (getting 90% and above in assignments) by actively engaging with course materials, lab sessions, and discussion forums. You can catch up with them by increasing your engagement with learning activities	2.36
	Your frequency of access to this course materials, lab sessions, and discussion forum is below the class average. Increasing your access and engagement with the materials will greatly enhance your knowledge and grade	2.09
Commitment and Consistency	By registering for this course, you have committed to improving your knowledge and skills in programming, so you should devote reasonable time every week to study for this course and perform your assignments	2.14
	Based on your activities on U-cursor, it appears that you were not very committed to this course. Increasing your engagement with this course materials, hands-on lab sessions, and assignments will help you to improve your programming skills and academic performance	3.47
	As you want to perform well in this course, you should show your commitment by accessing and studying your course materials and starting to work on your assignments early on rather than scrambling at the last minute	3.10
	By embracing this platform's course materials, assessments, and discussions, you can actively participate in your learning journey, resulting in higher grades and a well-rounded understanding of the subject matter	4.12

(*continued*)

Table 1. (*continued*)

Category	Messages	Persuasiveness Mean
	While merely attending classes may yield average results, utilizing the platform's tools and activities empowers you to take charge of your education, deepen your understanding, and ultimately achieve remarkable academic achievements	3.22
	Your activities on this platform are investments in your future success. By embracing the platform's opportunities for discussion, practice, and additional learning materials, you will be able to bridge any knowledge gaps and elevate your performance to new heights	4.34
	Do not settle for mediocre grades when you can unlock your full potential. By dedicating time to explore the course materials, discussion forum, and hands-on lab sessions for this course, you can supplement your classroom learning, reinforce key concepts, and significantly boost your academic performance	2.79
Self-Monitoring	We noticed that your activities for this course dropped since the previous week. By dedicating time to explore the course materials, the discussion forum, and the hands-on lab sessions for this course, you can enhance your understanding, improve your grades, and develop valuable skills that will set you apart from your peers	3.61
	Your average activity level for this course is low. Utilize more course materials on this platform and engage in discussion forums and hands-on lab sessions to enhance your understanding and grades	2.53

(*continued*)

Table 1. (*continued*)

Category	Messages	Persuasiveness Mean
	Think of your grades as a reflection of your past efforts. By improving your learning activities for this course, you will be able to bridge any knowledge gaps and elevate your performance to new heights	4.25
	You are doing great! Remember, each slight increase in engagement with your learning activities brings you closer to bridging any knowledge gaps and elevating your performance	4.58
	Remember, every assignment completed, every concept grasped, and every challenge overcome is a testament to your dedication and resilience. Keep going!"	4.64

3.2 Participants Recruitment

Our study received ethics approval from the University of Chile and the University of Saskatchewan ethics board. Forty-eight participants who used LMS for a course were recruited from the University of Chile. Our participants were at least 18 years old at the time of data collection; 12 were female (25%), 30 were male (62.5%) and other genders were 6 (12.5%). The average age of the participants is 19.6 years. The participants did not receive any compensation for participation. The persuasive messages were translated and validated in Spanish by experts to ensure that they conveyed the desired meanings. An online survey about the messages was created on a LMS and the participants were asked to rate the 18 messages based on their ability to encourage them to engage in their learning content on the LMS. After reading each message, the participants answered a questionnaire that measured their perceived persuasiveness of the message. The participants rated the perceived persuasiveness of the messages on four factors: motivational, effective, appropriate, and convincing.

3.3 Analysis

The main aims of this study are: 1) to examine the persuasiveness of the messages in encouraging students to improve their engagement in learning activities. 2) to investigate whether overall persuasiveness and the factors differ among messages employing various persuasive strategies. To achieve this, we used well-known statistical analysis techniques. The various steps taken to analyze our data include:

1. We calculated the average score for the four factors of persuasiveness for each message and we computed the overall persuasiveness based on the average of the four factors.

2. We evaluate the overall and constructs level reliability of our data using the intraclass correlation coefficient (ICC) [20, 21] to validate the messages.
3. To evaluate and compare the persuasiveness of the messages based on strategies, we ran the Friedman test, which is the non-parametric counterpart to the repeated-measures ANOVA (Analysis of Variance); i.e., it is commonly used in performing repeated-measures ANOVA on the same samples when data is not normally distributed. The latter was verified through the Shapiro-Wilk test.

4 Results

Based on the mean of persuasiveness factors, the last two messages on the social comparison and the first message on commitment and consistency were removed from further analysis as their average rating for each factor was below the median of the rating scale. Thus, a total of 15 messages qualified for further analysis. Figure 1 presents the overall persuasiveness of messages according to the strategies employed in their design. The intraclass correlation coefficient (ICC) was computed in SPSS using a 95% confidence interval, average measure, absolute agreement, and 2-way random-effects model with 48 raters across 15 messages. An excellent degree of reliability was found for overall and construct-level persuasiveness measurements. The average measure ICC for overall persuasiveness was .988 with a 95% confidence interval from .983 to .992 ($F_{(59,2773)}$ = 82.50, $p < .0001$). For constructs' reliability, the average measure of ICC for the persuasiveness of social comparison messages was .983 with a 95% confidence interval from .968 to 993 ($F_{(15, 705)}$ = 58.81, $p < .001$). The average measure of ICC for the persuasiveness of commitment and consistency messages was .981 with a 95% confidence interval from .969 to 991 ($F_{(23, 1081)}$ = 57.91, $p < .001$). The average measure of ICC for the persuasiveness of self-monitoring messages was .992 with a 95% confidence interval from .986 to 996 ($F_{(19, 893)}$ = 121.66, $p < .001$).

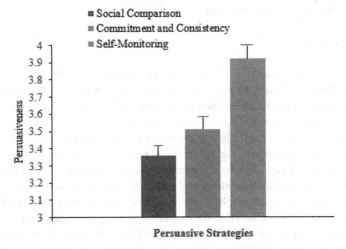

Fig. 1. Overall Persuasiveness of the Messages according to PT Strategies

Comparing the Persuasiveness of Messages Based on the Strategies. The results of the Friedman test show significant main effects of strategy type ($\chi^2(2) = 78.0, p < .0001$) for overall persuasiveness. This means that there are significant differences between the messages in the three strategies with respect to persuasiveness. The results of Post hoc analysis with Wilcoxon signed-rank tests and Bonferroni correction for multiple comparisons revealed significant differences in messages in terms of overall persuasiveness between Commitment and Social Comparison ($p < .043$), Self-Monitoring and Social Comparison ($p < .0001$), and Self-Monitoring and Commitment ($p < .0001$).

Furthermore, Friedman test results reveal significant main effects of strategy type for the persuasiveness factors of the messages:

- motivational ($\chi^2(2) = 62.09$, $p < .001$), effective ($\chi^2(2) = 31.70$, $p < .001$), appropriate ($\chi^2(2) = 61.46$, $p < .001$), and convincing ($\chi^2(2) = 73.08$, $p < .001$).
- The results of a post-hoc test using Bonferroni correction for multiple comparisons revealed significant differences in messages in terms of persuasiveness factor motivational between Self-Monitoring and Social Comparison ($p < .0001$), and Self-Monitoring and Commitment ($p < .0001$). The difference between Commitment and Social Comparison was not significant ($p = .248$).
- For the effective factor, significant differences exist between Self-Monitoring and Social Comparison ($p < .0001$), and Self-Monitoring and Commitment ($p < .0001$). The difference between Commitment and Social Comparison was not significant ($p = .1000$).
- For the appropriate factor, significant differences exist between Self-Monitoring and Social Comparison ($p < .0001$), and Self-Monitoring and Commitment ($p < .0001$). The difference between Commitment and Social Comparison was not significant ($p = .554$).
- For the Convincing factor, significant differences exist between Self-Monitoring and Social Comparison ($p < .0001$), Self-Monitoring and Commitment ($p < .0001$), and Commitment and Social Comparison ($p < .0001$).

5   Discussion

In this section, we discuss the implications of our findings, the effectiveness of each strategy, and the broader significance of our research in promoting positive behavioural changes among students. Our study findings reveal the effectiveness of the three strategies in influencing university students' engagement. The findings showed that the difference in the persuasiveness of the messages among the strategies is statistically significant. The messages based on self-monitoring are the most persuasive, followed by commitment and then social comparison. This means that empowering students with tracking of their academic progress, time management, and goal achievement, will play a crucial role in improving their motivation levels. This is consistent with some previous studies that found self-monitoring to be an effective persuasive strategy for behaviour change [14, 15]. The personalized nature of self-monitoring will allow students to set and track their own goals, which in turn, will boost their self-determination and intrinsic motivation.

Commitment and consistency emerged as another impressive persuasive strategy. Our finding is in line with prior research demonstrating the efficacy of commitment and consistency principles in fostering behavioural change [18]. Commitment and consistency increases users' sense of obligation and responsibility to perform the behaviour, which can strengthen their motivation and accountability. Messages encouraging students to make small initial commitments to their academic goals and consistently adhere to them demonstrated a remarkable impact. This suggests that by nurturing a sense of personal commitment, students were more likely to persist in their academic endeavours.

The persuasiveness of PT strategies is context and domain-dependent. Previous studies have shown that social comparison can be a powerful persuasive strategy for behaviour change [22, 23]. The social comparison messages that demonstrate learning progress and academic success of peers or embrace desirable study habits are particularly effective in motivating students to improve their engagement. However, the social comparison messages used in this study are the least persuasive among the three strategies. One possible explanation for this finding is that students will be more inspired by self-monitoring and commitment messages than social comparison messages.

The findings of this research hold several practical implications for educational institutions and those seeking to influence student behaviour positively. By tailoring persuasive messages to incorporate self-monitoring, commitment and consistency, and social comparison, educators and policymakers can develop more effective interventions to enhance student engagement and academic performance. The implications of these findings for persuasive technology design and research are as follows:

1. Persuasive technology designers should consider using self-monitoring and commitment strategies to design persuasive messages that aim to improve students' online learning outcomes. These strategies can increase students' motivation, feedback, accountability, and self-regulation for online learning.
2. Persuasive technology researchers should further investigate the factors that influence the effectiveness of different persuasive technology strategies on different target audiences and behaviours. These factors may include users' characteristics, preferences, goals, contexts, and emotions.

This research demonstrated that persuasive messages can be a useful and effective approach for improving students' engagement in online learning. By using these three strategies to design and deliver persuasive messages, online learning systems can influence students' attitudes and behaviours toward education and learning. This can enhance students' motivation, engagement, interaction, feedback, and support for online learning. This can also improve students' learning processes and outcomes regarding knowledge acquisition, skill development, and performance achievement.

6 Limitations

One limitation of our research study is that it was not conducted over a long-term period, preventing a comprehensive assessment of the effects of the persuasive strategies employed. Additionally, the selection of participants from a highly specific pool introduces a limitation as it may limit the generalizability of our findings to broader populations. Another limitation is the relatively small sample size of 48 participants, which may constrain the statistical power and generalizability of our results. Furthermore, our study relies on self-reported measures of persuasiveness for motivational appeal, which may not fully capture the actual effectiveness of the messages, particularly when implemented over an extended period within persuasive education systems. Although self-reporting is the predominant way of assessing persuasiveness, we acknowledge that the actual persuasiveness of the messages may differ when used for a long time in persuasive education systems.

7 Future Work

The main directions for future work are:

1. Designing and conducting experiments with the messages to determine their real effectiveness.
2. Conducting more studies with larger and more diverse samples of online learners from different institutions and contexts. This can increase the external validity and generalizability of the findings.
3. Exploring different message delivery modes (e.g., audio, video, multimedia) and different message exposure times (e.g., multiple times, spaced intervals). This can examine the optimal design and timing of the messages for different target audiences and behaviours.
4. Using more subjective and qualitative measures of online learning outcomes (e.g. interaction logs and physiological measures). This can capture the students' perceptions, experiences, and reflections on their online learning behaviours and outcomes.

8 Conclusion

This research constructed and evaluated persuasive messages based on three persuasive technology strategies: social comparison, commitment and consistency, and self-monitoring. The target behaviour was to improve students' engagement with their learning management system (LMS). The research conducted a survey among university students to measure their perceived persuasiveness in terms of four factors: motivational, effective, appropriate, and convincing. The research found that some of the messages could effectively promote students' engagement in education and learning. Also, the findings from the research revealed that self-monitoring was the most persuasive strategy, followed by commitment and consistency, and then social comparison. The messages could be adapted to create a significant positive effect on students' attitudes and intentions to adopt target behaviour.

Acknowledgments. This work was supported by NSERC through the Vanier Canada Graduate Scholarship CGV-175722 and Michael Smith Foreign Study Supplement of the first author, Vicerrectoría de Investigación y Desarrollo (VID), Universidad de Chile, grant: Ayuda de Viaje VID 2023 of the second author and the Discovery Grant program RGPIN-2021-03521 of the third author.

Disclosure of Interests. The authors have no competing interests to declare that are relevant to the content of this article.

References

1. Fogg, B.J.: Persuasive Technology: Using Computers to Change What We Think and Do. Morgan Kaufmann Publishers (2003). https://doi.org/10.1016/B978-1-55860-643-2.X5000-8
2. Goh, T.T., Seet, B.C., Chen, N.S.: The impact of persuasive SMS on students' self-regulated learning. Br. J. Educ. Technol. **43**, 624–640 (2012). https://doi.org/10.1111/j.1467-8535.2011.01236.x
3. Anagnostopoulou, E., et al.: From mobility patterns to behavioural change: leveraging travel behaviour and personality profiles to nudge for sustainable transportation. J. Intell. Inf. Syst. **54**, 157–178 (2020). https://doi.org/10.1007/S10844-018-0528-1
4. Orji, R., Moffatt, K.: Persuasive technology for health and wellness: state-of-the-art and emerging trends. Health Inform. J. **24**, 66–91 (2018). https://doi.org/10.1177/1460458216650979
5. Hirsh, J.B., Kang, S.K., Bodenhausen, G.V.: Personalized persuasion: tailoring persuasive appeals to recipients' personality traits. Psychol. Sci. **23**, 578–581 (2012). https://doi.org/10.1177/0956797611436349
6. Oinas-Kukkonen, H., Harjumaa, M.: Persuasive systems design: key issues, process model and system features. Commun. Assoc. Inf. Syst. **24**, 485–500 (2009). https://doi.org/10.17705/1cais.02428
7. Cialdini, R.: Influence: The Psychology of Persuasion. William Morrow and Company, New York (1984)
8. Festinger, L.: A theory of social comparison processes. Hum. Relat. **7**, 117–140 (1954). https://doi.org/10.1177/001872675400700202
9. Cialdini, R.B., Goldstein, N.J.: Social influence: compliance and conformity. Annu. Rev. Psychol. **55**, 591–621 (2004). https://doi.org/10.1146/annurev.psych.55.090902.142015
10. Orji, F.A., Vassileva, J.: A comparative evaluation of the effect of social comparison, competition, and social learning in persuasive technology on learning. In: Cristea, A.I., Troussas, C. (eds.) ITS 2021. LNCS, vol. 12677, pp. 369–375. Springer, Cham (2021). https://doi.org/10.1007/978-3-030-80421-3_41
11. Guéguen, N., Jacob, C.: Fund-raising on the web: the effect of an electronic foot-in-the-door on donation. Cyberpsychol. Behav. **4**, 705–709 (2001). https://doi.org/10.1089/109493101753376650
12. Katzev, R., Wang, T.: Can commitment change behavior? A case study of environmental actions. J. Soc. Behav. Personal. **9**, 13–26 (1994)
13. Munson, S.A., Consolvo, S.: Exploring goal-setting, rewards, self-monitoring, and sharing to motivate physical activity. In: 2012 6th International Conference on Pervasive Computing Technologies for Healthcare Workshop, PervasiveHealth 2012, pp. 25–32 (2012). https://doi.org/10.4108/icst.pervasivehealth.2012.248691
14. Orji, R., Nacke, L.E., Di Marco, C.: Towards personality-driven persuasive health games and gamified systems. In: Conference on Human Factors in Computing Systems – Proceedings, pp. 1015–1027. ACM Press, New York (2017). https://doi.org/10.1145/3025453.3025577
15. Şahin, M., Yurdugül, H.: Learners' needs in online learning environments and third generation learning management systems (LMS 3.0). Technol. Knowl. Learn. **27**, 33–48 (2022). https://doi.org/10.1007/s10758-020-09479-x
16. Kaptein, M., De Ruyter, B., Markopoulos, P., Aarts, E.: Adaptive persuasive systems: a study of tailored persuasive text messages to reduce snacking. ACM Trans. Interact. Intell. Syst. **2**(2), 1–25 (2012). https://doi.org/10.1145/2209310.2209313
17. Sallis, A., Harper, H., Sanders, M.: Effect of persuasive messages on national health service organ donor registrations: a pragmatic quasi-randomised controlled trial with one million UK road taxpayers. Trials **19** (2018). https://doi.org/10.1186/s13063-018-2855-5

18. Josekutty Thomas, R., Masthoff, J., Oren, N.: Adapting healthy eating messages to personality. In: de Vries, P.W., Oinas-Kukkonen, H., Siemons, L., Beerlage-de Jong, N., van Gemert-Pijnen, L. (eds.) PERSUASIVE 2017. LNCS, vol. 10171, pp. 119–132. Springer, Cham (2017). https://doi.org/10.1007/978-3-319-55134-0_10
19. Orji, F.A., Vassileva, J., Greer, J.: Evaluating a persuasive intervention for engagement in a large university class. Int. J. Artif. Intell. Educ. **31**(4), 700–725 (2021). https://doi.org/10.1007/S40593-021-00260-4
20. Shrout, P.E., Fleiss, J.L.: Intraclass correlations: uses in assessing rater reliability. Psychol. Bull. **86**, 420–428 (1979). https://doi.org/10.1037//0033-2909.86.2.420
21. Koo, T.K., Li, M.Y.: A guideline of selecting and reporting intraclass correlation coefficients for reliability research. J. Chiropr. Med. **15**, 155 (2016). https://doi.org/10.1016/J.JCM.2016.02.012
22. Orji, F.A., Greer, J., Vassileva, J.: Exploring the effectiveness of socially-oriented persuasive strategies in education. In: Oinas-Kukkonen, H., Win, K., Karapanos, E., Karppinen, P., Kyza, E. (eds.) PERSUASIVE 2019. LNCS, vol. 11433, pp. 297–309. Springer, Cham (2019). https://doi.org/10.1007/978-3-030-17287-9_24
23. Christy, K.R., Fox, J.: Leaderboards in a virtual classroom: a test of stereotype threat and social comparison explanations for women's math performance. Comput. Educ. **78**, 66–77 (2014). https://doi.org/10.1016/j.compedu.2014.05.005

Receptivity: Fostering Respectful Conversations in the Digital Public Sphere

Fatemeh Sarshartehrani⊙, Aaron Ansell⊙, Rafael Patrick⊙,
Denis Gracanin$^{(\boxtimes)}$⊙, and Sylvester Johnson⊙

Virginia Tech, Blacksburg, VA 24061, USA
{fatemehst,aansell,rncp,gracanin,saj240}@vt.edu

Abstract. The rapid advancement of digital technologies has reshaped educational environments, presenting both opportunities and challenges in enhancing student engagement and communication. Despite these technological strides, a critical gap remains in facilitating effective and respectful discourse, escalated by the rise of toxic discourse. This paper introduces the Receptivity, a novel educational technology concept designed to bridge this gap through an innovative, color-coded feedback system that allows teachers and students to communicate emotional and cognitive states in real time, fostering a respectful and inclusive conversational climate. Grounded in human-centered design principles and the concept of reflexivity in student engagement, Receptivity emphasizes the importance of immediate, non-intrusive feedback mechanisms. Our methodology employs an iterative development process, incorporating direct user engagement and expert consultations to refine the app's functionalities. Preliminary testing indicates that Receptivity enhances the quality of educational discussions, promoting a more adaptive learning environment. This work contributes to the discourse on digital behavior change interventions and student engagement, offering a fresh perspective on leveraging technology to improve educational outcomes. The paper discusses the implications of Receptivity for teaching and learning practices, highlighting its potential to transform educational experiences through enhanced communication and reflexivity.

Keywords: Human-Centered Design · Adaptive Instruction · Real-Time Feedback · Digital Discourse · Inclusive Communication · Student Engagement

1 Introduction

In the Digital age, educational technologies have evolved to address challenges in learning environments, yet gaps in promoting effective and respectful communication persist. The Receptivity app emerges as a novel solution, specifically designed to bridge these gaps by introducing an innovative approach to feedback and engagement in educational settings. Unlike existing technologies that passively facilitate communication, Receptivity actively empowers users—teachers

R. A. Sottilare and J. Schwarz (Eds.): HCII 2024, LNCS 14727, pp. 59–71, 2024.
https://doi.org/10.1007/978-3-031-60609-0_5

and students alike—to express their emotional and cognitive states through a simple, yet powerful, color-coded signaling system. This system addresses the urgent need for immediate, non-intrusive feedback mechanisms.

The rise of toxic discourse, both online and offline, presents a significant challenge in educational contexts, often leading to confrontational exchanges that undermine the quality of discussion. This issue is particularly detrimental where respectful dialogue is essential for effective learning and teaching. Recognizing this, our research focuses on mitigating the negative impacts of such toxic discourse, promoting a healthier conversational climate that is conducive to learning. The Receptivity app's design is informed by the gaps identified in current educational technologies, offering a real-time feedback system that is absent elsewhere.

Parallel to our focus on mitigating toxic discourse in educational environments, Warner et al. [17] investigate the efficacy of proactive content moderation through a mobile keyboard application across different communication platforms. Their study highlights the impact of design decisions—timing, friction, and AI model presentation—on user engagement and moderation effectiveness. This work emphasizes the potential of integrated digital tools not only to reduce toxicity but also to serve as educational instruments by prompting users to reflect on their communication choices.

The Receptivity app not only aims to reduce toxicity but also serves as an educational tool, encouraging users to reflect on their communication choices, thereby enhancing the inclusivity and productivity of discussions. This is significant as it fosters respectful interactions, reduces conflict, and encourages meaningful participation across a diverse spectrum of users, including those from minority communities who may be disproportionately affected by toxic discourse.

This work contributes to the existing body of research by providing a practical application of theories in student engagement and digital communication enhancement. It extends the concept of 'intentional causality' in learning, where students actively shape their engagement through the app, and addresses a gap in current educational technology by presenting a novel, real-time feedback system that enhances the reflexivity of learners. By aligning with recent advances in digital behavior change interventions and the human-centered design approach, Receptivity offers a fresh perspective on fostering engagement through technology in educational settings.

2   Related Work

In the realm of educational research, student engagement has long been acknowledged as a pivotal element in determining both the effectiveness of teaching methodologies and the depth of student learning. Traditional views of engagement have primarily centered on the visible, interactive aspects of the learning environment, emphasizing collaborative classroom dynamics as key drivers of student involvement [10]. This perspective, while invaluable, often overlooks the subtle, internal cognitive mechanisms that influence a student's engagement trajectory.

Recent shifts in educational theory, however, have brought a newfound focus on the concept of reflexivity in student engagement. This shift, largely influenced by the work of Peter Kahn [9], pivots from an external, interaction-based view to an introspective, self-driven perspective. Kahn's concept of "intentional causality" posits that true engagement is a product of a student's deliberate and conscious efforts to align their personal cognitive processes with their educational environment. Applying Margaret Archer's theories of reflexivity, Kahn is optimistic about measures that cultivate the right kind of student "self-talk," which encourages students to engage in a mental dialogue that positively influences their learning behaviors [1,9,18]. This theoretical evolution underscores the importance of internal cognitive and emotional processes in shaping student engagement, thus offering a more holistic understanding of what it means to be truly engaged in learning.

Building upon the concept of reflexivity in student engagement, the work of Rivera Muñoz et al. [13] introduces a compelling exploration into adaptive learning technologies. They demonstrate the transformative potential of these technologies for personalizing the educational journey in higher education. By adapting learning paths, resources, and activities to meet the individual needs and preferences of students, adaptive learning systems fill a crucial gap in providing immediate, adaptive feedback mechanisms. This personalization not only caters to the diverse learning styles and paces of students but also aligns with the introspective, self-driven perspective of student engagement, further enhancing the quality of learning outcomes.

The integration of digital technologies into educational settings marks a pedagogical revolution, paralleling theoretical advancements in educational research. The exploration of wearable technologies, such as those studied by Bauer et al. [3], and mobile applications designed to enhance cognitive skills, as shown by Asiri et al. [2], underscore the capacity of technology to create adaptive, personalized learning environments. This technological progression facilitates not only enhanced learning outcomes but also improved emotional well-being among learners, as indicated by Chen et al. [5] through the use of biometric data tracking.

Recent studies also emphasize the importance of personalizing feedback mechanisms in educational technologies. For instance, research exploring feedback strategies in the context of fraction arithmetic tasks has revealed that learner characteristics, including gender, knowledge level, and motivation, significantly influence how students respond to different types of feedback [14]. This finding supports the development of adaptive educational technologies that consider learner profiles to optimize engagement and learning outcomes.

Building on this foundation, the significance of addressing mental health through technological interventions becomes apparent. Research on mobile technologies for supporting mental health in youths [12] emphasizes the potential of these applications to improve access to mental health resources, particularly for underrepresented groups. This aligns with the Receptivity app's objectives, demonstrating the importance of incorporating emotional and cognitive considerations into educational technology to promote a healthier, more inclusive conversational climate.

Similarly, the utilization of peer-to-peer feedback mechanisms in residency programs shows the value of incorporating diverse feedback sources to enhance learning outcomes. These studies reveal that peer feedback, despite potential barriers such as insufficient training and perceived negative consequences, provides unique and valuable insights into the learning process, complementing traditional assessment methods [6].

Additionally, the impact of nonverbal communication on educational experiences further enriches our understanding of engagement and learning environments. Studies highlight how nonverbal cues, from teacher gestures to the physical setup of classrooms, significantly influence student perceptions and engagement [4,8]. Integrating insights from this research into educational practices can substantially enhance teacher-student interactions, making the case for a multifaceted approach to education that leverages both digital technology and an understanding of nonverbal communication to foster more effective and engaging learning experiences.

The research by Evmenova et al. [7] on designing a smartwatch application for young adults with intellectual and developmental disabilities (IDD) presents an innovative approach to fostering inclusivity in post-secondary education. By focusing on wearable technology, the study not only addresses the specific needs of students with IDD but also enhances their learning participation and independence. This aligns with the broader educational goal of utilizing technology to create adaptive and responsive learning environments, thereby supporting all students' educational experiences effectively. This integration emphasizes the critical role of technology in enhancing educational inclusivity and supporting diverse learner needs.

Our paper seeks to contribute to this innovative intersection of reflexivity-focused educational theory and digital technology. We present a detailed exploration of the engagement effects of a novel technological intervention—a Color-Coded Non-Verbal Feedback (CNVF) system. This system is designed to elicit and incorporate students' reflexive responses concerning their comprehension levels and preferences in lecture pacing and content delivery. By doing so, we aim to operationalize the theoretical concept of reflexivity in a practical, technology-driven context.

This research not only aims to enrich the theoretical discourse on student engagement but also endeavors to empirically demonstrate the transformative potential of integrating reflexive pedagogical approaches with cutting-edge digital tools. Through this, the study aspires to contribute significantly to the evolving narrative of student engagement in the digital age, offering insights and practical implications for educators, technologists, and theorists alike.

3   Methodology

3.1   Design Principles

The development of Receptivity was deeply informed by human-centered design principles, a framework that emphasizes the importance of tailoring technological solutions to the needs, capabilities, and contexts of end-users [11]. In the case of Receptivity, the primary users are teachers and students, groups that demand a tool that not only facilitates but also enriches the quality of conversations in digital and educational environments. The human-centered design approach adopted in our project guides every aspect of application development, from conceptualization to deployment.

We followed an iterative development process that involved direct engagement with end-users to ensure that the app's features met their actual needs. Our initial conceptualization was followed by prototyping, and we continuously sought user feedback and expert consultations to inform each iteration. This process was vital in refining the app's color-coded feedback system, which is a distinctive feature designed to enable participants to express their readiness to engage in discussions, thereby promoting a more inclusive conversation environment.

At the core of our design process was an application that transcends the mere functionality of facilitating discussions, aiming instead to enhance the inclusivity and productivity of these conversations. This goal was pursued by closely adhering to the design principles articulated by Donald Norman, which advocate for an empathetic and user-oriented approach to technology development [15]. However, we ensured that each aspect of Receptivity supported a practical purpose in addition to enriching the user experience in meaningful ways, rather than just incorporating these principles into a checklist.

To address the real-world challenges faced by users in engaging in respectful and constructive dialogue, especially on contentious topics, we focused on features that address the principle of usefulness. The app's color-coded signaling system, for example, was designed to meet the practical need for a simple yet effective way for participants to express their readiness to engage in various levels of discourse, thus facilitating a more inclusive environment for conversation.

As part of the overall goal of ensuring user inclusion, both usability and accessibility were considered equally important. Recognizing the diversity of our user base in terms of age, technological proficiency, and physical abilities, we prioritized an intuitive interface design. This was achieved through iterative testing and refinement, ensuring that navigation and interaction with the app remain straightforward for all users, including those with disabilities. The design process was imbued with an understanding that technology should adapt to its users, not the other way around.

The desirability of the app was also a key consideration, reflecting our belief that engagement with technology is significantly influenced by the user's emotional response to it [16]. Investing in a visually pleasing design and ensuring that the user experience is emotionally satisfying will foster a positive relationship between the app and its users, encouraging sustained engagement.

In ensuring the findability of features within the app, we addressed the need for an interface that allows users to quickly locate and use the functionalities they need without frustration or confusion. This aspect of design is crucial for maintaining the flow of conversation and ensuring that the technology serves as a seamless facilitator of discourse.

In the development of Receptivity, we ensured the application's credibility and trustworthiness. Given the sensitivity of privacy and data security, our approach was to engineer an application that inherently respects user privacy and autonomy. Central to achieving this was the decision to minimize the collection of personal data. Users are only required to choose a username, a process that empowers them to control their identity within the app. To further enhance the privacy of our users, we assign a unique user ID to each account, an approach that preserves user anonymity while enabling us to maintain the integrity of the app's functionality.

Understanding the potential need for users to participate in discussions without revealing their identity, we incorporated a feature that allows users to anonymize their responses. This functionality is pivotal in providing a safe space for open and honest signaling, particularly on sensitive or divisive topics. By design, this approach places Receptivity at the forefront of privacy-conscious applications, ensuring that the app is a trusted tool for fostering respectful and inclusive conversations.

Our application was designed with a human-centered approach to not only achieve its objectives but also align with users' needs and values.

3.2 Development Process

The development of Receptivity was characterized by an iterative, human-centered approach, heavily reliant on continuous engagement with teachers and students. This process was not linear but cyclical, allowing for constant refinements based on users' feedback and expert consultations. Each stage of development served as a foundation for the next, ensuring that the application not only met its intended goals but also addressed the needs of its users in real-world scenarios.

The Receptivity project began with a conceptualization phase, where the idea originated from a recognized need to improve the quality of discourse in educational settings. This phase involved literature reviews, and preliminary discussions with educators and students to identify the main challenges and opportunities for enhancing classroom and digital discussions. The goal was to gather a broad spectrum of perspectives on the existing discourse dynamics and the role technology could play in facilitating more respectful and constructive interactions. Through this exploratory process, we identified the critical need for a simple, intuitive mechanism that would allow participants to signal their readiness to engage in discussions at varying levels of intensity and sensitivity (Fig. 1).

Fig. 1. Early-stage interface of the Receptivity app, depicting the initial design of the Host and Audience Views with the color-coded feedback mechanism.

With a clear understanding of our users' needs, we moved into the prototyping phase. Early versions of Receptivity were developed, focusing on the core functionality of the color-coded signaling system. These prototypes were subject to internal testing, where the project team could interact with the app. This stage was crucial for identifying any technical issues early on and ensuring that the basic concept was sound and functional.

As prototypes became more refined, we initiated a series of feedback sessions involving teachers and students. These were not one-off meetings but part of a continuous feedback loop that allowed us to gather insights and suggestions for improvements directly from those who would ultimately use the app. Teachers provided invaluable input on how the app could be integrated into classroom settings to enhance engagement and inclusivity and the students offered perspective on the usability of the app and its potential to ease discussions in class. This feedback was instrumental in refining the app's features and interface, ensuring that Receptivity was both effective in its objectives and user-friendly.

Parallel to students' and teachers' feedback sessions, we engaged in expert consultations. These collaborations involved specialists in education, technology, and human-computer interaction. The experts contributed their knowledge and experience, challenging our assumptions and helping us to navigate the complex landscape of digital communication tools. This collaborative effort was vital in ensuring that Receptivity's development was not only informed by its immediate users but also grounded in the latest research and best practices in technology design and educational theory.

The development process of Receptivity was inherently iterative. Each round of feedback and testing led to refinements and sometimes significant changes to the application's design and functionality. This iterative approach allowed us to adapt flexibly to new insights and challenges as they arose, ensuring that the final product was not only technologically robust but also resonant with the needs and expectations of its users.

3.3 Testing Protocol

The Receptivity app's development was characterized by a structured testing protocol that aimed to evaluate its functionality, usability, and educational impact. This protocol was designed as a multi-phase approach, incorporating both alpha and beta testing cycles, to systematically identify areas for improvement and ensure the app's effectiveness in enhancing communication within educational settings. Despite the comprehensive nature of these initial testing phases, plans for a more detailed user study are in place to further explore the app's contributions to educational technology.

The alpha testing phase served as the app's first line of scrutiny, focusing primarily on internal evaluations with the development team and selected experts in technology and education. This phase aimed to identify technical issues, bugs, and potential usability challenges. Feedback obtained during alpha testing was critical for making significant refinements to Receptivity's design and functionality. Key areas of focus included the responsiveness of the color-coded feedback system, the app's stability across different devices, and the intuitiveness of the user interface. Modifications made in response to alpha testing feedback set the foundation for the subsequent beta testing phase.

Beta testing expanded the evaluation of Receptivity to include a broader and more diverse audience of end-users, specifically educators and students. This phase was instrumental in assessing the app's real-world usability and its impact on educational communication. Methods used during beta testing included distributed test releases to select classrooms, structured feedback sessions, and usability surveys. The feedback gathered from these sessions was invaluable for identifying user-specific needs and preferences, enabling further refinement of the app's interface and features. Beta testing also provided initial insights into how the color-coded feedback mechanism facilitated classroom interaction and engagement.

Alongside user testing, the Receptivity development team engaged with interdisciplinary experts, including scholars in mobile app development, communication, cultural studies, and educational technology. This expert feedback was solicited through a combination of structured interviews, demo sessions, and review meetings. Experts provided critical assessments of the app's design coherence, educational value, and technological innovation. Their contributions were pivotal in validating the app's conceptual underpinnings and ensuring that it adhered to high standards of academic and technological rigor.

The insights garnered from the alpha and beta testing phases, complemented by expert feedback, have laid a solid foundation for the next stages of Receptivity's development. Plans are underway for a comprehensive user study that will employ both quantitative and qualitative research methods to evaluate the app's effectiveness more systematically. This study will aim to measure the app's impact on student engagement, classroom dynamics, and educational outcomes. The development team intends to use the findings from this study to make data-driven improvements to Receptivity, ensuring that it continues to meet the evolving needs of its users in educational settings.

4   Implementation

4.1   App Overview

The Receptivity application presents a user-centric design aimed at enhancing interaction and engagement in educational settings. The interface showcases a color-coded feedback system, central to its design, which allows students to instantly communicate their comprehension and engagement levels with educators. The system uses three colors: red to signal a need to pause for questions, yellow for a request to repeat or slow down the content, and green to indicate that the audience is ready to proceed. This design translates complex emotional and cognitive feedback into a simple, universally understandable format.

4.2   Host and Audience Interface

The application features two main views: the Host view for educators and the Audience view for students.

In the Host view, educators can see aggregate feedback from the audience, which is quantified in real-time percentages. This aids educators in making data-driven decisions about lecture pacing and content depth. Educators can also monitor a list of feedback from their audience, displayed as a list with color-coded indicators against each participant's name. This feedback, represented by red, yellow, and green colors, reflects the students' requests to pause for questions, slow down, or proceed with the lecture, respectively. The Host view also provides the option to switch to 'Discussion Mode' or 'Poll Feedback', adding layers of interaction within a teaching session.

The Audience view allows individual students to provide their feedback, which contributes to the collective input displayed on the Host's screen. The interface is intentionally minimalistic to ensure clarity and ease of use during active engagement. Students participate in the feedback process by selecting their responses, which are then aggregated in the Host view and Audience view. Students can also anonymize their responses. This ensures a safe space for students to express their needs without the pressure of public identification. The option to anonymize responses even further safeguards students' privacy and encourages genuine feedback.

Receptivity enhances classroom engagement by offering features such as a simplified entry process for users with a straightforward interface for joining or hosting sessions via only session IDs and a username of their choice. Another feature worth mentioning is the participant list. In both Host and Audience views, this list provides a transparent overview of the class's feedback in real-time, fostering a sense of community and collective learning. Furthermore, users can Hide/Unhide Group Feedback which allows them to either view the group's overall feedback or focus on individual inputs. This option not only caters to diverse instructional styles and preferences but also empowers students to provide their feedback independently, without the influence or pressure of seeing others' responses. This encourages genuine individual input, free from potential bias or conformity, and promotes an environment where every student's voice can be heard and considered on its merit (Fig. 2).

Fig. 2. Receptivity Application Interface showcasing the Lecture Mode of Host and Audience Views with real-time color-coded feedback system.

4.3 Technical Details and Accessibility

From a technical standpoint, Receptivity has been meticulously engineered to be both responsive and intuitive. The app's infrastructure supports real-time data collection and analysis, ensuring immediate reflection of feedback in the lecture dynamic. Accessibility considerations have been paramount; the app's contrast and color choices have been optimized for visibility, including for those with color vision deficiencies. Furthermore, the touch targets are designed to be easily selectable, catering to users with varied motor skills, and the app's functionality remains consistent across different devices and screen sizes.

5 Discussion

This study presents the Receptivity as an innovative concept designed to enhance communication and engagement in educational settings through a novel color-coded feedback system. The development and implementation of Receptivity app were informed by a human-centered design approach, focusing on addressing the challenges of toxic discourse and the need for immediate, non-intrusive feedback mechanisms in educational environments. The iterative design process, grounded in direct engagement with teachers and students, ensured that the app was responsive to the real-world needs of its users, leading to the creation of a tool that not only facilitates but actively enriches the quality of educational discussions.

Receptivity addresses a significant gap in current educational technologies by offering a real-time feedback system that empowers users to express their emotional and cognitive states effectively. This system directly tackles the issues of toxic discourse and the lack of immediate feedback mechanisms, which have been identified as substantial barriers to effective communication in educational settings. By enabling participants to signal their readiness to engage in discussions at various levels, Receptivity fosters a more inclusive and respectful conversational climate. This approach aligns with the findings of Warner et al. [17], who highlighted the importance of design decisions in enhancing user engagement and moderation effectiveness.

The theoretical foundation of Receptivity is built upon the concept of reflexivity in student engagement, which suggests that true engagement arises from a student's intentional alignment of their cognitive processes with their educational environment. The color-coded feedback system operationalizes this concept by allowing students to actively shape their engagement, promoting a form of "intentional causality" in learning. This mechanism not only facilitates a more adaptive learning environment but also encourages students to reflect on their own learning processes, thereby deepening their engagement and participation in the educational discourse.

The implementation of Receptivity in educational settings has significant implications for teaching and learning practices. By providing educators with real-time, aggregate feedback on student comprehension and engagement levels, the app enables more informed decision-making regarding lecture pacing and content depth. This dynamic interaction between teachers and students can lead to more tailored and effective educational experiences, ultimately enhancing learning outcomes. Moreover, the ability of students to anonymize their responses encourages honest and open feedback, contributing to a safer and more inclusive learning environment.

While the development and initial testing of Receptivity have shown promising results, there are limitations to the current study that must be acknowledged. The testing protocol, though comprehensive, has yet to include a structured user study that quantitatively and qualitatively assesses the app's impact on educational outcomes. Future research should focus on conducting such a study to provide empirical evidence of Receptivity's effectiveness in enhancing student engagement and educational communication. Additionally, exploring the app's scalability and adaptability across different educational contexts and cultures would offer valuable insights into its broader applicability and potential for global impact.

6   Conclusion

Receptivity represents a significant advancement in the integration of digital technologies in education, offering a practical solution to the challenges of engagement and communication in learning environments. By leveraging the principles of human-centered design and the theoretical framework of reflexivity, the app holds the potential to transform educational practices, fostering environments that are more inclusive, respectful, and conducive to learning. As the educational landscape continues to evolve, tools like Receptivity will play a crucial role in shaping the future of student engagement and participation in the digital age.

References

1. Archer, M.S.: Making Our Way Through the World: Human Reflexivity and Social Mobility. Cambridge University Press, Cambridge (2007)
2. Asiri, Y.A., Millard, D.E., Weal, M.J.: Assessing the impact of engagement and real-time feedback in a mobile behavior change intervention for supporting critical thinking in engineering research projects. IEEE Trans. Learn. Technol. **14**(4), 445–459 (2021)
3. Bauer, M., Bräuer, C., Schuldt, J., Niemann, M., Krömker, H.: Application of wearable technology for the acquisition of learning motivation in an adaptive E-learning platform. In: Ahram, T.Z. (ed.) AHFE 2018. AISC, vol. 795, pp. 29–40. Springer, Cham (2019). https://doi.org/10.1007/978-3-319-94619-1_4
4. Beebe, S.A.: The role of nonverbal communication in education: research and theoretical perspectives. In: Annual Meeting of the Speech Communication Association. ERIC (1980)
5. Chen, J., et al.: FishBuddy: promoting student engagement in self-paced learning through wearable sensing. In: 2017 IEEE International Conference on Smart Computing (SMARTCOMP), pp. 1–9. IEEE (2017)
6. de la Cruz, M.S.D., Kopec, M.T., Wimsatt, L.A.: Resident perceptions of giving and receiving peer-to-peer feedback. J. Grad. Med. Educ. **7**(2), 208–213 (2015)
7. Evmenova, A.S., Graff, H.J., Genaro Motti, V., Giwa-Lawal, K., Zheng, H.: Designing a wearable technology intervention to support young adults with intellectual and developmental disabilities in inclusive postsecondary academic environments. J. Spec. Educ. Technol. **34**(2), 92–105 (2019)
8. Galloway, C.M.: An Analysis of Theories and Research in Nonverbal Communication. ERIC Clearinghouse on Teacher Education, Washington, D.C. (1972)
9. Kahn, P.E.: Theorising student engagement in higher education. Br. Edu. Res. J. **40**(6), 1005–1018 (2014)
10. Kuh, G.D., et al.: Excerpt from high-impact educational practices: what they are, who has access to them, and why they matter. Assoc. Am. Coll. Univ. **14**(3), 28–29 (2008)
11. Kulyk, O., Kosara, R., Urquiza, J., Wassink, I.: Human-centered aspects. In: Kerren, A., Ebert, A., Meyer, J. (eds.) Human-Centered Visualization Environments. LNCS, vol. 4417, pp. 13–75. Springer, Heidelberg (2007). https://doi.org/10.1007/978-3-540-71949-6_2
12. Litke, S.G., et al.: Mobile technologies for supporting mental health in youths: scoping review of effectiveness, limitations, and inclusivity. JMIR Mental Health **10**(1), e46949 (2023)
13. Muñoz, J.L.R., et al.: Systematic review of adaptive learning technology for learning in higher education. Eurasian J. Educ. Res. **98**(98), 221–233 (2022)

14. Narciss, S., et al.: Exploring feedback and student characteristics relevant for personalizing feedback strategies. Comput. Educ. **71**, 56–76 (2014)
15. Norman, D.: The Design of Everyday Things: Revised and Expanded Edition. Basic Books (2013)
16. O'Brien, H.L., Toms, E.G.: What is user engagement? A conceptual framework for defining user engagement with technology. J. Am. Soc. Inform. Sci. Technol. **59**(6), 938–955 (2008)
17. Warner, M., Strohmayer, A., Higgs, M., Rafiq, H., Yang, L., Coventry, L.: Key to kindness: reducing toxicity in online discourse through proactive content moderation in a mobile keyboard. arXiv preprint arXiv:2401.10627 (2024)
18. Wolters, C.A., Rosenthal, H.: The relation between students' motivational beliefs and their use of motivational regulation strategies. Int. J. Educ. Res. **33**(7–8), 801–820 (2000)

Navigating the Skies: Enhancing Military Helicopter Pilot Training Through Learning Analytics

Dirk Thijssen[✉], Pieter de Marez Oyens, and Jelke van der Pal

Royal Netherlands Aerospace Centre, Anthony Fokkerweg 2, 1059 CM Amsterdam, The Netherlands
{Dirk.Thijssen,Pieter.de.Marez.Oyens,Jelke.van.der.Pal}@nlr.nl

Abstract. Ensuring safety and proficiency in aviation requires effective training and maintenance of pilot skills. Pilots must maintain currency, implying regular flight, though some critical elements (e.g. emergency procedures) are rarely practiced in-flight. Simulator sessions are employed for safe practice and training. Simulators provide more opportunities for pilots to experience sophisticated training without safety risks and dependencies on for example weather and logistics. While human assessors currently evaluate pilot performance, a simulator can provide rich data for assessments, utilizing learning analytics for individual insights. Simulator data (and potentially aircraft data) is not regularly used yet in training to gain insights in performance. Furthermore, current training planning lacks consideration for individual needs, relying on fixed intervals and syllabi. This study aims to investigate how simulator data can be used to assess performance and if it provides valid statements about pilot performance. Thereafter, the data is used to develop a skill retention model accommodating personal differences. The research involves Chinook helicopter pilots from the Royal Netherlands Air Force, using simulator data to create tailored performance metrics for maneuvers. The ongoing study investigates the integration of simulator data into training assessments and aims to contribute valuable insights into pilot performance and skill retention. This paper presents the preliminary findings in the development of learning analytics within this context.

Keywords: Performance Assessment · Evidence-Based Training · Pilot Performance

1 Introduction

1.1 Related Work

Defence organizations are increasingly more focused on (operational) readiness, driven by a growing sense of urgency due to recent geo-political events, compelling them to be fully prepared for potential conflicts. In 2022, NATO leaders agreed on a new NATO Force Model [1, 21, 22]. This new model proposes a larger pool of high readiness forces

R. A. Sottilare and J. Schwarz (Eds.): HCII 2024, LNCS 14727, pp. 72–88, 2024.
https://doi.org/10.1007/978-3-031-60609-0_6

to improve NATO's ability to react in quick response. NATO members are commissioned to improve their operating methods in order to adhere to the new NATO Force Model. A critical component is the readiness of personnel. Readiness extends across every level of the organization, involving everyone from the military pilot, trained for specific goals and proficiency levels, to the commander, who requires thorough assessment reports and extensive data to make informed decisions regarding deployments and missions. Readiness can be subdivided into personnel, training, equipment on hand (i.e. amount) and equipment status (i.e. usability). Readiness can be assessed on an individual (e.g. pilot), team (e.g. unit) and collective level (e.g. joint forces) [2]. However, several divisions of readiness are available and can change per country and operational command.

Several studies from the RAND Corporation indicated that the current readiness assessment system falls short in effectively measuring the force's capacity to fulfill forthcoming mission requirements [2, 3, 23]. Investment in training assets is mentioned to help address the gaps in readiness assessment, specifically [2]:

- Mission Training through Distributed Simulation (MTDS)
- More simulators and new synthetic threat environments
- Aggregated force readiness measurement
- Adaptive, proficiency-driven training

The research investigates the utilization of data, particularly simulator data, for adaptive, proficiency-driven training. This field of research is known as learning analytics [4]. It involves the development of performance metrics to evaluate pilot performance. These metrics, in addition to instructor observations, help to outline trainee performance, leading to a more comprehensive assessment of readiness on an individual level.

Following this, the study explores the possibility of modeling performance over time based on the results of performance metric, which is called skill retention [5]. The integration of performance metrics and learning process modeling aims to optimize training by selecting and scheduling training tasks based on the individual needs of the trainee. This improves effectiveness of training and resource efficiency to meet the required proficiency level. Thereafter, results from a whole unit or force can be aggregated and analyzed from a commander's perspective, providing a comprehensive view of the readiness of the operating commands.

This paper presents the proposed method and initial development of performance metrics, along with preliminary results derived from the training session data. However, it does not delve into the modeling of skill retention. The performance metrics are crucial for modeling skill retention, and additional repeated measures are necessary. As data collection is ongoing, results will be published at a later stage. The research involves Chinook helicopter pilots from the Royal Netherlands Air Force (RNLAF), using simulator data to create tailored performance metrics for specific maneuvers. In the following section, we describe the Learning Analytics that may be applied to achieve this enhanced level of retention training and readiness.

1.2 Learning Analytics and Performance Metrics

Learning analytics is defined as the measurement, collection, analysis, and reporting of data on learners within their specific contexts. The primary objective is to comprehend and optimize the learning process and the learning environment [4].

The application of learning analytics further distinguishes itself based on the levels at which it operates. At one level, the focus is on individual (or team) learning processes, where performance is analyzed, providing valuable insights for self-evaluation, feedback, and performance assessment. Another level involves the analysis of data from large groups, offering valuable insights at the organizational level. Management can leverage these insights for improving learning trajectories and decision-making. In terms of military organizations, the management application is layered from managers of operational units responsible for training up to the commanders who decides on deployment and missions.

While the distinction between application at an individual or group level is well-known, another distinction is proposed between learning analytics within a training session and between multiple training sessions. Within the specific time frame of a single training session, learning analytics leverages data to provide insights into the performance of an individual. By extrapolating this functionality over multiple training sessions, aspects like skill retention can be addressed. While both types rely on similar data, distinct algorithms and analyses are required to effectively capture and assess performance trends over the course of time. For initial training, this could span weeks or months, while in continuation training, it may extend over months, years or even decades.

Types of Learning Analytics. A training session occurs with its own context, both foreseen (e.g. training scenario) and unforeseen (e.g. weather). Assessment of performance is highly dependent on this context and therefore learning analytics should be adjusted and fine-tuned for every context [6]. Military operators determine their actions based on the context, where a specific action may be correct in one context but improper in a slightly different one. The distinctions between contexts can involve subtle nuances but result in significant differences in desired behavior, particularly in critical situations. Regarding assessment, an instructor possesses the experience, knowledge and ability to recognize these differences based on their expertise and experience. An analytical model in learning analytics is less sensitive for the context and therefore it is important to provide sufficient context when interpreting the results. Models frequently tend to either overgeneralize or undergeneralize. Hence, it is crucial to determine the intended goal for using a performance metric and ensure its validity in that particular context. While there is currently no consistent framework for various types of learning analytics in military training, four distinct categories can be identified from the context of formal education: descriptive, diagnostic, predictive, and prescriptive analytics [4]. When applied to the training of a military operator, the definition slightly shifts from formal education.

Descriptive Analytics. Descriptive analytics offers insights into what takes place in the training scenario (e.g. visualizations of aircraft positioning relative to other objects). Descriptive analytics can be utilized in real-time and during an after-action review. Nevertheless, the insights presented by descriptive analytics still need interpretation from the user since they solely describe what occurs and do not assess or assign value to the observed behavior. Descriptive analytics aids instructors in gaining awareness of aspects that may be challenging to observe directly. During an after-action review, it serves to replay events and reconstruct potential causes of both success and mistakes. Its primary application is within a training session.

Diagnostic Analytics. Diagnostic analytics offer insights into why something happened [4]. Unlike descriptive analytics, diagnostic analytics involves a degree of assessment. The conclusiveness of the assessment varies for each analytic. The least conclusive form involves indicating abnormal behavior, such as a pilot deviating from procedures. This indication does not categorically state correctness or incorrectness; instead, it highlights an event that is potentially interesting for reflection. An example of a diagnostic analytic with high conclusiveness is a model that grades the execution of a turn based on the prescribed procedures. These analytics can help an instructor focus attention on intriguing events or on competencies that are complex to assess (letting the analytics handle simpler competencies). For effective use, high conclusiveness must be granted validity. The examples are mainly useful within a training session.

A diagnostic analytic can be applied in a context beyond specific actions, such as assessing the execution of a turn. This involves the assessment of competencies. Vatral et al. [7] developed a method in which diagnostic analytics, grading specific actions or performance indicators, are utilized to grade competencies. The results of multiple specific metrics relevant to a competency are combined to generate a competency score. The competency framework can be multi-layered, where competencies from a lower level can also be combined to create a higher-level competency score. The methodology expresses the score of a metric between 0 and 1, and this score is further categorized into three states (below, at, or above expectation) for a metric or competency score. Bayesian modeling is employed to combine these scores into a higher-level competency score [7]. This Bayesian scoring method is applied within a training session. However, there is also potential to use such approach between training sessions for long-term competency assessment.

Another approach for long-term competency assessment between training sessions is the use of the Elo rating system [8]. The Elo rating system, originating from competitive gaming, quantifies the skill level of a trainee. Each participant is assigned a numerical rating for each competency, and after each training session, the ratings are adjusted based on the outcome. Every training task or scenario is also provided by a numerical rating per relevant competency, which is the complexity level. The updating mechanism of the Elo rating system involves a dynamic adjustment of players' skill ratings after each match. The extent of the adjustment is determined by the outcome of the training session and the difference between the trainee's skill rating and the complexity of the training tasks. A lower-rated trainee gains more from performing well in higher-rated training tasks, while a higher-rated trainee experiences a smaller rating increase when performing well against lower-rated training tasks. The difference between the skill rating of the trainee and the complexity level of the task is utilized to calculate a probability of the win chance, expressed as a percentage, indicating the likelihood of a good performance by the trainee given the task's complexity. The adjustment, determined by a logistic function, ensures a dynamic and continuous reflection of changes in trainee's skill levels over time. The updating mechanism of the Elo rating system is visualized in Fig. 1. For a positive and effective learning curve, it is advisable to choose training tasks that align with the player's skill rating. This dynamic adjustment allows for the continuous assessment and monitoring of players' long-term competency, reflecting their evolving skill levels as they engage in more training sessions over time [8].

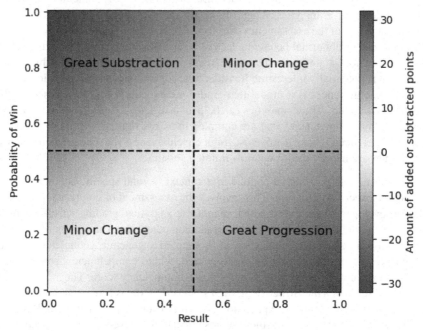

Fig. 1. A visualization of the updating mechanism of the Elo rating system.

Predictive Analytics. Predictive analytics are used to make estimates of future behavior. It combines historic data to identify patterns in the data and with the use of mathematical models and algorithms relationships between various variables can be captured in order to forecast trends [4]. These analytics are mainly useful between training sessions, for example in the prediction of skill retention. Anticipating when proficiency might diminish is advantageous for scheduling subsequent training sessions [9]. Various methodologies are employed to predict skill retention. The Predictive Performance Equation (PPE) adopts a theory-based approach, establishing a mathematical model based on: (1) the power law of learning, indicating that performance improves with practice; (2) the power law of forgetting, which suggests that performance declines over time since practice occurred; and (3) the spacing effect, emphasizing the benefits of distributed practice over time [9]. Another approach involves the utilization of machine learning algorithms, particularly those involving reinforcement learning and regression [10]. These approaches can also be integrated into hybrid models, where they complement each other [10].

Prescriptive Analytics. Prescriptive analytics provide recommendations and advice by utilizing historical data, such as scores, to identify training needs and relate them to various training options [4]. These analytics are frequently integrated into recommender systems, which can vary in their objectives and the techniques used to generate recommendations [11]. Recommendation systems can be applied across various time frames, as illustrated by the following examples. Firstly, a recommendation on very short notice involves the real-time adaptation of training scenarios based on the performance and mental state of the trainee. Performance, assessed by evaluator and/or other analytics, and mental state, determined by measures of brain activity, are used as inputs to adjust the scenario's complexity. For instance, a fighter pilot may encounter more or fewer adversary aircraft based on their performance [12]. Secondly, historical performance data can be utilized to construct the next scenario in a training. In the Pilot Training Next research program, historical assessment data and the scenarios constructed by instructors based on that data are provided to an Artificial Intelligence (AI) model. The model suggests the next training scenario by emulating the instructor's scenario construction process, considering the individual needs of each trainee, derived from the data [13]. The third example occurs on a similar timeline to the Pilot Training Next example. By employing the Elo rating system, competency scores are generated and updated over time. Competency scores that lag behind in development are identified, and subsequent training activities are recommended to address those underperforming competencies at a complexity level that matches the trainee's skill level. To facilitate this, an Experience Index is formulated, detailing which competencies are trained by each training task and at what complexity level [8].

Measures in Military Exercises. The utilization of analytics or measures is not entirely novel in the military domain. During military exercises, measures are frequently employed to obtain a more objective assessment of task execution. Two types of measures are defined: Measures of Effectiveness (MoE) and Measures of Performance (MoP). An MoE describes the desired outcomes, such as the mission objective, while an MoP characterizes the performance of an action irrespective of the overarching mission goal. These measures are established beforehand and evaluated afterward. Evaluation can be based on operator observations and opinions during debriefs, but outcomes can also be determined based on data [14]. The types of analytics mentioned in the previous section, primarily descriptive and diagnostic analytics, could also serve as input for determining the outcomes for MoE and MoP.

2   Methodology for Performance Metrics and Competency Scores

For modelling and optimizing retention training, our proposed method makes use of the methodology and results of Vatral et al. [7], which described the Generalized Intelligent Framework for Tutoring (GIFT). This research uses the Generalized Intelligent Framework for Tutoring (GIFT). GIFT is pioneering in the development of AI-based Augmented and Intelligent Tutoring Systems (IAAR-ITS) environments [7, 19, 20]. The research employs a competency-based approach, breaking down complex tasks into competencies and subtasks through cognitive task analysis, resulting in a layered competency profile. Each subsequent layer in the hierarchy further decomposes the complex task, specifying subtasks until the level of observable behavior and actions is reached. The bottom layer of the profile is made measurable through performance metrics [7, 20].

Subsequently, performance metrics developed for the bottom layer of the profile, representing observable behavior. The exact method for developing the analytics is not generic, as creating analysis models is context-dependent. The performance metrics are described in detail in Sect. 3.4. The layered competency profile allows scores at the bottom to be aggregated to higher layers. Figure 2 depicts the interactions between the different layers precisely, illustrating how a subordinate component can contribute to multiple superior components. The multi-layered competency profile that is shown is only a placeholder and will be defined at a later stage. A distinct visualization can be made for each maneuver or task [7, 20]. The orange boxes indicate that assessments scores of instructors can also be included in this method. This was not part of the initial methodology [7, 20].

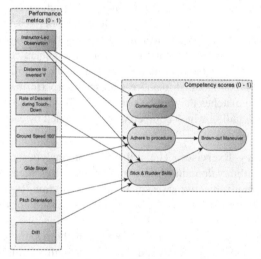

Fig. 2. An overview of the interaction between the performance metrics (left box) and the multi-layered competency profile (right box) for the brown-out maneuver.

A performance metric can thus contribute to multiple competencies [7, 20]. Watz et al. describe a criterion-based approach for developing learning analytics, defining rules and cut-off points with experts to convert measurements into assessments. An effective analysis model can be created with sufficient data and a well-defined concept of success. Watz et al. also investigate aggregating underlying components to higher-level ones, combining various measurements, both subjective and objective, into a performance score through a weighted average. Initial weights are assigned by experts and can later be refined using machine learning techniques [6]. However, Vatral et al. use a different method to aggregate scores of underlying components to higher-level ones [7, 20].

The result of a performance metric at the lowest level of the framework is a continuous number between zero and one, with three labels specified: (1) below expectation for a score below 0.4, (2) at expectation for a score between 0.4 and 0.9, and (3) above expectation for a score above 0.9. The continuous number can be based on a percentage or score. A continuous score between zero and one is used because Bayesian network models are employed to generate assessments for higher levels in the hierarchy. Bayesian networks are probabilistic models, used for predictions, representing relationships between variables. They consist of nodes representing variables and directed arrows indicating causal relationships between variables. Each node has a probability distribution representing the likelihood of each possible outcome of that variable, given the values of the variables it is linked to. These probability distributions are updated using Bayesian inference, using new data to calculate the likelihood of each possible explanation and determining which explanation is most likely. For each competency, the probability of the state (i.e. below-, at- and above-expectation) is calculated [7]. The Bayesian network model in this research consists of four models:

1. Three state competency model: the likelihood of a competence status given the result of a performance metric (i.e. below-, at- or above-expectation).
2. Transition model: determining the likelihood of a competence status changing after a training event (e.g., from below- to at-expectation).
3. Conditional model: assessing the likelihood of the higher-level competency having the same status as the underlying competencies.
4. Prior model: evaluating the probability of the team having a certain status based on prior knowledge and experience.

The method enables personalized training sessions by updating predictions within each session, optimizing training duration to maximize learning gains while minimizing diminishing returns. Additionally, it integrates into a larger ecosystem for evidence-based, data-driven trainee assessments across various scenarios and environments, ensuring consistent and reliable trend analysis regardless of training location or instructor.

3   Method

3.1   Experiment Design and Data Collection

Experiment Design. The experiment involved two groups of pilots: one conducting sorties every three months and the other every six months. The data collection period spans two years. These time intervals were selected to allow sufficient time for skill decay to occur. The discrepancy in time between the two conditions is implemented to explore differences in pilot performance after a specified duration. It is presumed that other flight experiences may impact the skill retention process. Therefore, in consultation with Subject Matter Experts (SME), three maneuvers are selected that are infrequently performed, or alternative procedures are employed to minimize the influence of other flight experiences. Additionally, the total number of flight hours and flight hours over the past year are recorded to assess whether these factors influence performance.

The three maneuvers performed are: brown-out landings in sandy environments, autorotation in case of engine failure, and instrument flying with cockpit instruments only. Variations were applied to the scenarios for each maneuver, but deviations from the proposed schedule resulted in scenarios being flown at different times or not at all.

The brown-out scenarios are the same for every training session. Four attempts are made in each training session. Participants are required to land to helicopter as close to a certain point as possible. This point is between two lamps within a formation of lamps, called inverted Y. Therefore, the brown-out maneuver only differs in timing (refresh training after three or six months).

Five scenarios are developed for the autorotation maneuver, conducted in an open field which provides plenty of landing spots. The scenarios vary in the location where they are conducted. Each training session involves one of these five scenarios. Additionally, two more complex scenarios are created, set above forests where fewer suitable landing spots are present, challenging pilots to make rapid and appropriate decisions. One complex scenario is included in training sessions four and eight for the three-month condition, and sessions two and four for the six-month condition. The simple scenario is repeated twice within a training session, while the complex scenario is executed only once.

Instrument flying is conducted in both a simple and complex scenario during every training session. The simple scenario alternates between two options for each session, while there are eight unique scenarios available for the complex scenario. Each type of scenario is performed once in each session. Besides that, the instrument flying differs in timing (refresh training after three or six months).

Data Collection. The data collection consists of saving simulator data and a questionnaire filled in by the participant. The simulator exports data in the DIS-protocol, which provides location, orientation and several speed parameters. The DIS-data is eventually converted to a CSV-file (see Sect. 3.4).

The questionnaire is filled out by the trainee after every training session. The first part is filled out before the training sessions starts. The following information is requested before the training session:

- The total amount of flight hours (rough approximation);
- The percentage of the total amount of flight hours on the CH-47F MYII CAAS (rough approximation);
- The total amount of flight hours in the last twelve months (rough approximation);
- Whether or not the participant is a staff pilot;
- Karolinska Sleepiness Scale (KSS) [17];
- A rating on self-efficacy on each maneuver between one and ten.

The following information is requested after the training session:

- A rating of perceived cognitive load on each maneuver between one and seven;
- A rating of perceived success on each maneuver between one and seven;
- A question whether the perceived cognitive load and/or success differed between the scenarios within a training session (including space to elaborate on their experience).

3.2 Apparatus

The research is conducted using the Transportable Flight Proficiency Simulator (TFPS). This high-fidelity simulator is utilized by CH-47 pilots of the RNLAF primarily for training (emergency) procedures. Efforts are underway to integrate the TFPS with other simulators, such as a second TFPS, AH-64 simulator, and Rear Crew Trainers (RCT). Consequently, tactical training will also be incorporated into the simulator's usage [15] (Fig. 3).

Fig. 3. The Transportable Flight Proficiency Simulator (TFPS) built by NAVAIR [16].

The pilots undertake a specially designed sortie for the research, featuring three maneuvers: brown-out, autorotation, and instrument flying. During the brown-out maneuver, pilots perform a landing in a sandy environment where the final phase of the landing lacks outside visuals due to swirling sand, potentially causing disorientation. While standard procedures typically involve using the autopilot, this research employs a more manual procedure. An autorotation, an emergency maneuver, is executed when the engines malfunction. Pilots initiate a steep dive to convert altitude into kinetic energy for the rotor head. As the ground approaches, this kinetic energy is utilized to reduce vertical speed and safely land the helicopter. Instrument flying restricts pilots to utilizing only the instruments within the cockpit, without relying on visual cues from outside. This practice is particularly valuable in low-visibility conditions such as nighttime or foggy weather. Arrival routes are flown using only cockpit instruments in this research setting.

3.3 Participants

Participant in this research were CH-47 pilots of the RNLAF. The complete population of active CH-47 pilots in the RNLAF is approximately sixty pilots, from which twenty pilots are only flying minimally due to other job responsibilities. In the present research 26 pilots are selected to participate. The participants are equally divided between the two conditions. As of the current report, eighteen training sessions have been completed. Specifically, one participant completed two sessions, while another completed three. Thirteen participants each completed one session. Unfortunately, data collection failed during three sessions.

3.4 Data Analysis

This paragraph describes the various activities performed for data analysis, all of which are conducted using Python.

DIS-Converter. The DIS protocol is primarily designed for connecting simulators and does not inherently support the development of learning analytics. To enable analysis, a software module was developed to convert DIS data into a CSV file format suitable for analysis. This module extracts key information such as timestamp, coordinates (i.e. x, y, z), orientation (i.e. pitch, roll, yaw), and various speed variables from each DIS message, organizing them into rows in the CSV file. This CSV file containing the extracted variables can then be utilized for the development of learning analytics.

Dashboard for Labeling. In each sortie, the three maneuvers are performed multiple times (i.e. attempt), and this data is recorded in one session per person. To create separate CSV files for each attempt, data extraction and labeling is necessary. While exploring automatic data labeling approaches, it was deemed that the effort to develop such software outweighed the time it would save. Instead, a dashboard was developed to visualize the altitude over time for each training session. The dashboard is depicted in Fig. 4. This visualization enabled the user to identify the different maneuvers and attempts based on their knowledge of the experiment. Users could then manually add starting and ending points to the timeline on the dashboard. After data collection, the recordings were processed through the dashboard, and the starting and ending points were used

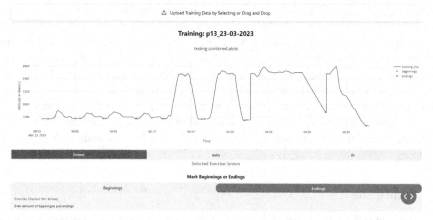

Fig. 4. Dashboard developed for data labeling, displaying time (x-axis) and altitude (y-axis) with starting (orange) and end points (green) to distinguish maneuvers. (Color figure online)

to generate distinct CSV files for each maneuver. Although the dashboard provided a solution for data labeling in a specific format with a specific goal, the lack of integrated data labeling tools or solutions in training devices remains a recurring issue [18].

Learning Analytics. Performance metrics were identified partly by consulting flight and training manuals, as they detail the execution of maneuvers and required speeds, among other factors. Also, the available data types, such as position and orientation of the helicopter, are considered. The design of the metrics was further refined through consultation with a simulator instructor, a qualified CH-47 pilot.

The performance metrics are listed below in Table 1. Each metric is accompanied by its relevance to specific maneuvers and a brief description of how it is computed. The performance metrics yield continuous scores, such as the distance in meters for the metric "distance to inverted Y." These scores are then normalized to a range between zero and one using predefined cut-off points, such as the distance at which a landing is deemed out-of-bounds. Subsequently, these normalized scores can be categorized as below, at, or above expectation levels. However, specific cut-off points are not disclosed due to classification reasons.

Table 1. The performance metrics with its relevance to the maneuver and a basic description on how they are computed.

Performance metric	Relevant maneuver	Computation
Distance to inverted Y	Brown-out	The pilots must land the helicopter at a specific location (the inverted y). The distance is calculated between the inverted y and the location the helicopter came to a standstill
Rate of Descent during Touchdown	Brown-out Autorotation	The moment of touchdown is determined. Thereafter, the vertical speed at that moment is compared to the cut-off points
Ground Speed 100'	Brown-out Autorotation	During landing the ground speed should decrease. The ground speed as reviewed at 100 ft. and 50 ft. in relation to the cut-off points
Glide Slope	Brown-out Autorotation	The glide slope in the last ten timesteps is compared to the cut-off points

(continued)

Table 1. (*continued*)

Performance metric	Relevant maneuver	Computation
Pitch Orientation	Brown-out Autorotation	The pitch in the last ten timesteps is compared to the cut-off points
Drift	Brown-out Autorotation	The roll and yaw in the last ten timesteps are compared to the cut-off points
Decision Making - Distance versus Time	Autorotation	The pilot must decide based on available landing sites between maximizing flight time or covering more distance with remaining kinetic energy after engine failure. Each option corresponds to a preferred speed range, determined by specific cut-off points
Correlation actual and suggested flight path of the approach	IFR	Every approach has a suggested flight path that must be followed. The correlation between the actual and suggested flight path is calculated indicating the adherence to the approach
Deviation from localizers	IFR	In an approach the pilot must adhere to a maximum and minimum altitude and wide based on so-called localizers. The absolute margins decrease when the runway is approached. The deviation/adherence from these minimum and maximum are calculated at several moments in the approach

4   Results

The paper presents preliminary results, with ongoing data collection and analysis. Although the analysis is still in progress, the initial version of three performance metrics, applied to the brown-out maneuver, is presented here. The algorithms have been applied to a dataset comprising fifteen training sessions. However, data from three sessions are corrupted, and efforts are underway to address this issue. Additionally, not all training sessions completed the planned four brown-out attempts. The preliminary results are presented in Table 2.

The results provide insight in the item discrimination of the performance metrics. Item discrimination is the measure of how well an item (e.g. test question, performance metric) can distinguish between those who perform well and those who do not. The

Table 2. Preliminary results for three brown-out maneuver performance metrics: scores categorized as below (0), at (1), and above (2) expectation. Mean scores are presented due to balanced category intervals, alongside standard deviation (SD) and response count (N).

		Mean	SD	N
Distance to Inverted Y	Attempt 1	0.25	0.62	12
	Attempt 2	0.33	0.65	12
	Attempt 3	0.11	0.33	9
	Attempt 4	0.63	0.92	8
Rate of Descent during Touchdown	Attempt 1	1.67	0.78	12
	Attempt 2	1.75	0.62	12
	Attempt 3	1.33	0.87	9
	Attempt 4	1.13	0.83	8
Pitch Orientation	Attempt 1	0.58	0.90	12
	Attempt 2	1.70	0.72	12
	Attempt 3	1.22	0.83	9
	Attempt 4	1.00	0.76	8

metric of 'distance to inverted Y' seems to have a poor item discrimination based on the low mean score. Approximately 80% of the attempts were labelled below expectation. This is also indicated by the relatively low standard deviation compared to the two other performance metrics.

5 Discussion

The preliminary results are limited; thus, it is premature to draw conclusions regarding the validity of the performance metrics or the pilots' performance. Nonetheless, this phase has provided valuable insights into the development and implementation of learning analytics in helicopter pilot training.

5.1 Design and Development Process of Performance Metrics

During development and computation of the performance metrics several challenges occurred. First, it was challenging to create a consistent algorithm to handle unexpected data, such as missing attempts or prematurely ended maneuvers (e.g., go-arounds). Additionally, obtaining conditional information necessary for computing performance metrics proved challenging. For instance, determining the exact moment of touchdown of the helicopter must be derived from simulator data, as there was no explicit label or variable provided. Establishing a consistent method to determine touchdown and similar information posed difficulties. Therefore, this version represents an initial attempt and may require refinement in subsequent iterations. These challenges align with the data labeling issue outlined by Bessey et al. [18]. Frequently, datasets lack standardized labeling, such as consistent start and stop times or event identification. Addressing this labeling problem enhances the data's utility and reduces the analysis workload. Possible solutions may include real-time labeling tools for simulator operators or classifier algorithms for automated data labeling, such as identifying maneuvers or helicopter touchdowns.

5.2 Item Discrimination and Validity

The preliminary results showed that item discrimination could be a potential problem regarding the validity of the performance metrics. In the continuation of this research it can be beneficial to statistically investigate the item discrimination, for example with the item-response theory.

Adjustments could enhance the usefulness of the performance metrics, such as improving item discrimination and validity. One planned adjustment is transitioning the outcome of the metric from a categorical to a continuous variable. For instance, consider the metric "distance to inverted Y," which measures the distance between the touchdown point and the intended landing spot. Currently, true/false statements check whether the distance belongs to the criteria of below, at and above expectation, resulting in a categorical variable. The true/false-statements reduces the variability massively in comparison to a continuous variable. To generate a continuous variable, the distance could be related to a minimum distance indicating perfect execution and a maximum distance indicating poor performance. This approach could yield a continuous variable ranging from zero to one, with values then categorized as below, at, or above expectation, aligning with suggestions by Vartral et al. (e.g. at expectation for a score between 0.4 and 0.9) [7, 20].

Furthermore, a review of the cut-off points could be considered. If a significant number of participants' attempts are consistently classified as below expectation for example, adjusting the cut-off points to enhance item discrimination might be advantageous. However, from an operational standpoint, altering established cut-off points could be undesirable. Doing so could potentially lead to either accommodating incompetent pilots if requirements are relaxed or losing competent pilots if requirements are made more stringent.

Additionally, it is important to assess whether the performance metrics accurately label performance compared to assessments made by instructors. In future phases of this research, training sessions will be organized wherein trainees are evaluated using the performance metrics alone, by an instructor, and by an instructor aided by the performance metrics. Examining the correlation between these three assessments can provide insights into the validity of the performance metrics.

5.3 Limitations and Future Research

One limitation of the research is the inconsistent attendance of participants, largely due to high operational demand, resulting in deviations from the proposed training schedule. This variability in the number and timing of training sessions significantly diverges from the initial plan. It is anticipated that this deviation may impact the statistical significance within tests and could potentially hinder the development of skill retention models.

One limitation of the research stems from the use of univariate analysis rather than multivariate analysis [7, 20]. The performance metrics in this study rely solely on DIS-data, which, while valuable, may be limited in capturing the full scope of activities. Future research should explore incorporating additional data types to enhance analysis.

Considerable knowledge and experience are being accumulated in regard to learning analytics, providing insights for developing new types and applications of performance metrics. It is crucial to establish a modular and expandable environment for these metrics, addressing aspects such as model and algorithm storage, version management, and the incorporation of feedback to refine results from previous iterations. Future research should focus on investigating the architecture of such an environment to support ongoing advancements in the field.

Acknowledgments. This study is part of the Adaptive Learning Ecosystem program, funded by the Dutch Ministry of Defense (grant number L2201).

Disclosure of Interests. The authors have no competing interests to declare that are relevant to the content of this article.

References

1. Wolting, J.: NAVO verandert huidige Force Model. Defensie Magazine - Landmacht 07 Artikel 8 (2022). https://magazines.defensie.nl/landmacht/2022/07/08_nieuwe-nato-force-model
2. Emmi, Y., et al.: How Training Infrastructure Can Improve Assessments of Air Force Readiness. RAND Corporation, Santa Monica (2023). https://www.rand.org/pubs/research_briefs/RBA992-1.html
3. Emmi, Y., et al.: Air Force Readiness Assessment: How Training Infrastructure Can Provide Better Information for Decisionmaking. RAND Corporation, Santa Monica (2023). https://www.rand.org/pubs/research_reports/RRA992-2.html
4. Society for Learning Analytics Research. https://www.solaresearch.org/about/what-is-learning-analytics/
5. Chittaro, L., Van der Pal, J., Oprins, E., Van Puyvelde, M., Taylor, H., Rankin, K.: Skill Fade and Competence Retention: A Contemporary Review. Brussels: North Atlantic Treaty Organization - Science and Technology Organization (2023)
6. Watz, E., Neubauer, P., Kegley, J., Bennet, W.: Managing learning and tracking performance across multiple mission sets. In: I/ITSEC (2018)
7. Vatral, C., Biswas, G., Naveeduddin, M., Goldberg, B.: Automated assessment of team performance using multimodal Bayesian learning analytics. In: I/ITSEC (2022)
8. Thijssen, D., Bosma, R.: Recommendation system in an integrated digital training environment for the 5th generation air force. In: I/ITSEC (2022)
9. Walsh, M.M., Gluck, K.A., Gunzelmann, G., Jastrzembski, T., Krusmark, M.: Evaluating the theoretic adequacy and applied potential of computational models of the spacing effect. Cogn. Sci. **42**(S3), 644–691 (2018). https://doi.org/10.1111/cogs.12602
10. Sense, F., et al.: Combining cognitive and machine learning models to mine CPR training histories for personalized predictions. Int. Educ. Data Min. Soc. (2021)
11. Uddin, I., Imran, A., Muhammad, K., Fayyaz, N., Sajjad, M.: A systematic mapping review on MOOC. IEEE Access (2021)

12. Tillema, G., Roza, M.: Data-driven and personalized training as a service infrastructure & technologies. In: I/ITSEC 2023, Orlando (2023)
13. Forrest, N.: Conceptualization and Application of Deep Learning and Applied Statistics for Flight Plan Recommendation. Air Force Institute of Technology, Ohio (2020)
14. Civil-Military Cooperation Centre of Excellence. Measures of effectiveness (MoE) and measures of performance (MoP). CIMIC Handbook (2020). https://www.handbook.cimic-coe.org/
15. Royal Netherlands Aerospace Centre. Case: Multi-Ship Multi-Type Helicopter Simulation Training Capability (n.d.). https://www.nlr.org/case/case-multi-ship/
16. Berry, T.: Transportable Flight Proficiency Simulator [Photograph]. Vertical Magazine: Bron (2021). https://verticalmag.com/press-releases/u-s-and-international-crew-members-train-on-flight-simulators-in-germany/
17. Akerstedt, T., Gillberg, M.: Subjective and objective sleepiness in the active individual. Int. J. Neurosci. **52**, 29–37 (1990)
18. Bessey, A., Waggenspack, L., Schreiber, B., Bennet, W., Jr.: Tackling the human performance data problem: a case for standardization. In: I/ITSEC (2022)
19. Goldberg, B., et al.: Forging competency and proficiency through the synthetic training environment with an experiential learning for readiness strategy. In: I/ITSEC (2021)
20. Vatral, C., Naveeduddin, M., Biswas, G., Goldberg, B.: GIFT external assessment engine for analyzing individual and team performance for dismounted battle drills. In: Ninth Annual GIFT Users Symposium (2021)
21. Twigt, A.: Task Force Defensienota 22 wil samen het verschil maken. De Vliegende Hollander (2023). https://magazines.defensie.nl/vliegendehollander/2023/09/06_successen-van-tf-22_slot
22. Monaghan, S., Wall, C., Morcos, P.: What Happened at NATO's Madrid Summit? [Critical Questions] (2022). http://tinyurl.com/444wthkm
23. Walsh, M., Taylor, W.W., Ausink, J.A.: Independent Review and Assessment of the Air Force Ready Aircrew Program: A Description of the Model Used for Sensitivity Analysis. RAND Corporation, Santa Monica (2019)

Augmented Intelligence for Instructional Systems in Simulation-Based Training

Joost van Oijen[✉]

Royal Netherlands Aerospace Centre, Amsterdam, The Netherlands
`Joost.van.Oijen@nlr.nl`

Abstract. Augmented Intelligence is a design pattern for a human-centered collaboration model of people and artificial intelligence (AI), where machines assist humans in tasks such as data analysis, information retrieval, decision-making, and task execution. In this study, the concept of Augmented Intelligence is applied within the context of an instructional system for simulation-based training. Here, the collaboration between human and machine is focused on the role of the instructor, which is to guide the learning process of one or more trainees toward some learning objective. We identify different levels of machine support to assist an instructor in this role during an adaptive training cycle. Additionally, two design aspects are discussed that contribute to increased levels of intelligence, namely the challenge of domain alignment to empower automation capabilities, and the benefits of simulation-based task environments to deliver AI-enabled approaches. Examples are discussed in the context of military training.

Keywords: Augmented Intelligence · Instructional system · Simulation · Training

1 Introduction

Simulation-based training is increasingly used as a method for training professionals on skills, abilities, and competencies in domains such as the military, aviation, or healthcare. Simulated task environments offer an interactive learning environment for trainees to experience representative real-world scenarios, while commonly being more cost-effective compared to live training. As in live training, simulation-based training is often instructor-led and supervised, where instructors prepare training scenarios, observe and assess trainee performance during training, provide feedback, and plan follow-up training needs and activities.

In current practice, instructors have little technological support for guiding simulation-based training processes. However, the increasing demand for data-driven, personalized and adaptive training has yielded promising research on supportive instructional technologies in areas such as learning analytics, human performance modelling, recommendation algorithms, and scenario generation. Fully automated approaches for adaptive training, such as Adaptive Instructional Systems (AIS), exclude the need for an instructor. However, these are often a bridge too far when considering complex domains

R. A. Sottilare and J. Schwarz (Eds.): HCII 2024, LNCS 14727, pp. 89–101, 2024.
https://doi.org/10.1007/978-3-031-60609-0_7

such as military training, which regularly involves learning complex knowledge, skills and abilities in dynamic task environments. The role of the human instructor cannot easily be disregarded as it brings rich knowledge about the task domain, individual learners, and instructional strategies that can be complex to model. Also, subjective insights may be based on years of experience, and an objective truth of 'adequate' trainee performance may not always be quantifiable by a machine. In this view, intermediate approaches seem more feasible, wherein machines support instructors for some instructional processes, while instructors retain control over others. In this paper we explore such different forms of instructional support using the concept of Augmented Intelligence.

Augmented Intelligence is "a design pattern for a human-centered partnership model of people and artificial intelligence (AI) working together to enhance cognitive performance, including learning, decision-making and new experiences." [1]. As a human-centered design, it aims to play an assistive role, combining strengths of both human and machine, while enhancing human intelligence rather than replacing it [2]. When applied to the context of instructor-led training, Augmented Intelligence can be used as a design pattern for implementing intelligent instructional systems to assist human instructors in their observation, assessment, decision-making, and task execution during adaptive training.

This paper presents a design pattern for developing Augmented Intelligence for instructional systems in the context of simulation-based training. The approach is based on progressive levels of machine support that can be considered for implementing an instructional system, illustrating some of the challenges for automation, as well as benefits for a human instructor. In addition, two design perspectives are highlighted that contribute to increased levels of intelligence. The first one addresses the problem of domain alignment, which relates to the grounding of (instructional) knowledge, required to tailor machine algorithms to operate in a particular task domain. The second one addresses the role of simulation-based task environments to enable AI approaches for enhanced machine support in instructional systems. Through advances in AI technologies and the need for data-driven solutions, (human-in-the-loop) simulations can play a central role in data provision and delivering AI-based approaches. Finally, we conclude the paper by summarizing the ideas presented with a future outlook.

2   Levels of Machine Support

In an instructional system, the role of an instructor is to guide the learning process of one or more trainees by tailoring instruction to optimize learning in the context of some learning objective. In the view of Augmented Intelligence, we consider the fulfillment of this role as a collaborative effort between human and machine, where the machine supports the human in its task. This concept is visualized in Fig. 1. In the figure, five levels of machine support are identified within an adaptive training cycle. These levels correspond to an instructor's ability to observe, measure and assess trainee performance, develop instructional plans, and instigate these through adaptations of the task environment.

A collaborative execution of the adaptive training process is shaped by a division of labor between the human instructor and the machine. The level of machine support is

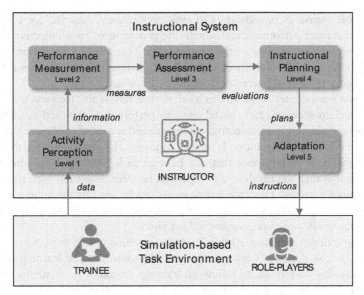

Fig. 1. Augmented Intelligence for an instructor role in an adaptive training cycle, conceptualized through different levels of machine support[1]

considered to be flexible, depending on the application. At one end, the machine provides no support and the instructor performs all levels by itself. At the other end, the machine performs all levels and the instructor would be out of the loop, in which case the machine acts as a true AIS. In between these extremes, the machine provides support for one or more levels, and communicates its results through some human machine interface (such as a dashboard or instructor operating station). The instructor can combine objective data obtained from automated processes with its own subject insights to implement remaining levels. Below we first describe each level in more detail, followed by an outline of illustrative related work.

2.1 Levels of Support

Level 1: Activity perception. At this level, the machine is able to observe the training environment and the trainee's behaviors and activities in that environment. By translating data that can be acquired from the simulation into meaningful, semantic information associated with the task domain, the human instructor can be supported in building situation awareness. When the machine can monitor distinct training tasks or activities as they are being performed, it enables the instructor to observe *what* tasks the trainee is undertaking, such that it can concentrate on relevant measurements to determine *how* the trainee is performing on those tasks.

Level 2: Performance measurement. At this level, the machine is able to extract and compute relevant performance metrics from the observed data. These metrics serve

---

[1] Note that the instructor role could also be seen as being fulfilled by the trainee itself, hereby suggesting self-guided training, as opposed to instructor-led training.

as quantifiable measures or indicators of task performance, required for consequent evaluation of trainee performance. Through the provision of such objective data, the human instructor can acquire data-driven insights that it can use as actionable information for making informed decisions: it can combine these insights with its own subjective insights in order to form of overall performance assessment.

Level 3: Performance assessment. At this level, the machine is able to assess and evaluate a trainee's performance on a task, based on some performance standard as a reference. A performance standard is an objective, predetermined notion of proficiency for a task that a trainee is expected to achieve. It can be represented by certain criteria, thresholds, or expert models of performance that can be used as a benchmark for comparison. An assessment is then the result of a comparison between measured performance and the performance standard. Objective assessments can be used by the human instructor (possibly augmented with its own subjective assessments) to update learner profiles, refine training needs, and plan instructional activities.

Level 4: Instructional planning. At this level, the machine is able to plan instructional activities, based on a trainee's assessed progression towards some learning objective. Instructional plans may relate to follow-up training schedules or scenarios, or to in-session adaptations such as direct interventions (e.g. scaffolding or feedback strategies to guide learning) or indirect interventions (e.g. change task complexity through environment adaptations). Personalized strategies may be used that take into account aspects such as learning preferences or measured learner engagement. Machine-based instructional plans can be used by the human instructor as recommendations or decision support on when and how to adapt training.

Level 5: Adaptation. At this level, the machine is able to realize instructional plans through instructions that configure, modify or interact with the training environment. In simulation-based training, this includes for instance the ability to generate or configure new training scenarios, to adapt the behavior of possible virtual role-players, or to provide direct feedback to the trainee. The human instructor can be provided with appropriate control interfaces to realize such changes in the training environment.

2.2 Related Work

Throughout all levels of support, increased technological progress is seen for automated approaches, driven by the increased demand for more data-driven, personalized, and adaptive training solutions. Simulation-based training is particularly suited as it provides a training environment that can be effectively created, observed, and controlled. However, fully end-to-end systems like AISs are still rare in simulation-based training, especially in complex domains such as military training. In this domain, related research and technology are more often dedicated to individual levels of support. Below, we touch upon related research in this context for illustrative purposes.

To support perception at *Level 1*, military standards such as Distributed Interactive Simulation (DIS) have been developed to represent raw simulation data as real-time, domain-specific information about military entities and events in a simulation environment [3]. Additionally, standards such as the Experience API (xAPI) have been utilized in simulation-based training to infer trainee activities from synthetic environments [4]. In line with activity recognition, model tracing techniques have been proposed to track activities and contexts in real-time, based on known hierarchical task models of the trainee [5], hereby enabling context-based measurements at *Level 2.*

Performance measurement at *Level 2* is a well-addressed topic. Specifically for simulation-based training, a systematic review of methodologies and best practices for computer-assisted performance measurement is provided in [6]. In specific domains such as fighter pilot training, performance measurement has been extensively researched to measure technical and non-technical skills, including task proficiency, teamwork, communication and situation awareness [7, 8]. Within this domain, measurement tools such as PETS [9] provide a framework to develop and deliver performance metrics, and has been used e.g. to support subject matter experts in evaluating training effectiveness [10] or to assess AI pilots to support training [11].

Closely related in research is performance assessment at *Level 3*, which can be automated when some 'absolute' measure of desired performance can be known. It is often recognized that fully automated assessment is not feasible in many domains and that human judgement cannot easily be replaced, giving rise to partial automation or assessment aids. Though, several studies on automated assessment exist. For instance, in [12], a system was evaluated that assessed performance based on observed examples of good and bad performance in tactical air engagement scenarios, showing a high degree of agreement with a human grader; and in [13], an automated assessment method for training simulators was investigated where assessment rules are learned from observing experts and students performing training tasks. A broad review on systems and trends for automated after action review in military training is given in [14].

After assessments are made, instructional planning at *Level 4* caters the planning of teaching activities for follow-up training or in-session interventions. For instance, in [15], methods are explored to tailor personalized training programs for maintaining currency, based on measured or predicted skill decay; in [16], a recommendation system has been proposed that recommends optimal training tasks based on training needs and measured trainee competency levels; and in [17], real-time difficulty adjustments are implemented through the adjustment of AI opponent behavior, based on measured trainee proficiency.

Finally, for adaptation at *Level 5*, (semi-)automated approaches have been explored to relieve instructors from the often resource intensive manual activity of scenario construction. For instance, in [18], generative techniques are leveraged to generate training scenarios aligned with learning objectives and individual learner characteristics; whereas in [19], semi-automated methods are used that allow an instructor to direct the generation process to reflect its own preferences. Implementing adaptive strategies are strongly guided by instructional design principles on how, when and what to adapt [20].

2.3 Concluding

The identification of different levels of machine support presented in this section guides the design of increased automation while considering the human role in the process. The levels of support that can be provided for a particular application depends not only on the available technology and algorithms to fulfill particular levels but also on the ability to the capture required domain and instructional knowledge associated with those levels in machine language. For instance, a fully automated instructional system requires a machine understanding of the task domain, instructional needs and means throughout the system. In the next section, we discuss this challenge. Finally, a successful deployment lies in the trustworthiness of the system to provide validated automated aids, as well as the human instructor's acceptance of and trust in the system.

3   Domain Alignment

In the previous section, different levels of machine support were identified for an adaptive training process. One of the challenges for implementing increased levels of support is the tailoring and alignment of machine algorithms to operate in a particular (instructional) task domain. This is also known as *domain alignment* and is the process of adapting or tailoring algorithms to perform well on a specific domain or task. For an instructional system, this relates to the problem of how to incorporate, ground, and align domain knowledge throughout the system, such that algorithms can be semantically aligned across the processes of an adaptive training cycle. The domain alignment problem for an adaptive training process is shown in Fig. 2. In the figure, domain alignment is conceptualized through top-down requirement drivers for domain-specific knowledge that is needed to implement different levels of support. This is explained using an example below, where the different knowledge concepts are shown in italic.

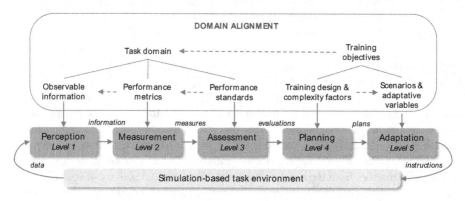

Fig. 2. Domain alignment in an adaptive training process. The solid arrows represent information flows; the dashed arrows represent top-down requirement drivers.

3.1   An Example

Consider a competency-based training for a fighter pilot, designed to train specific technical or non-technical skills through whole-task or part-task training (e.g. specific mission types or tactical engagements). The *training objectives* specify a set of competencies or skills that can be trained through different training tasks in a simulated environment. Job-specific competencies, skills and tasks may originate from a pilot's competency profile or training needs analysis. The training objectives drive the scope of the *task domain*, which encompasses the (expert) knowledge about the tasks to be trained. These are the tasks that (1) trainees should have the opportunity to master during training, and (2) instructors should be able to observe, measure and assess. The ability to assess a training task requires a *performance standard* as a reference of desired performance. Associated are *performance metrics* that should be measurable to compare against this standard. These, in turn, drive requirements for the kind of *information* that should be observable from the simulation data in order to compute them.

When the trainee has been assessed on a task and has shown sufficient mastery, the *training design* may be updated to prioritize other training tasks for follow-up training. Alternatively, the same task may be trained but different *complexity*. The training design drives the requirements for the kind of *scenarios* that should be supported in the simulated task environment for the trainee to experience. The need for in-session adaptations further drives the requirements for *adaptive variables* to be supported.

The example illustrates two key aspects of domain alignment. First, it shows how requirements for domain knowledge can be derived from a top-down analysis, starting from instructional objectives, all the way down to what is needed from a simulation environment in terms of data acquisition, scenario elements, and adaptation options. Second, it shows the dependencies between knowledge concepts across different levels of support. When these concepts can be computationally grounded and interconnected, this promotes alignment between supporting algorithms throughout the system.

3.2 Approaches

Related work that involves domain alignment can be distinguished between bottom-up and top-down oriented approaches. Bottom-up approaches focus more on data strategies, infrastructures, architectures, standards, and tools to manage, represent, and communicate domain-related knowledge within learning ecosystems. For instance, in [21], the authors discuss the need for data-driven learning analytics and explore architectures and infrastructures for data management that help organizations transform performance data into actionable insights. Further, data strategies have been proposed that focus on standards to promote uniform integration of learning technologies. For instance, ADL's Total Learning Architecture (TLA) defines data standards for concepts such as observable activity data from learners; training session meta-data; competency definitions; and learner profiles [22]. Finally, standardization efforts are undertaken within the AIS community on the definition and interoperability of components in an AIS system, including components to manage domain knowledge, track learner data, plan instruction, and provide user interfaces [23, 24]. Such initiatives are supported by frameworks such as GIFT that provide tools, methods, and standards for developing AISs [25].

Top-down approaches focus more on the collaborative process between instructional designers and engineers on how to model instructional design concepts. For instance, the need for a human-centered design approach is argued to orchestrate so-called actionable learning analytics [26]. It focuses on three principles, namely (1) the use of learning design to derive needs for algorithmic learning analytic (LA) solutions, (2) grounding educational theories in LA solutions, and (3) facilitating stakeholder involvement in the design process to reflect the needs and values of instructors and trainees. Advances in AI technologies also benefit human-centered approaches. For instance, large language models (LLM) can play a role in knowledge acquisition, elicitation and organization of acquired knowledge in computational forms [27]. Further, their potential is reviewed to support personalization and teacher activities related to generating automatic feedback, personalized learning tasks, learning content (such as scenarios), and recommenders [28].

In conclusion, effective domain alignment empowers automation capabilities in instructional systems. The grounding of domain-specific knowledge and its semantic alignment throughout the system not only requires suitable data strategies and infrastructures, but also necessitates a collaborative effort between instructional designers and engineers, matching instructional needs with technological solutions.

4   Simulation-Based Task Environments

Simulation-based training can deliver tailored training programs for trainees in specific task domains. Compared to live task environments, simulated task environments offer several benefits for leveraging AI-enabled approaches in instructional systems, particularly for developing learner models of performance and expert models of behaviors. Figure 3 illustrates this notion.

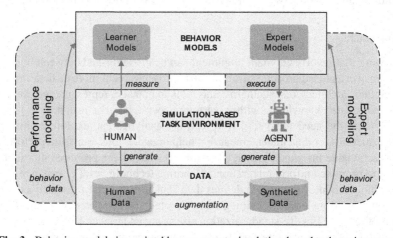

Fig. 3. Behavior models in a mixed human-agent, simulation-based task environment.

In the figure, the simulation environment is shown as a mixed human-agent environment that can be populated with human actors and agent actors. For human actors (e.g. trainees), human-in-the-loop simulations can effectively measure and capture learner data, which is needed to develop human learner models of performance. For agent actors, simulations can be used to learn and deploy expert models of behaviors for various purposes (as will be described). For both type of actors, the simulation mediates in the generation of behavior data, where this data can consequently be used for data-driven AI approaches in the development of either learner or expert models. Below, we elaborate on these principles, referencing illustrative related research.

4.1   Learner Models of Performance

Human performance models are used in instructional systems to measure, track or predict skill development of learners over time. Human-in-the-loop simulations can acquire the learner data required to develop such (personalized) models. To give some examples, in [29], learner data is collected from trainees across a series of team training sessions in order to measure team performance progression over time. Bayesian inference is used to propagate measures from domain-specific performance metrics to higher-level teamwork competencies. In [15], learner data is collected to develop a predictive model of skill decay, which is used for time-based scheduling of future training sessions to maintain personnel currency of complex skills. In related research, agents are used to simulate human learning on a task. The envisioned application was to employ simulated learners to computationally derive an optimal part-task training design for humans on the task [30].

4.2   Expert Models of Behavior

Expert models in simulation-based task environments represent computational models of performance. Traditionally expert models have often been hand-crafted using rule-based techniques. More recently, through advances in areas such as deep reinforcement learning, expert behaviors can also be learned by agents in complex task environments [31]. Expert models of behavior are used for a variety of purposes in instructional systems. For instance, in [32], expert models of fighter pilot behaviors are trained using reinforcement learning from human feedback (RLHF), where tacit expert preferences are encoded in human-readable form and used for automated assessment or demonstration learning. Alternatively, expert models are used to represent embodied agents as virtual role-players. For instance, in [33], agents are trained to support adaptive team training and influence learning engagement in simulation-based training. Embodied agents can also assume a tutoring role, exhibiting teaching activities such as through prompting or using other social learning methods [34].

4.3   The Role of Data

To support the development of data-driven AI approaches for behavior models, such as described above, the simulation plays a central role in collecting and generating (learner) behavior data. The simulation can be agnostic to whether such data is generated by human actors or agent actors. Regarding human learner data, an issue often faced is that the amount of (historical) data that can be obtained from regular training sessions is insufficient for training AI algorithms. To address this, researchers have resorted to alternative approaches. One approach is an organizational one that focuses on human data collection strategies. This includes, for instance, the organization of dedicated data collection sessions with students or experts [35], or outsourcing to a broader audience using (semi-)public online (serious) games to harvest player behaviors [15]. An alternative approach is a technological one and uses synthetic data to address the data scarcity problem. For instance, data augmentation methods have been developed to generate new representative behavior data from limited human data samples [36, 37]. Backwards, synthetic agents have been used to generate synthetic data as representative human data, for instance to investigate data-driven behaviour modelling for demonstration learning [38], or to develop activity recognition algorithms for human behaviors [39].

In conclusion, it is seen that simulation-based task environments enable the development of data-driven AI approaches, mediating in human data collection, synthetic data generation, and providing interactive machine learning environments. Their role in supporting performance modelling and embodied tutoring in simulation-based training further empowers automation capabilities in instructional systems.

5 Conclusion

In this paper, we positioned the concept of Augmented Intelligence for instructional systems as a design pattern for increased instructor support through automation in simulation-based training. We started by identifying different levels of machine intelligence to support the role of an instructor during an adaptive training cycle. These levels range from more low-level support, such as measuring trainee performance and adapting the task environment, to more high-level support, such as performance assessment and planning instructional action, in correspondence to training objectives. The design pattern provides guidance on implementing increased levels of support while recognizing the human role in the process.

Next, we discussed two design perspectives that foster increased levels of support. The first one highlights the need for effective domain alignment, relating to the grounding of domain knowledge to allow machine understanding of the task domain and instructional needs. Key is an alignment of bottom-up approaches that lay the infrastructural foundation for managing knowledge throughout the system, with top-down approaches that focus on the role of instructional designers in the design process.

The second perspective highlights the role of simulation-based task environments to drive AI-enabled approaches for developing behavior models. Human performance models support the measurement or prediction of trainee performance, whereas expert models support training through capabilities of automated assessment, virtual role-players, demonstration learning, or social learning strategies. Simulation-based task environments facilitate AI-based approaches for these models through data collection, data generation, and delivering interactive machine learning environments within the respective task domain.

When projecting on future developments, advances in AI will be instrumental in enhancing Augmented Intelligence for instructional systems. Recent trends in generative AI, large language models, or human-directed reinforcement learning show innovative approaches to cater to effective domain alignment, empowering instructional designers to collaborate in the design process, concerning delivering fit-for-purpose training environments, scenarios, interpretable learning analytics, and personalized instruction. In parallel, continued advances in immersive and digital twin technologies are blurring the line between simulated and real-world task environments, such that instructional aids for simulation-based training will become more aligned with, and accessible to mixed reality training environments.

References

1. Gartner: Gartner Glossary (2023). www.gartner.com/en/information-technology/glossary/augmented-intelligence
2. Sadiku, M.N., Musa, S.M.: Augmented intelligence. In: Sadiku, M.N., Musa, S.M. (eds.) A Primer on Multiple Intelligences, pp. 191–199. Springer, Cham (2021). https://doi.org/10.1007/978-3-030-77584-1_15
3. IEEE Standard for Distributed Interactive Simulation–Application Protocols. IEEE Std 1278.1-2012 (Revision of IEEE Std 1278.1-1995). 1–747 (2012). https://doi.org/10.1109/IEEESTD.2012.6387564
4. Hernandez, M., et al.: Enhancing the total learning architecture for experiential learning. In: Interservice/Industry Training, Simulation, and Education Conference (I/ITSEC), Orlando, FL (2022)
5. van Oijen, J.: Human behavior models for adaptive training in mixed human-agent training environments. In: Interservice/Industry Training, Simulation and Education Conference (I/ITSEC). I/ITSEC (2022)
6. Salas, E., Rosen, M.A., Held, J.D., Weissmuller, J.J.: Performance measurement in simulation-based training: a review and best practices. Simul. Gaming **40**, 328–376 (2009)
7. Smith, E., Borgvall, J., Lif, P.: Team and collective performance measurement. Report of NATO Research and Technology Organization, RTO-TR-HFM-121-Part-II (2007)
8. Mansikka, H., Virtanen, K., Harris, D., Jalava, M.: Measurement of team performance in air combat – have we been underperforming? Theor. Issues Ergon. Sci. **22**, 338–359 (2021). https://doi.org/10.1080/1463922X.2020.1779382
9. Portrey, A.M., Keck, L.B., Schreiber, B.T.: Challenges in developing a performance measurement system for the global virtual environment. Lockheed Martin Systems Management MESA AZ (2006)
10. Rowe, L.J., Prost, J., Schreiber, B., Bennett, W., Jr.: Assessing high-fidelity training capabilities using subjective and objective tools. In: 2008 Interservice/Industry Training, Simulation, and Education Conference (I/ITSEC) Proceedings (2008)
11. Freeman, J., Watz, E., Bennett, W.: Assessing and selecting AI pilots for tactical and training skill. In: NATO MSG-177 (2020)
12. Stevens-Adams, S.M., Basilico, J.D., Abbott, R.G., Gieseler, C.J., Forsythe, C.: Performance assessment to enhance training effectiveness. In: Interservice/Industry Training, Simulation, and Education Conference (I/ITSEC), Orlando, FL (2010)
13. de Penning, L., Kappé, B., Boot, E.: Automated performance assessment and adaptive training for training simulators with SimSCORM. In: Interservice/Industry Training, Simulation, and Education Conference (I/ITSEC), Orlando, FL (2009)
14. Johnson, C., Gonzalez, A.J.: Automated after action review: state-of-the-art review and trends. J. Defense Model. Simul. **5**, 108–121 (2008)
15. van der Pal, J., Toubman, A.: An adaptive instructional system for the retention of complex skills. In: Sottilare, R.A., Schwarz, J. (eds.) HCII 2020. LNCS, vol. 12214, pp. 411–421. Springer, Cham (2020). https://doi.org/10.1007/978-3-030-50788-6_30
16. Thijssen, D., Bosma, R.: Recommendation system in an integrated digital training environment for the 5th generation air force. In: Interservice/Industry Training, Simulation and Education Conference (I/ITSEC). I/ITSEC (2022)
17. Demediuk, S., Raffe, W.L., Li, X.: An adaptive training framework for increasing player proficiency in games and simulations. In: Proceedings of the 2016 Annual Symposium on Computer-Human Interaction in Play Companion Extended Abstracts, pp. 125–131 (2016)
18. Rowe, J., Smith, A., Pokorny, B., Mott, B., Lester, J.: Toward automated scenario generation with deep reinforcement learning in GIFT. In: Proceedings of the Sixth Annual GIFT User Symposium, pp. 65–74 (2018)

19. Luo, L., Yin, H., Cai, W., Lees, M., Zhou, S.: Interactive scenario generation for mission-based virtual training. Comput. Animation Virtual Worlds **24**, 345–354 (2013)
20. Rebensky, S., Perry, S., Bennett, W.: How, when, and what to adapt: effective adaptive training through game-based development technology. In: Interservice/Industry Training, Simulation and Education Conference (I/ITSEC). I/ITSEC (2022)
21. Watz, E., Neubauer, P., Shires, R., May, J.: Precision learning through data intelligence. In: Sottilare, R.A., Schwarz, J. (eds.) Adaptive Instructional Systems, pp. 174–187. Springer, Cham (2023). https://doi.org/10.1007/978-3-031-34735-1_13
22. Smith, B., Milham, L.: Total learning architecture (TLA) data pillars and their applicability to adaptive instructional systems. In: Stephanidis, C., et al. (eds.) HCII 2021. LNCS, vol. 13096, pp. 90–106. Springer, Cham (2021). https://doi.org/10.1007/978-3-030-90328-2_6
23. Sottilare, R.: Exploring methods to promote interoperability in adaptive instructional systems. In: Sottilare, R.A., Schwarz, J. (eds.) Adaptive Instructional Systems, pp. 227–238. Springer, Cham (2019). https://doi.org/10.1007/978-3-030-22341-0_19
24. Sottilare, R.: Understanding the AIS problem space. In: Proceedings of the 2nd Adaptive Instructional Systems (AIS) Standards Workshop (2019)
25. Sottilare, R., Brawner, K.: Component interaction within the generalized intelligent framework for tutoring (GIFT) as a model for adaptive instructional system standards. In: The Adaptive Instructional System (AIS) Standards Workshop of the 14th International Conference of the Intelligent Tutoring Systems (ITS) Conference, Montreal, Quebec, Canada (2018)
26. Dimitriadis, Y., Martínez-Maldonado, R., Wiley, K.: Human-centered design principles for actionable learning analytics. In: Tsiatsos, T., Demetriadis, S., Mikropoulos, A., Dagdilelis, V. (eds.) Research on E-Learning and ICT in Education: Technological, Pedagogical and Instructional Perspectives, pp. 277–296. Springer, Cham (2021). https://doi.org/10.1007/978-3-030-64363-8_15
27. Allen, B.P., Stork, L., Groth, P.: Knowledge engineering using large language models. Trans. Graph Data Knowl. **1**, 3:1–3:19 (2023). https://doi.org/10.4230/TGDK.1.1.3
28. Mazzullo, E., Bulut, O., Wongvorachan, T., Tan, B.: Learning analytics in the era of large language models. Analytics **2**, 877–898 (2023)
29. Vatral, C., Biswas, G., Mohammed, N., Goldberg, B.S.: Automated assessment of team performance using multimodal Bayesian learning analytics. In: Interservice/Industry Training, Simulation and Education Conference (I/ITSEC). I/ITSEC (2022)
30. van Oijen, J., Roessingh, J.J., Poppinga, G., García, V.: Learning analytics of playing space fortress with reinforcement learning. In: Sottilare, R.A., Schwarz, J. (eds.) HCII 2019. LNCS, vol. 11597, pp. 363–378. Springer, Cham (2019). https://doi.org/10.1007/978-3-030-22341-0_29
31. Berner, C., et al.: Dota 2 with large scale deep reinforcement learning. arXiv preprint arXiv: 1912.06680 (2019)
32. Bewley, T., Lawry, J., Richards, A.: Learning interpretable models of aircraft handling behaviour by reinforcement learning from human feedback. In: AIAA SCITECH 2024 Forum, p. 1380 (2024)
33. Sottilare, R., McGroarty, C., Ballinger, C., Aris, T.: Investigating the effect of realistic agents on team learning in adaptive simulation-based training environments using GIFT. In: Sinatra, A.M. (ed.) Proceedings of the 11th Annual Generalized Intelligent Framework for Tutoring (GIFT) Users Symposium. U.S. Army Combat Capabilities Development Command – Soldier Center (2023)
34. Goldberg, B., Cannon-Bowers, J.: Feedback source modality effects on training outcomes in a serious game: pedagogical agents make a difference. Comput. Hum. Behav. **52**, 1–11 (2015)

35. Samuel, K., et al.: AI enabled maneuver identification via the maneuver identification challenge. In: Interservice/Industry Training, Simulation and Education Conference (I/ITSEC). I/ITSEC (2022)
36. Ucuzova, E., Kurtulmaz, E., Gökalp Yavuz, F., Karacan, H., Şahın, N.E.: Synthetic CAN-BUS data generation for driver behavior modeling. In: 2021 29th Signal Processing and Communications Applications Conference (SIU), pp. 1–4 (2021)
37. Brandt, B., Dasgupta, P.: Synthetically generating human-like data for sequential decision making tasks via reward-shaped imitation learning. CoRR. abs/2304.07280 (2023). https://doi.org/10.48550/ARXIV.2304.07280
38. Kamrani, F., Luotsinen, L.J., Løvlid, R.A.: Learning objective agent behavior using a data-driven modeling approach. In: 2016 IEEE International Conference on Systems, Man, and Cybernetics (SMC), pp. 002175–002181 (2016)
39. Romero, A., Carvalho, P., Côrte-Real, L., Pereira, A.: Synthesizing human activity for data generation. J. Imaging **9**, 204 (2023)

Adaptive Learning Experiences

Student Perceptions of Adaptive Learning Modules for General Chemistry

Angelo Cinque[1,2], Jennifer Miller[1,3], Tamra Legron-Rodriguez[1],
James R. Paradiso[4(✉)], and Nicole Lapeyrouse[1]

[1] Department of Chemistry, University of Central Florida, Orlando, FL 32816, USA
{tamra.legron-rodriguez,nicole.lapeyrouse}@ucf.edu
[2] Department of Health Sciences, University of Central Florida, Orlando, FL 32816, USA
[3] College of Medicine, University of Central Florida, Orlando, FL 32816, USA
[4] Center for Distributed Learning, University of Central Florida, Orlando, FL 32816, USA
James.Paradiso@ucf.edu

Abstract. Due to the sheer breadth of content covered and large enrollment numbers in fundamental chemistry courses at the University of Central Florida (UCF), instructors and course designers often look for ways to accommodate a high degree of learner variability—in terms of interest, background, and content knowledge—to enhance the potential for student success. One way to accomplish this is through incorporating an adaptive instructional system (AIS), such as Canvas Mastery Paths (MP), into the course design and implementation processes: a method ultimately aimed at improving student learning outcomes through the development and delivery of flexible, robustly aligned content–assessment sequences. In the current study, four (4) MP course modules on the topics of *Measurements*, *Atomic Theory*, *Quantum Mechanics*, and *Molecular Polarity* were created, with two optional surveys administered at the completion of the modules to gather insight into students' sentiments toward and perceived learning impact from MP. Overall, students felt like they had a better understanding of the course material and that MP had improved their attitude toward general chemistry.

Keywords: Canvas Mastery Paths · adaptive learning · chemistry education · mixed-methods research

1 Introduction

Since the advent of adaptive instructional systems (AISs), instructors have been able to better meet the distinct needs of students by way of AIS's inherent capacity to dynamically modify the learning experience and provide immediate, tailored feedback on formative activities [1–3]. AISs help lessen students' perturbations, keep learners engaged with the material, and enhance their ability to absorb course content efficiently—features which are especially relevant for students learning chemistry [1, 4].

A. Cinque and J. Miller—Contributed equally to this work.

© The Author(s), under exclusive license to Springer Nature Switzerland AG 2024
R. A. Sottilare and J. Schwarz (Eds.): HCII 2024, LNCS 14727, pp. 105–115, 2024.
https://doi.org/10.1007/978-3-031-60609-0_8

Students who take general (fundamental) chemistry courses often possess varying levels of requisite knowledge for the course, which can be challenging for instructors to accommodate within the context of a 'live' semester/in real-time [1]. However, an AIS's ability to calculate individual students' level of subject matter understanding and create personalized educational experiences from that information reduces the ill effects of any pre-existing, content-specific knowledge gaps on the part of the learners [1, 5]. The unique attributes of AISs, therefore, make these systems especially suitable for chemistry instructors with large course enrollments (N > 100) by allowing them to give assignments that permit students to evaluate their own performance through instant feedback at the question and content levels [4].

While a wide variety of AISs exist in the marketplace, the Canvas learning management system was chosen for this study, as it provides educators and institutions a robust platform to deliver online course materials and assess student learning through built-in features, like Mastery Paths (MP). MP is designed to provide pathways for each student—depending on their subject-matter proficiency—so that those who exhibit low performance in a given area are provided automated, scaffolded instruction to step them through the learning materials, while students with a strong grasp of the content are able to move forward to the next topic in the learning path at an accelerated pace [6, 7]. MP also provides an added layer of learning and interaction data that offers the course instructors insights into student mastery and trajectory through the content and assessment pathways.

Cognitive Load Theory. While developing the MP learning modules, cognitive load theory (CLT) was used as a framework to design and develop the content. CLT refers to the level of demand required by one's working memory when encountering new information. To enhance learning, the transfer of items from working memory to long-term memory must be at peak efficiency. In this study, three types of cognitive loads were considered: intrinsic cognitive load, extraneous cognitive load, and germane cognitive load [8].

Intrinsic Cognitive Load Refers to the Inherent Complexity of Processing New Information. Since the goal for this course design was to simplify rather than complicate the transmission of new information, the researchers looked for ways to reduce intrinsic load by utilizing AIS features that allow for the creation of learning scaffolds, while 'chunking' the content contained within those scaffolds to facilitate cognitive processing [9–11]. 'Chunking' is an instructional method that can be described as breaking content in small, digestible units of information to prevent students from being mentally overwhelmed [10]. 'Scaffolding' is another technique used by the researchers to reduce the chances of overburdening students' working memory. These strategies were applied to aid students in reaching their learning goals. Through scaffolding student learning pathways through the AIS (MP), students were able to solve content-related problem sets in a manageable way, while gaining more competence and confidence to complete future pathways without as much prompting and scaffolding.

Extraneous cognitive load refers to unnecessary or distracting information in the working memory [11], and the research team recognizes that extraneous cognitive load is something to be reduced, as it distracts from a student's working memory and hinders the processing of new information.

Germane cognitive load refers to the mental resources required for the construction of new schema or the long-term storage of knowledge [12, 13], so to maximize the deep processing of new information, the researchers made sure to interleave any novel concepts with those learned prior.

Research Questions. This mixed-methods study investigates the following research questions (RQ):

1. What was the students' perception of Canvas Mastery Paths based on their experiences with the modules?
2. What impact did the Mastery Path activities have on students' perceived learning of and attitude toward chemistry?
3. What aspects of Mastery Path activities did students find the most useful?

2   Methods

Institutional Review Board.   The surveys were administered through the Canvas Learning Management System and approved by the University of Central Florida Institutional Review Board (STUDY00004338). The descriptions of the survey design and participant numbers are detailed below.

2.1   Study Setting

This study was conducted at The University of Central Florida (UCF), one of the largest public universities by enrollment in the nation. UCF is a Hispanic-Serving Institution with a diverse student body. Currently, the university's student population is 43.5% White, 29.2% Hispanic/Latino, 9.2% Black, 7.5% Asian, 4.6% International, 4.6% Multiracial, 1.2% Not Specified, 0.1% Native Hawaiian/Other Pacific Islander, and 0.1% American Indian/Alaska Native. University enrollment by gender is reported as binary: 45.2% Male, 54.7% Female, and 0.06% not specified.

Data was collected over two semesters, Fall 2022 and Spring 2023, in a large general chemistry course, referred to at UCF as Chemistry Fundamentals I. Both course offerings were taught by the same instructor and were structured similarly. The course was taught using a flipped classroom format, and students were provided with various low- and high-stakes assessments (Table 1). Course meeting times included three 50-min in-person meeting times each week and mandatory enrollment to a discussion course. To facilitate active learning in these large enrollment courses, Fall 2022 (n = 373) and Spring 2023 (n = 445) semesters, undergraduate learning assistants were integrated into the classroom activities to help reduce the student-to-instructor ratio. The discussion consisted of a recitation session and the completion of virtual lab experiments (accounting for 10% of the student's grade). Prior to coming to class, students would watch online lecture videos and complete a low-stakes quiz based on the material covered in the video (accounting for 10% of their grade, with their two lowest attempts dropped). During the in-person meeting times, students would work in small groups to complete a weekly worksheet (participation accounting for 5% of their grade). In addition, students had high-stakes summative weekly quizzes (15% of their grade with their lowest attempt dropped), exams (40% of their grade with the lowest exam replaced by the final), and a cumulative final exam (20% of their grade). In addition, students had access to additional student success resources, including supplemental instruction, weekly recitation sessions held by a graduate student, optional MP learning modules, and weekly office hours held by the course instructor.

Table 1. Overview of the course grading scheme

Grade distribution (%)	0%	5%	10%	10%	15%	20%	40%
Course Activity	MP learning modules	Group worksheets and in-class participation	Recitation sessions and virtual labs	Practice quizzes	Summative quizzes	Final exam	Unit exams

2.2 Participants

Participants were students enrolled in Chemistry Fundamentals I, in Fall 2022 and Spring 2023 semesters. Two types of surveys were administered to students enrolled in the courses. Module surveys were offered at the end of each of the four MP modules during both Fall 2022 and Spring 2023 (n = 173) to gauge student perception of the individual MP modules. An end-of-semester survey was distributed to students enrolled in the Spring 2023 course that interacted with an MP module (n = 69) to gather overall perception of MP. The MP was not graded, and student participation was voluntary with the option to discontinue working on the MP modules at any time throughout the semester. Student demographic information from the two courses was obtained from university data. Race/ethnicity and gender data are from the time of admission. Compiled demographic of students enrolled in the two semesters in terms of race/ethnicity: 43% White, 33% Hispanic/Latino, 9% Black, 8% Asian, 6% Multiracial, and 1% Not Specified/Other and gender: 34% Male and 66% Female. The authors recognize the terms used for race/ethnicity and gender may not represent the identity of the participants; the terms reflect available institutional data.

2.3 Mastery Path (MP) Modules

The four chemistry topics selected for the MP modules chosen for this study were *Measurements, Atomic Theory, Quantum Mechanics*, and *Molecular Polarity*. These topics were selected based on feedback from students in prior semesters. The modules were designed as ungraded practice to help prepare students for higher-stakes assessments.

2.4 Surveys

Module Surveys. A survey was distributed at the end of the four MP modules and was available to students who completed the entire MP. These module surveys contained free-response questions (*What information in the module did you find useful?* and *What would you change about the module?*) and one ranking question about perceived time spent on each module (*How long do you feel it took to complete the module?*). The student-facing language described the MP modules as "Mastery Challenges" and is reflected below in the survey questions and responses. The data from the free response questions were analyzed through thematic analysis, with both coders independently identifying recurring participant responses. The two coders met and formed a codebook for each question. These codebooks were then used to code all subsequent responses with a 20–25% overlap between coders. Gwett's AC1 score [14] was used to determine interrater reliability (IRR), and the researchers set a threshold of 0.85. For any reliability falling below the threshold, coders met to discuss differences, and those responses were reanalyzed. The IRR ranged from 0.88–0.99.

End-of-Semester Survey. The end-of-semester survey was distributed in the Spring 2023 semester to gather feedback from participants about their overall experience with MP. This survey contained twelve Likert-scale questions (Fig. 1) with a scale of 1 (strongly disagree) to 5 (strongly agree). The survey was adapted from prior literature regarding online homework systems [4].

3   Results

3.1   What Information in the Mastery Challenge Did You Find Useful?

The final codebook included five codes (see Table 2) related to the information students found the most useful in the MP. Students found many aspects of MP helpful, with the most frequent themes being *Practice Problems and Examples* and *Content-Related Topics*. The frequency of responses can be found in Fig. 1. These codes were consistent for each of the four individual MPs. These modules also helped students reflect on their understanding of their strengths, weaknesses, and areas for improvement, with one student saying, "I found that it made me seriously consider the conversion of measurements in my formulas… I am now taking more time to carefully look at the wording of problems so as to not output the incorrect response." In addition to the practice, students expressed the usefulness of the supplementary resources they received from their completed assessments. Students also voiced their appreciation of the MP layout and overall module structure.

Table 2. Codebook for "What information in the Mastery Challenge did you find useful?"

Code	Definition	Example code
Practice Problems and Examples	A mention or reference to practice problems, examples, and explanations found in MP	"The extra practice on the ice cube and sandbox questions were incredibly useful"
Content Related Topics	Students' comments mentioning content related to topics	"the pneumonic for khdbdcm was really helpful" "I thought that the tables where most useful, especially when referring to conversions"
Metacognition	Reflects students' own understanding of their strengths, weaknesses, and areas for improvement	"I found that it made me seriously consider the conversion of measurements in my formulas… I am now taking more time to carefully look at the wording of problems so as to not output the incorrect response"
Assessments	Any mention of an MP quiz or test	"I liked the practice tests" "The quizzes help me"
Layout/Overall Module Structure	Students' comments regarding the MP layout and overall module structure	"I found the layout of the mastery challenge helpful" "I liked how it is organized so that we have to complete each part before we procced on to the next part"

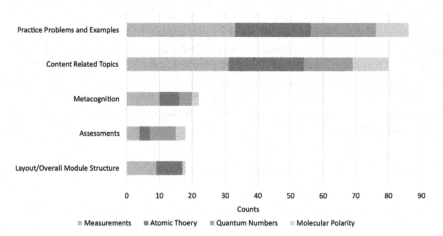

Fig. 1. Top 5 most prevalent codes for: "What information in the Mastery Challenge did you find useful?". AC1 score of 0.8875.

3.2 What Would You Change About the Mastery Challenge Module?

The final codebook included five codes (see Table 3) related to what students would change in the MP. The most frequent code was *Nothing to Change*, with one student quoted as saying, "Nothing, I honestly believe it's perfect for some good practice." The next most prominent codes were *Feedback*, *Layout/Overall Module Structure*, *Additional Practice*, and *More Examples* (Fig. 2). While most students did not want to change anything about the MP, some of them mentioned aspects of MP that they would like to see improved. For instance, the second most frequent code for this question shows that students would like to see more feedback when they got a question wrong. Another common theme related to changes to the layout/overall module structure, with some mentioning the length of content was too long. Although the practice problem and example code were the predominant codes that students found useful about MP (Fig. 1), two codes emerged, showing that students would like even more worked-out examples and additional practice with the material (Fig. 2).

Table 3. Codebook for "What would you change about the Mastery Challenge module?"

Code	Definition	Example code
Nothing to change	Student says that nothing about MP needs to be changed	"I would change nothing" "Nothing, I honestly believe it's perfect for some good practice"
Feedback	Mentioned of wanting feedback or explanation to incorrectly answered problems	"If there were step by step solutions to the problems that we get wrong" "An explanation to the questions wrong"
Layout/Overall Module Structure	Students' comments regarding the MP layout and overall module structure	"The length, it was a little longer and a little more reading than I thought"
Additional Practice	Students mention additional practice problems or reference to more quiz questions	"More sig fig problems on the quizzes with multistep" "I would include more practice problems"
More Examples	Students mention desire for more worked out examples	"more worked out examples" "I think it'd be helpful to have corresponding worked problems available after taking some of the quizzes. So students can review examples of questions they struggled with and find out how they went wrong on their own"

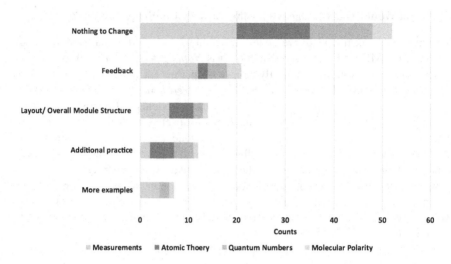

Fig. 2. Top 5 most prevalent code for "What would you change about the Mastery Challenge module?". The AC1 score for this data was 0.9947.

3.3 End-of-Semester Survey

Analysis of the Likert scale questions in the end-of-semester survey indicates that 79.7% (n = 55) of students strongly agreed or agreed that they had an overall positive experience with MP, and 50.7% (n = 35) of students believed that MP helped improve their attitude toward chemistry. In addition, 42.0% (n = 29) of students strongly agreed or agreed that by completing MP modules, they received higher scores on their exams, and 65.2% (n = 45) believed that completing MP made them think more about chemistry. Regarding students' feelings towards MP, 60% of respondents strongly disagreed or disagreed with the statement that these modules were a waste of time. Overall, MP led to improved

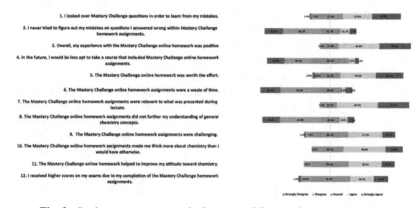

Fig. 3. Student responses to end-of-semester Likert-scale survey questions

attitudes toward chemistry, and students perceived that they received higher scores due to completing these modules.

In addition, 80.9% (n = 55) of students strongly agreed or agreed that the homework assignments in MP were relevant to the content covered during course lectures, and 69.6% (n = 48) agreed that the assignments were worth their effort (Fig. 3). This exhibits one of the hallmarks of MP functionality—the instructor's ability to customize the content and questions delivery. MP also integrates seamlessly with the course curriculum, which is one of the primary reasons students said they would not change anything about the MP modules. Figure 1 supports this finding, showing that a majority of students stated that they found the *Practice Problems and Examples* and *Content-Related Topics* to be the two most useful aspects of the MP platform.

4 Discussion

4.1 RQ1: What Was the Students' Perception of Canvas Mastery Paths Based on Their Experiences with the Modules?

Analysis of the Likert scale questions in the end-of-semester survey (Fig. 1) indicated that 79.7% (n = 55) of students either strongly agreed or agreed that their experience with MP was an overall positive one, despite 65.2% (n = 45) of students saying that they found the MP homework assignments to be challenging.

AISs have been shown to improve the course performance of students enrolled in general chemistry, however, a common grievance students have with these platforms is the time spent to achieve the desired results [4]. This was not seen as much of an issue with MP in this study, as 63.8% (n = 44) of students expressed that the MP homework assignments were not a waste of their time (Fig. 3), and the majority of students spent, on average, between 0–2 h per module (Fig. 4). Time spent was also not one of the prevailing criticisms that students stated they had with the platform, with one student describing MP as "easy to go through in a reasonable amount of time" (Fig. 4). However, this could also be due to these MP modules not being graded and being voluntary.

4.2 RQ2: What Impact Did the Mastery Path Activities have on Students' Perceived Learning of and Attitude Toward Chemistry?

Students who utilized the MP modules generally felt they had a better grasp of the course material and that MP had improved their attitude toward general chemistry. MP made students feel "more comfortable with the material," with 50.7% (n = 35) saying the platform helped to elicit a more positive attitude towards general chemistry (Fig. 3). Students additionally found that the online homework assignments made them think more about chemistry and led them to have an improved attitude towards chemistry and achieve higher exam scores. Apropos of students' learning comprehension, one student said, "I feel like I strengthened my understanding of the important topics of this unit." Although 63.8% (n = 44) of the students revealed that the MP homework assignments helped to further their understanding of general chemistry concepts, only 42.7% (n = 29) of students felt that this enhanced comprehension led to better course performance (Fig. 3).

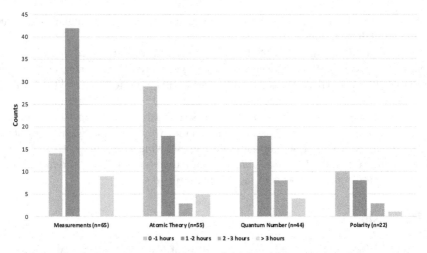

Fig. 4. Student responses for perceived time spent on MP modules

4.3 RQ3: What Aspects of Mastery Path Activities Did Students Find the Most Useful?

Students found the additional practice, the module content, and the mode in which this content was presented to be the most helpful. Students also implied that they were able to use MP to help identify their strengths and weaknesses, with 63.8% (n = 44) of them saying they were able to use the MP questions to learn from their mistakes (Fig. 1 and 3). To better understand the student's perspective, one student is quoted as saying: "I figured out the topics that I need to get more help on before the exam in the upcoming weeks." For some students, this translated to improved course performance, as 42.7% (n = 29) of students agreed or strongly agreed that they received higher exam scores because of the platform (Fig. 3). While students found the MP modules to be a helpful resource, they stated that they would have liked to see more feedback and explanations when they got a question wrong. Overall, students found the MP to be a useful tool, and 69.6% (n = 48) thought it was worth the effort to complete, with a large majority stating that the practice problems and examples and content-related topics were the two most useful aspects of the MP platform.

5 Conclusion (and Future Work)

In a higher education market saturated with educational technology designed to offer personalized learning experiences to students, any combination of interoperable tools can be curated, combined, and deployed within a very short time frame with the intent to support student learning. This study, however, highlights the benefits of using a singular tool that doubles as a learning management system (LMS, Canvas in this case) and adaptive learning platform, reducing potential technical (single platform) and financial (no added cost) barriers for students. While MP does not operate from sophisticated machine learning algorithms, the tool allows instructional designers and instructors to build pedagogically-sound, empirically-based educational experiences that promote student mastery through the use of automated content and assessment pathways that bridge learning gaps between the most and least adept student populations—creating efficiencies for those who possess adequate prior knowledge and introducing scaffolds for those who need further support to reach subject matter proficiency—fostering learning equity and student achievement.

To expand on this study, the research team plans to investigate if students who progress further into the MP have a better understanding of the chemistry content. The group additionally plans to examine if students who participated in MP show increased performance on topics covered in the course assessments.

References

1. Liu, M., et al.: Investigating the effect of an adaptive learning intervention on students' learning. Educ. Tech. Res. Dev. **65**(6), 1605–1625 (2017)
2. Kelly, D.: Adaptive versus learner control in a multiple intelligence learning environment. J. Educ. Multimed. Hypermedia **17**(3), 307–336 (2008)
3. Koedinger, K.R., Aleven, V.: Exploring the assistance dilemma in experiments with cognitive tutors. Educ. Psychol. Rev. **19**(3), 239–264 (2007)
4. Richards-Babb, M., et al.: General chemistry student attitudes and success with use of online homework: traditional-responsive versus adaptive-responsive. J. Chem. Educ. **95**(5), 691–699 (2018)
5. Walkington, C.A.: Using adaptive learning technologies to personalize instruction to student interests: the impact of relevant contexts on performance and learning outcomes. J. Educ. Psychol. **105**(4), 932 (2013)
6. Kerr, P.: Adaptive learning. ELT J. **70**(1), 88–93 (2015)
7. What is MasteryPaths? [Cited 2024 1/27]. https://community.canvaslms.com/t5/Canvas-Basics-Guide/What-is-MasteryPaths/ta-p/404483
8. How do I use Mastery Paths in course modules? [Cited 2024 1/27]. https://community.canvaslms.com/t5/Instructor-Guide/How-do-I-use-Mastery-Paths-in-course-modules/ta-p/906
9. Sweller, J.: Cognitive load during problem solving: effects on learning. Cogn. Sci. **12**(2), 257–285 (1988)
10. Sweller, J.: Cognitive load theory, learning difficulty, and instructional design. Learn. Instr. **4**(4), 295–312 (1994)
11. Sweller, J.: Cognitive load theory and educational technology. Educ. Tech. Res. Dev. **68**(1), 1–16 (2020)
12. Sweller, J.: Cognitive load theory. In: Mestre, J.P., Ross, B.H. (eds.) Psychology of Learning and Motivation, pp 37–76. Academic Press (2011)
13. Sweller, J., van Merrienboer, J.J.G., Paas, F.G.W.C.: Cognitive architecture and instructional design. Educ. Psychol. Rev. **10**(3), 251–296 (1998)
14. Gwet, K.: Handbook of Inter-Rater Reliability, pp. 223–246. STATAXIS Publishing Company, Gaithersburg (2001)

Analytical Approaches for Examining Learners' Emerging Self-regulated Learning Complex Behaviors with an Intelligent Tutoring System

Daryn A. Dever$^{(\boxtimes)}$ (iD), Megan D. Wiedbusch (iD), and Roger Azevedo (iD)

University of Central Florida, Orlando, FL, USA
daryn.dever@ucf.edu

Abstract. Self-regulated learning (SRL), or the ability for a learner to monitor and change their cognitive, affective, metacognitive, and motivational processes, is a critical skill to enact, especially while learning about difficult topics within an intelligent tutoring system (ITS). Learners' enactment of SRL behaviors during learning with ITSs has been extensively studied within the human-computer interaction field but few studies have examined the extent to which learners' SRL behaviors quantitatively demonstrate a functional system (i.e., equilibrium of repetitive and novel behaviors). However, current analytical approaches do not evaluate how the functionality of learners' SRL behaviors unfolds as time on task progresses. This paper reviews two analytical approaches, both based within categorical auto-recurrence quantification analysis (aRQA), for examining how learners' SRL complex behaviors emerge during learning with an ITS. The first approach, binned categorical aRQA, segments learners' SRL behaviors into bins and performs categorical aRQA on the SRL behaviors enacted within those bins to produce metrics of complexity that demonstrate how learners' functionality of their SRL systems change over time. The second approach, cumulative categorical aRQA, continuously calculates complexity metrics as learners enact SRL behaviors to identify the evolution of learners' functional SRL. These two approaches allow researchers to identify how the functionality of SRL behaviors change over time in relationship to the occurrences within the ITS environment. From this discussion, we provide actionable implications for contributing to how learners' SRL functionality can be visualized and scaffolded during learning with an ITS.

Keywords: Self-regulated Learning · Categorical auto-Recurrence Quantification Analysis · Intelligent Tutoring System · Complexity

1 Introduction

Self-regulated learning (SRL) is critical for successful learning with intelligent tutoring systems (ITSs) as learners are required to define the task and set goals to accomplish the task, enact cognitive and metacognitive strategies to achieve set goals, and modify strategies to complete the task more efficiently [1–5]. In deploying SRL behaviors (i.e., enactment of cognitive and metacognitive SRL strategies), learners can enhance

© The Author(s), under exclusive license to Springer Nature Switzerland AG 2024
R. A. Sottilare and J. Schwarz (Eds.): HCII 2024, LNCS 14727, pp. 116–129, 2024.
https://doi.org/10.1007/978-3-031-60609-0_9

their understanding of instructional materials throughout the ITS, thereby increasing their domain knowledge [6–9]. However, learners typically have difficulty accurately and effectively engaging in SRL [10] due to several factors such as learners' limited knowledge in the strategies they could deploy [5, 11]. As such, several studies have aimed to understand how learners should be engaging in SRL during learning with an ITS (e.g., [7, 12, 13]) but most of these studies investigate how learners transition across specific strategies without evaluating the overall emergent behavior of SRL. This traditional micro-level approach limits our understanding of how learners should be enacting macro-level SRL behaviors throughout the entire task. To address this limitation, the current study describes two analytical approaches to examining SRL through a Complex Systems Theory (CST) approach. These approaches can be used to evaluate the extent to which the behaviors of a system (i.e., SRL behaviors) display an emerging balance between novel and repetitive behavioral patterns. These methods have to potential to significantly expand our understanding of how learners should be deploying SRL behaviors over time while learning with an ITS, provide contributions for developing learner models informing the scaffolds within these environments, and extend beyond ITS environments and SRL into other aspects of human-computer interaction.

2 Examining Complexity in Self-regulated Learning

Self-regulated learning (SRL) is a critical skill that refers to learners' ability to monitor and change their own cognitive, affective, metacognitive, and motivational processes during a learning task [4, 5]. Across the several models and frameworks of SRL [5, 14–17], SRL has been touted as cyclical and recursive phases in which learners set goals and plans, deploy strategies to achieve those plans, and reflect on the effectiveness of those strategies to then modify future plans and strategies to achieve the overall goal. Underlying these models are the assumptions that SRL is multidimensional where there are multiple components which need to interact to produce SRL behaviors and temporally unfolding in which SRL evolves over time [7]. From the traditional models of SRL, we posit that SRL is a complex system as defined by Complex Systems Theory (CST), providing us the opportunity to incorporate concepts (e.g., far-from-equilibrium) and quantitative methodologies (e.g., auto-recursive quantification analysis) for interpreting how learners have and should deploy SRL behaviors over time while learning.

2.1 Complex Systems Theory

Complexity science is an interdisciplinary framework that is typically applied within the more naturalistic sciences, such as thermodynamics [18], combustion [19], physics [20], animal behavior [21], etc. to explain how systems operate [22, 23]. Within complexity science, Complex Systems Theory (CST) is defined as the study of abstract principles which explain the organization and behavior of a system [22, 24, 25]. CST states that a system can be considered complex if it characterizes self-organization [26], interaction dominance [27], and emergence [28] in which the behavior of a system cannot be attributed to a single component, rather it is the interaction and feedback/feedforward processes of these components that give rise to nonlinear system behavior [25, 29].

Past studies have argued that SRL can be considered a complex system defined through CST as SRL behaviors are not dependent on a single controller to dictate behaviors, SRL comprises of multiple components that must interact with each other, and the overall SRL behavior exhibited by learners cannot be attributed to a single component [30, 31]. For example, a component of SRL is a single SRL strategy such as planning. However, for SRL to occur, planning must be interacting or paired with other components of SRL such as control processes, the presence of goal setting, or enactment of other SRL strategies. As such, one may not say that SRL occurs when a single strategy is present, rather SRL behaviors arise when several SRL strategies are enacted to monitor, control, and change one's learning processes (see [32]). However, questions arise as to how to determine when learners' SRL systems demonstrate behaviors that are desirable for learning outcomes. In other words, how should learners' SRL systems demonstrate behaviors that are conducive to learning and how should the complexity of these behaviors change over time?

2.2 Far-From-Equilibrium

Within CST are several concepts which allow researchers to identify the extent to which system-level behaviors demonstrate healthy, functional systems. One such concept is far-from-equilibrium which defines the health of a system by its fluctuation between stable and chaotic states [33]. Specifically, a healthy functional system will demonstrate a balanced deployment of stable and chaotic states where the more removed from a stable state the system becomes, a functional system will periodically revert back to its initial stability. Conversely, an unhealthy dysfunctional system will exist consistently in either extremely stable or extremely chaotic states [34]. The extent to which a system can be labeled as either functional or dysfunctional is dependent upon how studies can capture and measure the complexity of learners' SRL systems.

2.3 Methods for Capturing Complexity in SRL

The application of a complex systems lens to SRL research is in the understanding of how intricate combinations of factors result in emergent phenomena [35–37]. In other words, how can micro-level variables (e.g., a single click) combine to produce unpredictable but commonly occurring outcomes at the macro-level (e.g., making a metacognitive judgements). Central to this approach is the understanding that SRL behaviors cannot be readily reduced to the description or analysis of the individual system component behaviors [38, 39]. This has implications for the methodological and analytical approaches that research can take to studying complex systems. Koopmans [39] outlines three major priorities of complex (dynamical) system research including the study of complex processes, the study of behavior stability and changes over time, and the study of qualitative transformations. Across these research priorities, we have seen multiple methodological approaches to address questions about the applicability of a complex system's lens to SRL research. Some of these methods include qualitative approaches (e.g., ethnography), while others are quantitative approaches aimed at addressing the structure of variable networks (e.g., social network analysis), the dynamics of communications (e.g., orbital decomposition), the dynamics of states (e.g., state space grids, transition matrices), time-series based predictions (e.g., fractional differencing), and recurrence of patterns in behaviors (e.g., recurrence quantification analysis). These approaches within educational research tend to focus primarily on behaviors of individuals (e.g., clicks within a learning environment; [30]) but can also be applied to social processes such as communication (e.g., [40]) or interactions between individuals (e.g., [41]). In this study, we utilize a type of recurrence quantification analysis (categorical aRQA) to examine the evolution of the recurrence of learner behavioral patterns as indicators of (un)healthy SRL systems.

3 Categorical Auto-recurrence Quantification Analysis

Auto-recurrence quantification analysis (aRQA) is a method that captures internal non-linear changes within both continuous as well as categorical time series [42]. For the purposes of this paper, we focus on categorical aRQA to identify repetitive patterns within learners' single time series (hence 'auto') with the overall goal of quantifying the extent to which a time series exemplifies a balance between novel and repetitive behaviors. By using categorical aRQA, categories of actions or the semantic meaning of behaviors can be analyzed, including the sequences in which learners engage in self-regulatory strategies during a learning task. Within categorical aRQA, learners' time series of events that occurred during a learning session can be plotted against itself. We

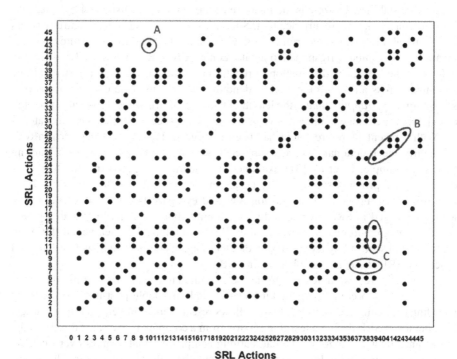

Fig. 1. Example categorical aRQA plot in which a learner's SRL actions within MetaTutor are mapped on the X and Y axes.

use Fig. 1 as our first example of an auto-recurrence plot. A recurrence plot is a structured two-dimensional matrix in which a time series is represented on both the X and Y axes. At the intersection of the X and Y axes within the matrix, a black dot represents the times at which the action the learner engages in on the X time series matches that of the action on the Y time series. Non-matching actions remain as a white indicator. Recurrence plots incorporate three main components: (Fig. 1A) recurrent point; (Fig. 1B) diagonal line representing a behavioral pattern; and (Fig. 1C) vertical and horizontal recurrent point structures representing repetitive behaviors.

From the structure of a recurrence plot, several metrics can be calculated including (but not limited to) recurrence rate, percent determinism, entropy, laminarity, and trapping time which all provide different insights into learners' time series data [43–45]. Each metric allows researchers insight into the complexity of system behaviors. *Recurrence rate* is the proportion of repetitious events across the full recurrence plot where the density of the recurrent points on the plot are quantified. Recurrence rate provides insight into how many similar states are present within the system and can be used to understand how recurrent, or repetitive, events within a time series changes over time. *Percent determinism* refers to the proportion of recurrent points on diagonal structures. Determinism is the degree of predictability a system demonstrates. If a system shows greater determinism through longer diagonal structures within the recurrence plot (i.e., greater percent determinism), the system behaviors demonstrate greater predictability, or stability, whereas shorter diagonal structures in the plot denote a greater degree of chaos, or instability. *Entropy* is the measure of the complexity displayed by deterministic structures within a recurrence plot. Similar to percent determinism, entropy refers to the probability that the recurrence plot will display diagonal line structures where a greater entropy value corresponds to greater complexity within a system. *Laminarity* is defined as the frequency distribution of repetitive events on vertical or horizontal line structures across all recurrence points, denoting the deterministic characteristics of a system. Finally, *trapping time* is the length of the vertical structures on a recurrence plot, calculating the average number of events that a system spends within a certain state.

Within recent literature, there has been a surge in the use of categorical aRQA to identify the functionality of learners' SRL systems as they engage in a learning task. For example, Li et al. [31] examined how learners engaged in SRL behaviors during clinical problem-solving using aRQA on log-file data. Another study by Dever et al. [30] used aRQA on learners' log files and eye gaze behaviors to identify how learners engaged in SRL processes during learning about microbiology with a game-based learning environment. Dever et al. [12] also examined how learners differing in the amount of scaffolding received by pedagogical agents influenced how they deployed functional SRL strategies calculated using aRQA.

Across all three examples, aRQA has been applied to each learner's full time series to evaluate system complexity, making metrics only available post-hoc, or after they have finished with their task. While this allows researchers to understand how learners have deployed SRL during a learning session from a complexity view, the disadvantage to this analytical methodology is that learning is a process that evolves over time. As such, analyses examining learning processes should be reflective of learning dynamics. While a few studies have used RQA methods to track regularity across time [46], only one study to the authors' knowledge has been published that attempts to evaluate how learners' aRQA metrics change over time within SRL. A study by Poquet et al. [41] examined how learners' recurrent SRL behavior changed over time and how learners could be profiled based on these fluctuations in behavior. This study used latent growth curve modelling on aRQA metrics that were calculated on undergraduate learners' event logs within an online learner management system after a period of one week across several weeks within a semester. While this study is the closest to understanding how learners' complexity metrics change over time, aRQA is still evaluated post-hoc, after a full week of learning activities and after a full semester. Because of this, the nuanced fluctuations *as* learners engage in learning processes such as SRL are not captured. We propose examining the fluctuations of these metrics by re-calculating aRQA in (near) real-time during ITS interactions to understand emerging SRL complexities. We propose two methods for calculating emerging SRL system complexity *during* a single learning task: (1) binning actions (e.g., Actions 1–10, 11–20, 21–30, etc.) to perform aRQA after a pre-determined number of actions have occurred or when a learning phase has been completed (i.e., binned categorical aRQA); and (2) re-calculating aRQA metrics when a new action is performed, inclusive of all prior actions (i.e., cumulative categorical aRQA).

4   Analytical Approaches for Examining Emerging Complexity

To compare our two analytical method approaches, we used a sample of learners who completed MetaTutor, an ITS about the human circulatory system (see [1]). Within MetaTutor, learners were required to interact with pedagogical agents who prompted them to engage in SRL strategies as they learned with instructional materials. There were several SRL strategies a learner could enact or be prompted to use including note taking, content evaluations, summarizing, inference making, prior knowledge activation, monitoring progress towards goals, judgments of learning, feelings of knowing, and

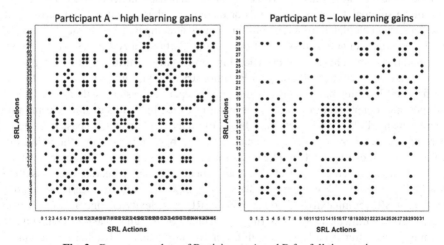

Fig. 2. Recurrence plots of Participants A and B for full time series.

planning. From the undergraduate students who learned with MetaTutor, two participants were selected based on their normalized change scores where the participant with the highest learning gain (Participant A) demonstrated the greatest increase in their human circulatory knowledge from pre- to post-test after interacting with MetaTutor where the other participant (Participant B) had the lowest learning gain (i.e., demonstrated negative learning gains; see Fig. 2). These participants were selected to serve as examples of how binned and cumulative categorical aRQA can be used to examine emerging complexities in learners' SRL processes, discuss research questions that can be used to better understand how SRL should unfold over time, and provide actionable implications for transforming how learner models interpret learners' SRL strategies and how scaffolds can be used to assist learners in deploying functional SRL.

4.1 Binned Categorical aRQA

A binned categorical aRQA approach fragments a time series into bins (see Fig. 3). For each of these bins, new aRQA metrics can be calculated and compared. This approach allows researchers to describe the complexity of a learner's SRL system in terms of how system changes because of phases (e.g., orientation vs. reflection phases) or events that occur (e.g., pedagogical agent prompting). For example, within SRL literature, learners should cyclically enact forethought, performance, and reflection phases throughout the learning task [47]. Actions which occur throughout these phases can be used within aRQA analyses to identify the functionality of the behavior displayed in each phase and how this transforms as learners move from one SRL phase to another. Another example may be the SRL actions themselves that are enacted within learning phases such as the processes required in scientific reasoning including information-gathering, hypothesis generation, and hypothesis testing [48]. In applying this binning method, researchers could better understand how learners deploy SRL strategies within each of these phases and how this changes over time.

In Fig. 3, we arbitrarily created bins of a static size, but future iterations of this method could determine bins based off environmental phases (i.e., SRL strategies enacted to achieve subgoal 1 versus subgoal 2) or theoretical phases as described above. The aRQA metrics calculated from each bin can be used to compare learners' degree of complexity and functionality across bins. For example, Participant A demonstrated visually functional system behaviors across each bin, potentially contributing to their greater learning gains. In contrast, Participant B demonstrated a visually functional system in the first and last bins but demonstrated a highly repetitive system within their second bin, possibly contributing to their lower learning gains. From this example, using this binning method allows researchers to begin to understand at which phase or state of learning learners demonstrate dysfunctional system behaviors and how this relates to learning outcomes to provide implications for scaffolding emerging SRL complexity.

By using a binned categorical aRQA analytical approach, researchers can ask: (1) how do interventions implemented at different time periods influence learners' demonstration of SRL complexity? (2) do learners' SRL complexity change as a result of a specific event within the ITS?; (3) during which phase of learning do learners with higher learning gains differ from learners with lower learning gains? Advantages to a binned categorical aRQA approach include an ability for researchers to understand how phases

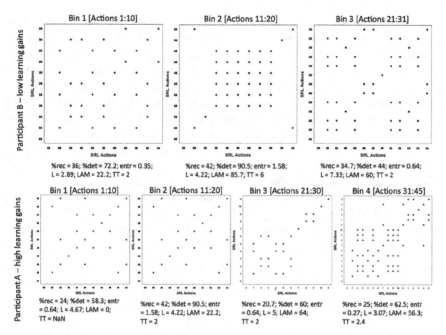

Fig. 3. Binned categorical aRQA recurrence plots and corresponding metrics.

of a learning process influence how learners demonstrate functional SRL behaviors. In this, we can evaluate the competencies of learners' SRL strategy use within a learning context while also evaluating learners' competence in transitioning across SRL phases.

However, this approach is heavily contextualized to the environment itself in which researchers will need to have already predetermined how the environment dictates how learners engage in the learning process. For example, a theoretically-based binned categorical aRQA approach assumes that for a certain number of actions learners are engaging in a single learning phase where the transition to the next learning phase is uniform across learners. While this could be challenging, the environmental affordances, such as the agency of learners, would need to be mapped to the theoretical model or framework for this approach to be successful. Additionally, this approach also ignores the assumption of parallel processing where learners could be engaging in multiple phases of learning simultaneously.

4.2 Cumulative Categorical aRQA

A cumulative categorical aRQA approach uses learners' first ten actions to calculate metrics, with each action thereafter being used to re-calculate aRQA. This approach can calculate learners' emerging complexity in their system behaviors where as the learner progresses through a task, the trends in aRQA metrics that are continuously re-calculated can show how a learner's behaviors demonstrate increasingly functional, dysfunctional, or unstable functionality.

As an example, we used a cumulative categorical aRQA analytical approach to understand how Participant A and B's recurrence rate and percent determinism changed over time. There are several interpretations researchers can glean from this approach. For example, Participant A (who had greater learning gains) demonstrated a more level approach to their overall recurrence rate and percent determinism scores than Participant B (who had lower learning gains) which can be confirmed with a growth modeling approach. Researchers can identify the rapidity in which stability of cumulative aRQA metrics are reached as well as the magnitude of change in these metrics. In comparing the change across these metrics between more successful and less successful learners, researchers can begin to understand how a functional system displays emerging complexity through this approach. With a cumulative categorical aRQA approach, we can ask: (1) to what extent do learners display an emerging SRL complexity?; (2) at what time do learners display a shift from nonfunctional to functional SRL behaviors?; (3) does the stability of learners' complexity of SRL behaviors indicate greater functional behaviors? (Fig. 4).

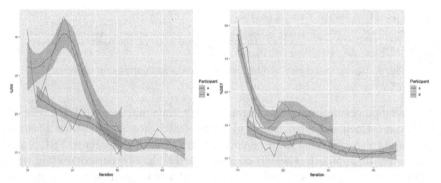

Fig. 4. Cumulative categorical aRQA %RR (left) and %DET (right) as time progresses on task for Participants A (high learning gains) and B (low learning gains).

The major advantage of using this cumulative categorical aRQA approach is that each re-calculation of aRQA metrics accounts for the full history of learners' actions which addresses the assumption that SRL unfolds over time. In addition, this approach can be used without contextualization to the environment in which SRL is occurring as it is the overall system behavior that is of interest, rather than the specific strategies or phases of learning.

5   Future Directions for Research

Our paper addresses several major conceptual, theoretical, methodological, analytical, and educational issues related to self-regulated learning and learning with intelligent tutoring systems [1, 49]. First, treating complexity in self-regulated learning refers to recognizing and managing the various factors, variables, and intricacies involved when individuals actively monitor and regulate their learning. The first significant step is to assume and understand conceptually that the complexity in SRL can arise from several sources, including cognitive factors such as the mental processes involved in learning, such as attention, memory, and problem-solving, which contribute to the complexity. Learners must navigate and regulate their cognitive processes effectively. Also included are metacognitive factors, which involve awareness and control of one's thinking processes that are quite demanding cognitively and may induce extraneous cognitive load on learners [5]. Often neglected are motivational factors that are likely to fluctuate throughout learning but are rarely captured given the over-reliance on self-report and lack of sophisticated tools to examine the temporal unfolding of motivational states over longer time dimensions (e.g., changes over minutes, hours, days) that impact a learners' evolving sense of self and impact factors like interest, perceived competence, and the perceived value of the learning task. Lastly, complexity also emerges when we con-sider the role of emotional states, such as dealing with confusion, frustration, and boredom that require emotion regulation strategies to continue to engage in productive learning, reasoning, and problem-solving. Treating complexity in self-regulated learning involves understanding, addressing, and optimizing these various factors. It includes the development of skills and strategies that enable learners to adapt to different situations and challenges. By acknowledging and addressing the complexity inherent in self-regulated learning, educators and learners can enhance the learning processes' effectiveness and efficiency, expanding the theoretical assumptions about the dynamics of SRL processes that unfold in (near-)real-time during learning-system interactions. Future research should empirically test assumptions about the dynamics of various SRL processes to determine when, how, and why changes will occur, by how much, if they recur (or not), give a myriad of affordances provided by intelligent tutoring systems.

Our study also highlights the significant benefits of using complex system theory analytical approaches such as auto-recurrence quantification analysis (aRQA) to measure the complexity of self-regulatory learning processes during learning with intelligent tutoring systems [7]. We also highlight several productive areas for future research. First, our analytical techniques allow researchers to capture dynamic intricate interactions between various components and thus present analyses of temporal patterns and recurrent structures in the self-regulatory learning process, providing insights into how different elements interact over time. As such, we need to better understand the nature of each SRL process and its subprocesses (e.g., metacognitive judgments such as content evaluations vs. judgment of learning [JOLs]; see [6, 50]) and the internal (e.g., learners' cognitive system) and external conditions (e.g., scaffolding provided by the intelligent tutoring system's pedagogical agents) under which they are triggered and how they are monitored and regulated by the learner and system. Second, they can capture, identify, and quantify nonlinear patterns of SRL processes, which are crucial for understanding the adaptive and evolving nature of self-regulatory processes in the learner. Future research must determine how to provide learners with intelligent, adaptive individual scaffolding based on the dynamics of each SRL process and their unique fluctuation over time. Third, we can detect phase transitions, such as when complex systems undergo phase transitions (e.g., planning, enactment to reflection), where there is a qualitative change in the system's behavior, indicating shifts in the learner's strategies or cognitive states. However, future research needs to capture how phases of SRL change not only temporally and sequentially but also whether they may last longer for certain learners when they return to previous phases to support learners' SRL and why. Fourth, a significant advantage of our analytical approach is the ability to quantify adaptability—i.e., learning is an adaptive process, and aRQA provides a means to quantify the adaptability of self-regulatory strategies. By analyzing recurrence patterns, researchers can gain insights into how learners adjust their strategies in response to challenges or changing task demands, such as reactions to pedagogical agents' scaffolding, the complexity of instructional materials, etc. This will require future research to focus on using multimodal trace data to capture, isolate, and examine SRL processes at a fine-grained level of precision. Fifth, we can assess system robustness and stability since complex systems theory emphasizes the importance of robustness and stability. As such, aRQA provides information on the system's stability, helping to identify periods of effective self-regulation and potential vulnerabilities in the learning process. Again, multimodal trace data will be key in pinpointing these periods and generating behavioral signatures that intelligent tutoring systems can use as triggers to pedagogically intervene. Sixth, our approach allows for a holistic under-standing of the complexity of SRL by allowing researchers to examine patterns of recurrence at multiple levels and across different dimensions of self-regulatory learning; analytical approaches such as aRQA allow researchers to consider the complex interplay of cognitive, metacognitive, motivational, and emotional factors. A major emphasis for future research is to decipher and determine if these patterns, both within and across SRL processes, are correlational or causational in nature. Lastly, our approach allows researchers to develop more sophisticated, intelligent tutoring systems that are capable of measuring, modeling, tracking, and therefore providing more effective individualized scaffolding and feedback due to their ability to make inferences about multimodal trace data about the complexity of self-regulatory learning process-es. This will enhance their predictive modeling and forecasting of quantitative and qualitative changes in learners' SRL knowledge, strategies, and skills. In summary, applying complex system theory analytical approaches like aRQA to study self-regulatory learning processes with intelligent tutoring systems provides a nuanced and comprehensive understanding of the dynamic, nonlinear, and adaptive nature of learning. These insights can inform future learning technologies, embed more sophisticated pedagogical interventions, and contribute to developing more effective intelligent tutoring systems [3, 51].

Acknowledgments. The research reported in this manuscript was supported by funding from the National Science Foundation (DRL#1916417, DRL#1661202 and DUE#1761178). The authors would also like to thank the members of the SMART Lab at the University of Central Florida.

Disclosure of Interests. The authors have no competing interests to declare that are relevant to the content of this article.

References

1. Azevedo, R., et al.: Lessons learned and future directions of MetaTutor: leveraging multi-channel data to scaffold self-regulated learning with an intelligent tutoring system. Front. Psychol. **13**, 813632 (2022)

2. Azevedo, R., Gašević, D.: Analyzing multimodal multichannel data about self-regulated learning with advanced learning technologies: issues and challenges. Comput. Hum. Behav. **96**, 207–210 (2019)

3. Azevedo, R., Wiedbusch, M.: Theories of metacognition and pedagogy applied in AIED systems. In: du Boulay (ed.) Handbook of Artificial Intelligence in Education, pp. 45–67. Springer, Dordrecht (2023)

4. Winne, P., Azevedo, R.: Metacognition and self-regulated learning. In: Sawyer, K. (ed.) The Cambridge Handbook of the Learning Sciences, 3rd edn, pp. 93–113. Cambridge University Press (2022)

5. Winne, P.H.: Cognition and metacognition within self-regulated learning. In: Schunk, D.H., Greene, J.A. (eds.) Educational Psychology Handbook Series. Handbook of Self-regulation of Learning and Performance, pp. 36–48. Routledge/Taylor & Francis Group (2018)

6. Azevedo, R., Dever, D.A.: Metacognition in multimedia learning. In: Mayer, R.E., Fiorella, L. (eds.) Cambridge Handbook of Multimedia Learning, 3rd edn., pp. 132–142. Cambridge University Press, Cambridge (2022)

7. Dever, D.A., Sonnenfeld, N.A., Wiedbusch, M.D., Schmorrow, S.G., Amon, M.J., Azevedo, R.: A complex systems approach to analyzing pedagogical agents' scaffolding of self-regulated learning within an intelligent tutoring system. Metacognition Learn. 1–33 (2023)

8. Dever, D.A., Wiedbusch, M.D., Romero, S.M., Azevedo, R.: Investigating pedagogical agents' scaffolding of self-regulated learning in relation to learners' subgoals. Br. J. Educ. Technol. (2024)

9. Taub, M., Azevedo, R.: How does prior knowledge influence eye fixations and sequences of cognitive and metacognitive SRL processes during learning with an Intelligent Tutoring System? Int. J. Artif. Intell. Educ. **29**(1), 1–28 (2019)

10. Munshi, A., Biswas, G.: Personalization in OELEs: developing a data-driven framework to model and scaffold SRL processes. In: Isotani, S., Millán, E., Ogan, A., Hastings, P., McLaren, B., Luckin, R. (eds.) Artificial Intelligence in Education. Lecture Notes in Computer Science, vol. 11626, pp. 354–358. Springer, Cham (2019). https://doi.org/10.1007/978-3-030-23207-8_65

11. Schunk, D.H., Greene, J.A.: Handbook of Self-regulation of Learning and Performance, 2nd edn. Routledge/Taylor & Francis Group, New York (2018)

12. Dever, D.A., Sonnenfeld, N.A., Wiedbusch, M.D., Azevedo, R.: Pedagogical agent support and its relationship to learners' self-regulated learning strategy use with an intelligent tutoring system. In: Rodrigo, M., Noburu, M., Cristea, A., Dimitrova, V. (eds.) International Conference on Artificial Intelligence in Education, pp. 332–343. Springer, Cham (2022). https://doi.org/10.1007/978-3-031-11644-5_27

13. Wiedbusch, M., Dever, D., Wortha, F., Cloude, E., Azevedo, R.: Revealing data feature differences between system- and learner-initiated self-regulated learning processes within hypermedia. In: Sottilare, R.A., Schwarz, J. (eds.) Human Computer Interaction International, pp. 481–495. Springer, Cham. (2021). https://doi.org/10.1007/978-3-030-77857-6_34

14. Mayer, R.E.: Computer games in education. Annu. Rev. Psychol. **70**, 531–549 (2019)

15. Nelson, T.O.: Metamemory: a theoretical framework and new findings. Psychol. Learn. Motiv. **26**, 125–173 (1990)

16. Pintrich, P.R.: The role of goal orientation in self-regulated learning. In: Boekaerts, M., Pintrich, P.R., Zeidner, M. (eds.) Handbook of Self-regulation, pp. 451–502. Academic Press (2000)

17. Zimmerman, B.J., Moylan, A.R.: Self-regulation: where metacognition and motivation intersect. In: Hacker, D.J., Dunlosky, J., Graesser, A.C. (eds.) Handbook of Metacognition in Education, pp. 299–315. Routledge (2009)

18. Chatterjee, A., Iannacchione, G.: The many faces of far-from-equilibrium thermodynamics: deterministic chaos, randomness, or emergent order? MRS Bull. **44**(2), 130–133 (2019)

19. Sujith, R.I., Unni, V.R.: Dynamical systems and complex systems theory to study unsteady combustion. Proc. Combust. Inst. **38**(3), 3445–3462 (2021)

20. Battiston, F., et al.: The physics of higher-order interactions in complex systems. Nat. Phys. **17**(10), 1093–1098 (2021)

21. Fisher, D.N., Pruitt, J.N.: Insights from the study of complex systems for the ecology and evolution of animal populations. Curr. Zool. **66**(1), 1–14 (2020)

22. Favela, L.H., Amon, M.J.: Reframing cognitive science as a complexity science. Cogn. Sci. **47**(4), e13280 (2023)

23. Mitchell, M.: Complexity: A Guided Tour. Oxford University Press, New York (2009)

24. Van Orden, G.C., Holden, J.G., Turvey, M.T.: Human cognition and 1/f scaling. J. Exp. Psychol. Gen. **134**(1), 117–123 (2005)

25. Wiener, N.: Cybernetics. Sci. Am. **179**, 14–19 (1948)

26. Kelso, J.A.S.: Dynamic Patterns: The Self-organization of Brain and Behavior. MIT Press, Cambridge (1995)

27. Strogatz, S.H.: Nonlinear Dynamics and Chaos: With Applications to Physics, Biology, Chemistry, and Engineering, 2nd edn. CRC Press, New York (2015)

28. Francescotti, R.M.: Emergence. Erkenntnis **67**, 47–63 (2007)

29. Favela, L.H.: Cognitive science as complexity science. Wiley Interdisc. Rev. Cogni. Sci. **11**, e1525 (2020)

30. Dever, D.A., Amon, M.J., Vrzakova, H., Wiedbusch, M.D., Cloude, E.B., Azevedo, R.: Capturing sequences of learners' self-regulatory interactions with instructional material during game-based learning using auto-recurrence quantification analysis. Front. Psychol. **13**, 813677 (2022)

31. Li, S., Zheng, J., Huang, X., Xie, C.: Self-regulated learning as a complex dynamical system: examining students' STEM learning in a simulation environment. Learn. Individ. Differ. **95**, 102144 (2022)

32. Fan, Y., et al.: Improving the measurement of self-regulated learning using multi-channel data. Metacogn. Learn. **17**, 1025–1055 (2022)

33. Prigogine, I., Stengers, I.: Order Out of Chaos: Man's New Dialogue with Nature. Flamingo, London (1985)

34. Larsson, J., Dahlin, B.: Educating far from equilibrium: chaos philosophy and the quest for complexity in education. Complicity Int. J. Complex. Educ. **9**, 1–14 (2012)

35. Hilpert, J.C., Marchand, G.C.: Complex systems research in educational psychology: aligning theory and method. Educ. Psychol. **53**(3), 185–202 (2018)

36. Marchand, G.C., Hilpert, J.C.: Complex systems approaches to educational research: introduction to the special issue. J. Exp. Educ. **88**(3), 351–357 (2020)

37. Marchand, G.C., Hilpert, J.C.: Contributions of complex systems approaches, perspectives, models, and methods in educational psychology. In: Schutz, P.A., Muis, K.R. (eds.) Handbook of Educational Psychology, 4th edn., pp. 139–161. Routledge, New York (2023)

38. Hilpert, J.C., Greene, J.A., Bernacki, M.: Leveraging complexity frameworks to refine theories of engagement: advancing self-regulated learning in the age of artificial intelligence. Br. J. Edu. Technol. **54**, 1204–1221 (2023)

39. Koopmans, M.: Education is a complex dynamical system: challenges for research. J. Exp. Educ. **88**(3), 358–374 (2020)

40. Gorman, J.C., Amazeen, P.G., Cooke, N.J.: Team coordination dynamics. Nonlinear Dyn. Psychol. Life Sci. **14**(3), 265–289 (2010)

41. Poquet, O., Jovanovic, J., Pardo, A.: Student profiles of change in a university course: a complex dynamical systems perspective. In: Hilliger, I., Khosravi, H., Rienties, B., Dawson, S. (eds.) LAK23: 13th International Learning Analytics and Knowledge Conference, pp. 197–207. ACM (2023)

42. Webber, C.L., Zbilut, J.P.: Recurrence quantification analysis of nonlinear dynamical systems. Tutorials Contemp. Nonlinear Methods Behav. Sci. **94**, 26–94 (2005)
43. Bhardwaj, R., Das, S.: Recurrence quantification analysis of a three level trophic chain model. Heliyon **5** (2019)
44. Meinecke, A.L., Handke, L., Mueller-Frommeyer, L.C., Kauffeld, S.: Capturing non-linear temporally embedded processes in organizations using recurrence quantification analysis. Eur. J. Work Organ. Psy. **29**(4), 483–500 (2020)
45. Wallot, S.: Recurrence quantification analysis of processes and products of discourse: a Tutorial in R. Discourse Process. **54**(5–6), 382–405 (2017)
46. Amon, M.J., Vrzakova, H., D'Mello, S.K.: Beyond dyadic coordination: multimodal behavioral irregularity in triads predicts facets of collaborative problem solving. Cogn. Sci. **43**(10), e12787 (2019)
47. Zimmerman, B.J., Schunk, D.H.: Self-regulated learning and performance: an introduction and an overview. In: Zimmerman, B.J., Schunk, D.H. (eds.) Handbook of Self-regulation of Learning and Performance, pp. 15–26. Routledge/Taylor & Francis Group (2011)
48. Klahr, D., Dunbar, K.: Dual space search during scientific reasoning. Cogn. Sci. **12**, 1–48 (1988)
49. D'Mello, S.K., Graesser, A.: Intelligent tutoring systems: how computers achieve learning gains that rival human tutors. In: Schutz, P.A., Muis, K.R. (eds.) Handbook of Educational Psychology, 4th edn, pp. 603–629. Routledge (2023)
50. Greene, J.A., Azevedo, R.: A macro-level analysis of SRL processes and their relations to the acquisition of a sophisticated mental model of a complex system. Contemp. Educ. Psychol. **34**(1), 18–29 (2009)
51. du Boulay, B.: Pedagogy, cognition, human rights, and social justice. Int. J. Artif. Intel. Educ. 1–6 (2023)

Moving Beyond Physiological Baselines: A New Method for Live Mental Workload Estimation

Torsten Gfesser$^{(\boxtimes)}$ ⓘ, Thomas E. F. Witte ⓘ, and Jessica Schwarz ⓘ

Fraunhofer FKIE, Fraunhoferstr. 20, 53343 Wachtberg, Germany
{torsten.gfesser,thomas.witte,
jessica.schwarz}@fkie.fraunhofer.de

Abstract. The analysis of physiological data can provide valuable information on the mental state of users interacting with a technical system, such as an intelligent tutoring system. By obtaining live estimations of mental workload a learning system can adapt, e.g., the level of difficulty of tasks to the learners needs. However, the analysis and interpretation of physiological data usually requires a baseline recording at a rested state prior to or after a task limiting their practical value. Additionally, the baseline of a physiological measure cannot be considered as a stable value but varies between days and even within a day interpersonally, so the validly calibrated data of a baseline become invalid over time limiting its value for long term use cases.

This paper proposes a new method for near real time mental workload estimation. A machine learning model which predicts the mental workload based on the heart rate variability (HRV) derives metrics without the necessity of baseline recordings. First, a machine learning model is trained on a dataset of previously collected physiological data and corresponding mental workload ratings. Subsequently, physiological measures are collected continuously from a participant throughout tasks. The model is then used to predict the participant's mental workload in real time based on the HRV data.

The results of our pilot study show first empirical support, that the proposed analysis technique is able to estimate mental workload in near real time with an accuracy of 90%.

As this technique does not depend on baseline recordings it has the potential to be specifically valuable in applied settings such as adaptive training systems or to monitor the mental health of workers in safety-critical industries. The method could also be extrapolated for the analysis of other physiological measures in future research.

Keywords: Mental Workload · Baseline · Real Time · Physiological · HRV · Artificial Intelligence

1 Introduction

Live estimation of mental states, such as mental workload, is an important requirement for the design of adaptive technical systems that adapt their behavior to the current state of the user. As an example, adaptive systems may use live detection of mental

R. A. Sottilare and J. Schwarz (Eds.): HCII 2024, LNCS 14727, pp. 130–146, 2024.
https://doi.org/10.1007/978-3-031-60609-0_10

workload to support the operator, if a critically high level of mental workload, has been detected. In the learning context such information could be used by an adaptive training system to specifically provoke states of high workload to train the learner how to cope with critical conditions. Measuring mental workload by physiological measures such as heart rate, heart rate variability, or pupil size is one of the most prominent approaches in this context. Benefits compared to e.g., subjective ratings are that most of these measures can be recorded continuously during a task without disturbing the user and data can be analyzed in near real time. However, physiological reactions can differ strongly between and even within individuals depending on the fitness level, caffeine consumption, and physical activity among others. A common approach to account for inter- and intraindividual differences is recording a baseline at the beginning or at the end of a task in a relaxed state and comparing recorded data during a task with the baseline value. However, in real-world applications baseline recordings are not always a suitable method as there is often not enough time for a baseline recording. Also, if baselines must be recorded regularly to be able to work with a technical system, this can be disturbing and often lowers the user acceptance of the system.

In this paper we introduce a new method for live mental workload estimation without baseline recordings, making it more applicable for real-world settings. We used the heart rate variability (HRV) as a physiological measure to develop this method, because it is considered as an established indicator of mental workload and stress.

Section 2 gives an overview on prior studies on HRV assessment. Section 3 introduces our new method for live analysis, also describing the data used for validation. Section 4 describes the AI Classifier, which was trained for the live analysis of mentally demanding tasks based on the calculated metrics from Sect. 3. This paper ends with a discussion, the limitations of the method as well as conclusions and future developments.

2 Heart Rate Variability as an Indicator of Mental Workload

Mental workload refers to the amount of mental effort and resources required to perform a specific task, encompassing cognitive processing, attention, and effort exerted by an individual while engaging in a task [1]. It is a measure of the cognitive and perceptual demands of a task, influenced by factors such as task complexity, time pressure, and environmental conditions [2]. Heart rate variability (HRV) analysis measuring the variation of time intervals between heartbeats is a valuable method for assessing autonomic function of the cardiovascular system. As such it is often used as a potential biomarker for various health conditions [3] and for predicting mental workload in various settings.

The HRV can be calculated using various methods, including time-domain and frequency-domain. In time-domain analysis, parameters such as SDNN (standard deviation of NN intervals) and RMSSD (root mean square of the differences between adjacent NN intervals) are commonly used [4]. Frequency-domain methods involve the use of spectral analysis, such as fast Fourier transform and autoregressive model, to calculate parameters like high frequency and low frequency components [5]. The Task Force of the European Society of Cardiology recommends the use of SDNN and RMSSD as widely adopted measures of HRV [4].

Literature shows that there is a negative correlation between the SDNN and subjective mental workload [6], which also applies to the RMSSD [7]. Delliaux et al. [8]

and Radüntz et al. [9] have focused on characterizing the impact of mental workload on cardiovascular function using HRV non-linear indexes and inherent timescales of cardiovascular biomarkers, providing insights into the relationship between HRV and mental workload. Their results indicate that mental workload significantly lowered the non-linear dynamics of RR interval [8] and that the assessment of mental workload using cardiovascular biomarkers' inherent timescales provide valuable insights into the physiological responses associated with varying levels of cognitive demand [9]. Also, Forte et al. [10] concluded in their systematic review about the relationship between HRV and cognitive functions, that HRV is closely linked to cognitive function. Veltman and Jansen [11] emphasized the differentiation of mental effort measures and its consequences for adaptive automation, highlighting the importance of HRV in assessing cognitive workload. The physiological basis of HRV as a reflection of autonomic nervous system activity and its role in emotion regulation further supports its relevance in adaptive systems, as highlighted by Witte et al. [12].

Veltman and Gaillard [13] concluded that HRV is a sensitive index for mental workload when tasks are highly demanding, emphasizing its relevance in assessing cognitive demand during complex tasks. Research by Cinaz et al. [14] and Shao et al. [15] has focused on the use of HRV in monitoring mental workload levels during office-work scenarios and human-robot interaction, respectively, highlighting the versatility of HRV in diverse domains and environments. Shao et al. [15] conducted a comparison analysis of different time-scale HRV signals, demonstrating the applicability of HRV in evaluating cognitive demand during interactive tasks. This supports the notion that HRV can discern fluctuating task demands and attenuate during mentally straining workloads, as stated by Nardolillo et al. [16].

The findings from these studies collectively highlight the potential of HRV as a valuable physiological marker for assessing mental workload during software tasks, providing insights into cognitive demand and adaptive responses in various task environments.

2.1 Physiological Baselines

One of the most common methods to account for intra- and interindividual differences in physiological measures, such as HRV, is to use baselines. This method is considered as the gold standard for analyzing HRV and other physiological measures, where the "tonic level measured immediately prior to stimulation is referred to as the baseline, the level of activity against which we compare the phasic response to a stimulus" [17]. A baseline can be obtained by taking the average of HRV values over a specified time interval, such as five minutes, under resting or non-stressful conditions before or after a task [18]. This baseline can then be used to compare it against HRV values collected during physical or mental tasks. The method of obtaining a baseline is visualized in Fig. 1.

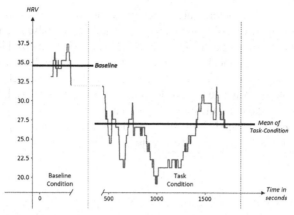

Fig. 1. Baseline for HRV obtained in the first five minutes (300 s) in comparison to the HRV mean of the task condition.

2.2   Challenges of Live HRV Analysis

While HRV is a valuable health metric, it has certain peculiarities that make it difficult to establish a definitive baseline that is valid over long periods. When it comes to baseline-measurements the respective physiological variable should be stable across the analyzed period. Regarding short-term measurements, Tarkiainen et al. [19] conducted a study on the stability over time of short-term HRV, indicating that most short-term HRV measures were highly stable over time in laboratory conditions. They further conclude, that their SDNN obtained during 40-min recordings was more stable than the SDNN obtained during 5-min periods and the SDNN showed large variability in consecutive recordings. The stability and variance of HRV appear to vary depending on the period of recording.

Baseline measurements compare a short period of HRV values to another period of HRV values, similar to a consecutive recording. However, baseline measurements are implying that the physiological variable is stable in variance across time, regardless of the period. This assumption is important because it enables the comparison of the participant's response to a stimulus and their initially recorded baseline level. Physiological measurements can result in chaotic timeseries [20] violating this assumption. Chaotic timeseries refers to a type of time series data that exhibits chaotic behavior, characterized by instability, unpredictability, and sensitivity to initial conditions [21]. Without stability, there may be no anchor point in the psychophysiological data that could serve as a good baseline.

HRV baselines can become less valid over time because the range of values being evaluated can change. This may be the case when comparing consecutive recordings of short-term HRV. Individual HRV values vary greatly. Age, gender, health status, and even psychotropic substances like caffeine intake can influence the HRV. Secondly, intrapersonal homeostasis is a factor of influence, causing natural fluctuations even without external stimuli [22]. That means that a fixed baseline can shift throughout the day based on the current physical state of the individuum. Comparing absolute values, like the baseline, across individuals or even comparing someone's current HRV to their own baseline, or any past measurements, can therefore be misleading. If the HRV changes too much without external factors, there is a high chance of false positives (false alarms e.g., false classification of tasks as mentally demanding) or false negatives (misses, missed detections of mentally demanding tasks). This case is visualized in Fig. 2 where according to the initial baseline, huge parts of the following values would be classified as significant mentally demanding. A valid and reliable baseline must thereby shift based on the current physiological change, that is not related to external factors. It has to migrate with it as a measure, like normalizations do. A change in the current physiological state must be recognized and the shift of the baseline and a shift due to physiological or cognitive impact factors has to be differentiated.

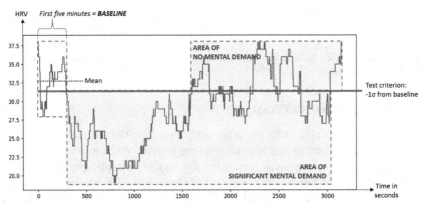

Fig. 2. HRV values from one participant over a period of around 55 min, where the first five minutes are serving as the baseline. The test criterion by which HRV is classified as mentally demanding is HRV values below the first standard deviation (-1σ) based on the baseline values.

The condition of the participant is usually unknown when the baseline is taken. If the subject is too excited at the beginning, then the HRV may only increase over time. If the subject is too relaxed at the beginning, then the HRV may initially lose altitude rapidly. This is called the law of initial value, where the magnitude of a response to a stimulus depends on the starting level of the measured variable [23]. For example, if a participant already has a very low HRV, which would also be measured as a baseline, then HRV-lowering stimuli may no longer be noticeable. However, if the participant has a very high HRV at the beginning, it may be that the entire test period is significantly below the initial value and therefore the effects of individual stimuli are no longer recognized.

Small or finer oscillations in the data are completely ignored when physiological baselines only compare the mean of a few conditions. Oscillations within a condition, based on single stimuli, will vanish through the calculation of means for whole experimental phases.

HRV is often used in retrospective analysis, where at least the values of the recorded timespan are completely available for state-of-the-art time series analysis techniques, such as removing possible trends or seasonality from the data. But without knowing the future data, an online or real time method, can only rely on the current data and data recorded in the past. Future data can't be forecasted because of the chaotic nature of the HRV time series, as stated previously in this section. So, all useful metrics must be calculated based on the latest and passed HRV values.

The stated challenges are mostly valid for detailed live analysis with only HRV as a parameter. Therefore, several approaches like the multifactorial RASMUS Framework from Schwarz and Fuchs [24], use a combination of multiple parameters to provide more robust user state analyses.

Hoover et al. [25] tried to detect changes in mental workload based on real-time monitoring of HRV. Their original intent was to determine whether a change in task caused a change in HRV measurement. Using sub-Gaussian functions, they were able to successfully detect change points based on a change in tasks. This provided insight that mentally demanding tasks can be identified by changes in HRV. Further, they conclude that their method can successfully detect changes that are quite subtle.

However, to the best of our knowledge literature does not suggest a profound method for the live classification of short or ultra-short fluctuations of a single physiological parameter, like the HRV.

3 Introducing a New Method for Live Mental Workload Estimation

In this section we present a new method to enable a more detailed analysis of HRV fluctuations that may be useful e.g., for an efficient live analysis of workload in adaptive system design or for the evaluation of applications and tools in and outside of a laboratory. Our method aims at eliminating the need for regular baselines and providing a metric that is stable over time periods for classifying mental demanding fluctuations in the HRV.

3.1 Re-analysis of Data from a Prior Study

A prior study conducted by Bruder and Schwarz [26] was used to develop the method of this paper by re-analyzing the data, where the HRV was calculated as a rolling 300 Heartbeat SDNN. A HRV baseline was calculated for each participant based on the first 120 s. During that time, the mental workload was kept low to moderate. This baseline was further used as a test criterion where a HRV value lower than one standard deviation from the baseline distribution will be classified as critically high and a value greater than one standard deviation as critically low mental workload when coinciding with a performance decrement. Figure 3 visualizes the concept and classification based on the data of one participant from the original study.

The initial study of the authors was followed by a validation study which confirmed that their used method is temporally valid and, moreover, could distinguish between three different conditions named baseline, high workload, and monotony. Workload assessment was based in this study on a combination of HRV and four other workload indicators (the number of tasks, number of mouse clicks, pupil diameter, and respiration rate) to compensate for inaccurate classifications of single indicators. The method proposed in this paper aims at developing this approach further by providing HRV-based live classifications of even short-term mentally demanding tasks, thus increasing the accuracy of this classifier.

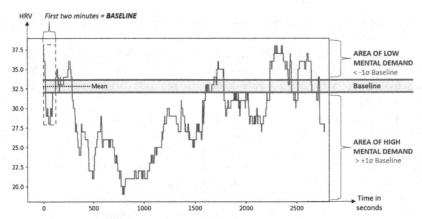

Fig. 3. The classification of low, none and high mental demand based on a two-minute baseline, where a HRV value lower than 1σ will be classified as critically high and a value greater than 1σ as critically low mental workload.

3.2 Experimental Setting

For the evaluation of the method provided in this paper, we utilized the existing data from the previous study conducted by Bruder and Schwarz [26] about the evaluation of diagnostic rules for real time assessment of mental workload within a dynamic adaptation framework. The framework was operationalized for an air traffic surveillance task.

The original study involved a sample of 15 participants (8 males, 7 female) aged between 20 and 51 years (M = 31.26 ± 8.27). A multisensory chest strap (Zephyr BioHarness 3) was used to collect data on HRV and respiration rate. Pupil diameter was recorded with an eye tracker placed underneath the monitor.

Participants began with a ten-minute training session where the examiner clarified task completion for each subtask. Following this, they engaged in a 45-min experimental test divided into three continuous phases, punctuated with a survey when a performance decrement on one of the tasks had been detected. In this survey, participants rated their perceived mental workload and other mental states. The experiment's duration therefore varied based on user performance, with additional workload ratings recorded both after training and at the experiment's conclusion to establish individual baselines [26].

3.3 Concept for a Live Analysis of Cognitive Workload

Steps for live analysis of the HRV data are normalizing the incoming HRV values and calculating additional metrics such as the slope of the ascending and descending HRV. Finally, the normalized HRV and metrics are used to be classified using a pre-trained machine learning model. The whole data flow is shown in Fig. 4. The steps specifically of the live classification will be explained in detail in the following section.

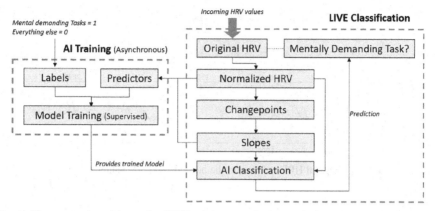

Fig. 4. The processing of incoming HRV values towards the prediction of mentally demanding tasks.

3.4 Live Normalization for Non-stationary Timeseries

It is important to normalize the HRV for comparability even of the same subject, because of the unpredictability of the future HRV values and regarding to the intraday differences.

In our method, we estimate a rolling normal distribution with a maximum likelihood estimation (MLE) based on the last 60 HRV values. The calculated mean from the distribution serves as the current plateau of the HRV, from which a new gradient will relatively be measured. Mathematically, the distribution means from the last 60 HRV values will be subtracted from the incoming HRV values, to get the relative value compared to the last minute in our case.

$$HRV = \{hrv_1, hrv_2, \ldots, hrv_n\}$$

$$Rolling\ Mean = \frac{1}{60}\sum_{i=n-59}^{n} HRV_i$$

$$HRV_{Normalized} = \{hrv_1 - Rolling\ Mean, hrv_2 - Rolling\ Mean, \ldots, hrv_n - Rolling\ Mean\}$$

This normalization forces the values to move around a center of 0 for each subject, which acts detrending, so it removes trends in the time series. Figure 5 plots the HRV values from one subject over a period of around 53 min with the normalized HRV values below.

3.5 Online Change Point Detection

The second step is to recognize a change in the incoming flow of HRV values. Whenever there is a significant change, we must mark the position of that change point.

A comparison of two single values, so the last past value and the newest, won´t be accurate in case of slightly fluctuating values. To address this problem, we always take the last twenty HRV values, where the first ten values are compared with a Wilcoxon rank sum test to the latest ten values, formulated as:

$$HRV_{Normalized} = \{hrv_1, hrv_2, \ldots, hrv_n\}$$

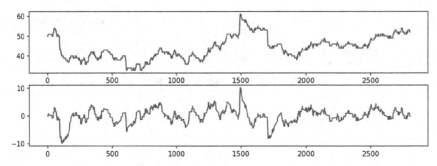

Fig. 5. Original HRV values and the normalized HRV values below.

$$HRV_{Group1} = \{hrv_{n-20}, hrv_{n-19}, \ldots, hrv_{n-10}\}$$

$$HRV_{Group2} = \{hrv_{n-9}, hrv_{n-8}, \ldots, hrv_n\}$$

$$W = WilcoxonRankSum(HRV_{Group1}, HRV_{Group2})$$

This is visualized in Fig. 6 are grouped colored in blue. In most cases, one or both groups do not meet the test assumptions of parametric inference statistical tests, such as the assumption of normal distribution. Therefore, the two groups are compared using a non-parametric Wilcoxon rank sum test with a fixed 5% alpha significance level.

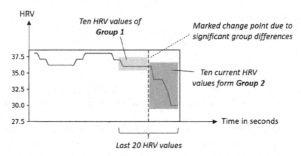

Fig. 6. Calculation of change points based on the previous 1–10 values (Group 1) and the newest 11–20 values (Group 2).

In our method, the test criterion of the rank sum test will always be tested twice as one-sided tests for lower and greater significant difference. If the latest ten values are significantly greater, than an ascending changepoint will be reported. If significantly lower, a descending changepoint will be reported.

$$W_{Lower} = WilcoxonRankSum(HRV_{Group1}, HRV_{Group2})$$

$$Descending\ Changepoint = \begin{cases} True\ if\ W_{Lower}\ is\ < 0.05\ \alpha \\ False\ otherwise \end{cases}$$

$$W_{Greater} = WilcoxonRankSum(HRV_{Group1}, HRV_{Group2})$$

$$Ascending\ Changepoint = \begin{cases} True\ if\ W_{Greater}\ is\ < 0.05\ \alpha \\ False\ otherwise \end{cases}$$

If there are two or more change points, each following the same direction, e.g., all lower or all higher, the series of all such change points is cached until the series is broken by a new change point with a different direction. The cached series then forms a single ascent or descent to be able to calculate a slope over the entire ascent or descent series.

After a change point has been detected, a dead time of 10 HRV values starts in order to avoid that the already significant range of values is compared again. Figure 7 visualizes the dead time, where the first group of ten values was significantly different compared to the second group of ten values. So, a new change point was identified between the first and second groups. To avoid a new comparison of the first group, no group comparison is carried out as long as ten new values are available. After that, for each new value, the groups are calculated as specified in the paragraphs above.

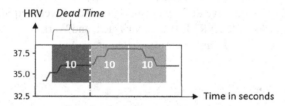

Fig. 7. Dead time in the change point detection.

3.6 Calculating the Slopes

The third step is to calculate the slopes between the change points. Every time a new change point is detected, we can calculate the slope between the previous change point and the new one, as well as the slope over an entire series of ascent or descent change points. Figure 8 visualizes the calculated slope between two descending change points.

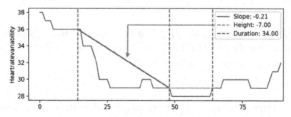

Fig. 8. Two visualized slopes between three change points. Red dotted lines are descending change points, green dotted lines are ascending change points. The red line is the ascending slope between two change points, followed by a yellow line indicating no slope. (Color figure online)

3.7 Feeding the Classifier

When a new slope has been calculated, we add the slope, its height and length to the previously normalized HRV values for the whole length of the calculated slope. For the length of the latest slope all normalized HRV values, the slope itself with its length and height will be passed to the classifier. For each normalized HRV value, the classifier will then predict if it is part of a mentally demanding sequence.

4 Artificial Intelligence Classifier Training

In this section, we describe how we trained the classifier, starting with the data we used, the model architecture, the training process, model evaluation, and finally further analysis.

4.1 Training Data and Preprocessing

For our study, we examined the video data of all participants from the study of Bruder and Schwarz [26] and annotated five different tasks. Participants were required to complete three surveillance tasks called NRTT, Unknown Track, and Warn/Engagement. Figure 9 shows the three different tasks and their associated areas.

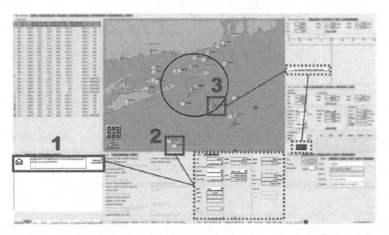

Fig. 9. Original Task User Interface, showing 1) a non-real-time track (NRTT) to process, 2) a new unknown track to classify and 3) a hostile track entering the self-protection zone that must be warned or further engaged if it proceeds to get closer.

The Non-Real-Time Tracks (NRTT) are displayed with information about a track that must be created manually, with specific information such as classification, speed, and direction, all of which must be entered into a form. Unknown tracks sometimes appeared as yellow symbols on the map, and the participant had to classify them within a form according to their position, speed, and direction. In the third task, hostile tracks, marked as red symbols on the map, were moving towards our position in the center of the map. When these tracks crossed the first line, the participant had to manually warn the track by clicking a warn button to the right. If an enemy track crossed Fig. 10 on the right.

We annotated the surveys as a separate task and task-independent parts of the video, such as the start of the experimental software, as other. Visualizes all of a participant's tasks and their processing times. The three monitoring tasks could occur simultaneously which made the first half of the experiment mentally more demanding than the second half.

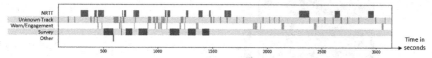

Fig. 10. Visualized time slots from one participant of all five different annotated classes.

We marked the three surveillance tasks as *mentally demanding tasks* which was coded as 1, whereas the periods surrounding these tasks were coded 0. The stated statistical model was defined as:

$$Mentally\ demanding\ Task\ \sim\ normalized\ HRV\ +\ Slope\ +\ Length\ of\ Slope\ +\ Height\ of\ Slope\ +\ \varepsilon$$

The calculation of the predictors as well as the criterion was based on the steps as described in paragraph 2.

We split the data into 70% training data and 30% test data for model validation afterwards. The training data consists of 70% of the dataset, which is around 6:52 h of training material, coded in 22.109 data rows. In contrast, the test data consists of 30% of the dataset, which is around 2:57 h of evaluation material, coded in 9.476 data rows that will only be used afterwards for the purpose of evaluation.

4.2 Model Architecture

We have chosen a modified version of the Decision Tree Classifier algorithm called Extra Trees Classifier written in python v.3 from the package scikit-learn [27] in v.1.3.2. This package implements the Extra Trees Classifier as a meta estimator that fits several randomized decision trees on various sub-samples of the dataset and uses averaging to improve the predictive accuracy and controls over-fitting, indicating its robust predictive capabilities in healthcare applications [28].

The hyperparameters were optimized using a grid search, resulting in a best fit with the parameters *n_estimators = 100* and *max_features = 3*.

With respect to possible class imbalance, we also calculated the class weights for the two possible states of the criterion, which were 0.839 for non-critical tasks and 1.236 for critical tasks. These class-weights were given to the following machine learning algorithm.

4.3 Training and Evaluation

Based on the test dataset, consisting of 30% (2:57 h) of the whole dataset, the classifier's discrimination accuracy is 90.78%, which is based on 8.603 right classified cases in contrast to 873 wrong classifications, as shown in the confusion matrix in Fig. 11.

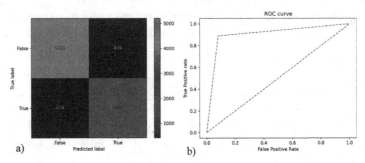

Fig. 11. a) Confusion Matrix based on test data. b) Receiver operating characteristic curve for the same test data.

A Receiver Operating Characteristic curve (ROC) is a graphical plot that illustrates the trade-off between the true positive rate (TPR) and the false positive rate (FPR) at different threshold settings, which is shown in Fig. 11 for our model evaluated with the test dataset. The calculated Receiver Operating Characteristic Area Under the Curve (ROC AUC) score indicates the ability of a model to distinguish between positive and negative examples across all possible classification thresholds. A ROC AUC score of 100% indicates a perfect model, while a score of 50% indicates a model that is no better than random guessing. Our trained model reaches an ROC AUC score of 94.02%, indicating that the model is very good in distinguishing between mentally demanding sequences in the HRV and sequences without. The ROC AUC score of our model is suggesting that the model is highly effective at identifying the true positive and true negative cases, also reflecting the results of the confusion matrix.

Feature importance measures the relative contribution of each feature to the classification accuracy of a machine learning model, which is a crucial aspect of understanding how a machine learning model works and can be used to improve its performance. It also shows how much information a specific feature adds to a model, which can help to decide whether the adding or deletion of a feature can optimize a model. In this study, we investigated the feature importance of the used predictors for predicting mental demanding sequences in the HRV. The four classes of features were the normalized HRV, the slope, its length and height. We found that the most important features are the normalized HRV the slope itself and the length of the slope, which are together accountable for 92.10% of the model's accuracy. This suggests that the height of an ascending or descending HRV, with a feature importance of 7.90%, is not as essential for predicting mental demanding tasks as the length of that slope with a feature importance of 26.96%. The slope itself, which is combining the length and height, has a feature importance valued with 26.49%. The normalized HRV has the highest feature importance with 38.65%.

Because of the partial redundancy of the predictors slope, length, and height, it may be sufficient to just use the slope and the normalized HRV values. In a further analysis, we trained a second model for testing with the normalized HRV and the slope as the only two predictors. Figure 12 shows the accuracies and ROC-AUC scores for different combinations of predictors.

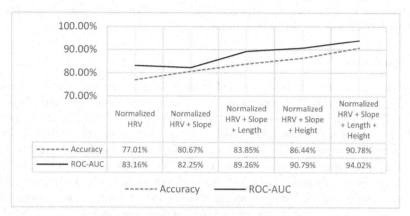

	Normalized HRV	Normalized HRV + Slope	Normalized HRV + Slope + Length	Normalized HRV + Slope + Height	Normalized HRV + Slope + Length + Height
Accuracy	77.01%	80.67%	83.85%	86.44%	90.78%
ROC-AUC	83.16%	82.25%	89.26%	90.79%	94.02%

Fig. 12. Accuracies and ROC-AUC scores for different combinations of predictors.

The further analysis indicates, that the normalized HRV explains most of the variance regarding mentally demanding tasks. Furthermore, the height of the slope seems to explain more variance than the length, but in the use of both features, variance of the height was lower as stated in the paragraphs before. In comparison to the model using just the normalized HRV + Slope the additional use of the predictors length or height increases the ROC-AUC score, therefore improving the model in identifying the true positive and true negative cases for up to ~8%.

In summary, our trained model detects mentally demanding tasks with a high accuracy of 90.78%. The ROC-AUC score of 94.02% shows that the model can distinguish positive and negative cases very well. The analysis shows that it appears possible to run the model with fewer predictors than features, thus reducing complexity.

5   Discussion

5.1   Benefits of Live HRV Analysis

Heart rate variability (HRV) analysis has long provided valuable insights into health and well-being. However, traditional methods often rely on establishing individual baselines, making them cumbersome and limiting.

Our live analysis reduces the complexity by tracking HRV fluctuations in near real time, enabling the identification of changes during mentally demanding tasks. This eliminates the need for lengthy baseline measurements, saving time in research settings and making the technology applicable outside of lab environments. Without spending time establishing baselines, a system could start adapting to new situations faster, potentially leading to quicker and more optimal responses. Adapting without baselines could allow the system to learn and improve continuously, incorporating new information and experiences without needing to re-establish a static starting point.

In conclusion, our live HRV method indicates that real time classifications of short-time mentally demanding tasks based on HRV is possible. Additionally, our method eliminates the need for baseline measurements what offers significant advantages for real time applications like adaptive systems.

5.2 Limitations

Several limitations need to be considered regarding the overall use of HRV for measuring cognitive workload in general, and for the described method without a baseline specifically. One major limitation is that HRV is altered by many confounding variables, like body movement, general stress level, and psychoactive substances like caffeine. Discriminating whether a HRV change is caused, for example by physiological activity, homeostasis, or cognitive workload by analyzing the raw data, is difficult [29]. Context information are usually needed as a co-variable to address that issue. Those data are not always possible to collect. The applicability of HRV as a sole parameter to measure CWL is restricted because of that. To address the issue of discriminating different cofounding factors, the experiment was performed in a highly controlled laboratory environment. That limits the possible extrapolation of the results to field studies, or productive systems.

Another limitation is an increased data noise because the tasks for the training of the model were manually annotated, based on video material captured during the experiment.

Also, the described method was only used for a binary classification of cognitive workload, where multiple mentally demanding tasks at the same time where merged together. Usually, multiple tasks have to be monitored and performed independently, and some tasks are interrupted by others. The allocation of cognitive workload load due to specific tasks in that scenario has to be further investigated in future work and is a limitation of the results.

6 Conclusion and Future Developments

This study has demonstrated the potential of using heart rate variability (HRV) as a marker of cognitive workload, showing the way for further exploration and practical applications.

While this study successfully estimated mentally demanding tasks with an accuracy above 90%, future research can delve deeper into understanding the real time changes in HRV within single tasks or specific task types. This level of granularity could reveal which tasks or specific segments within tasks are most demanding, allowing for targeted interventions or workload balancing. Additionally, investigating the impact of multitasking on HRV could provide valuable insights into its unique cognitive demands.

Integrating other physiological metrics beyond HRV, such as pupil dilation, could potentially enhance the accuracy and comprehensiveness of workload assessment. Combining multiple physiological measures might create a more robust and nuanced understanding of cognitive state. While other physiological metrics, like pupil dilation, share some similar connections to mental and emotional states as HRV, further research is needed to establish other metrics like the HRV as a reliable standalone measure for the intended method. It's important to note that replacing HRV with another physiological measure requires careful consideration of their specific functionalities and limitations within the given method's context.

Determining the optimal level of accuracy required for different applications is crucial. For instance, adaptive systems requiring real time adjustments might demand high accuracy, while workload evaluation might be tolerant of some margin of error. Tailoring the model's complexity and resource requirements to specific use cases will optimize its practicality and efficiency.

Future research could focus on differentiating mental and physical activities solely based on HRV data. This could enable applications like monitoring mental fatigue during physical exercise or distinguishing between cognitive stress and physical exertion in real time.

This study represents a step forward in utilizing HRV in live analysis to assess cognitive workload. Future research along the proposed avenues can refine and broaden this understanding, leading to impactful applications across various domains.

References

1. Davis, D.H.J., Oliver, M., Byrne, A.J.: A novel method of measuring the mental workload of anaesthetists during simulated practice. Br. J. Anaesth. **103**(5), 665–669 (2009)
2. Galy, E., Paxion, J., Berthelon, C.: Measuring mental workload with the NASA-TLX needs to examine each dimension rather than relying on the global score: an example with driving. Ergonomics **61**(4), 517–527 (2018)
3. Mccraty, R., Shaffer, F.: Heart rate variability: new perspectives on physiological mechanisms, assessment of self-regulatory capacity, and health risk. Global Adv. Health Med. **4**(1), 46–61 (2015)
4. Kemper, K.J., Hamilton, C., Atkinson, M.: Heart rate variability: impact of differences in outlier identification and management strategies on common measures in three clinical populations. Pediatr. Res. **62**(3), 337–342 (2007)
5. Gąsior, J.S., et al.: Normative values for heart rate variability parameters in school-aged children: simple approach considering differences in average heart rate. Front. Physiol. **9**, 342109 (2018)
6. Li, W., Li, R., Xie, X., Chang, Y.: Evaluating mental workload during multitasking in simulated flight. Brain Behav. **12**(4), e2489 (2022)
7. John, A.R., et al.: Unravelling the physiological correlates of mental workload variations in tracking and collision prediction tasks: implications for air traffic controllers. IEEE Trans. Neural Syst. Rehabil. Eng. **30**, 770–781 (2021)
8. Delliaux, S., Delaforge, A., Deharo, J.C., Chaumet, G.: Mental workload alters heart rate variability, lowering non-linear dynamics. Front. Physiol. **10**, 565 (2019)
9. Radüntz, T., Mühlhausen, T., Freyer, M., Fürstenau, N., Meffert, B.: Cardiovascular biomarkers' inherent timescales in mental workload assessment during simulated air traffic control tasks. Appl. Psychophysiol. Biofeedback **46**, 43–59 (2020)
10. Forte, G., Favieri, F., Casagrande, M.: Heart rate variability and cognitive function: a systematic review. Front. Neurosci. **13**, 436204 (2019)
11. Veltman, J.A., Jansen, C.: Differentiation of mental effort measures: consequences for adaptive automation (2003)

12. De Witte, N.A., Sütterlin, S., Braet, C., Mueller, S.C.: Getting to the heart of emotion regulation in youth: the role of interoceptive sensitivity, heart rate variability, and parental psychopathology. PloS One **11**, e0164615 (2016)
13. Veltman, J.A., Gaillard, A.W.K.: Indices of mental workload in a complex task environment. Neuropsychobiology **28**, 72–75 (1993)
14. Cinaz, B., Arnrich, B., Marca, R.L., Tröster, G.: Monitoring of mental workload levels during an everyday life office-work scenario. Pers. Ubiquit. Comput. **17**(2), 229–239 (2013)
15. Shao, S., Wang, T., Li, Y., Song, C., Jiang, Y., Yao, C.: Comparison analysis of different time-scale heart rate variability signals for mental workload assessment in human-robot interaction. Wireless Commun. Mob. Comput. **2021**, 1–12 (2021)
16. Nardolillo, A.M., Baghdadi, A., Cavuoto, L.A.: Heart rate variability during a simulated assembly task; influence of age and gender (2017)
17. Stern, R.M., Ray, W.J., Quigley, K.S.: Quigley, Psychophysiological Recording, vol. 59 (2001)
18. Ernst, G.: Methodological issues. In: Heart Rate Variability, pp. 51–118. Springer, London (2014)
19. Tarkiainen, T.H.: Stability over time of short-term heart rate variability. Clin. Auton. Res. **15**, 394–399 (2005)
20. Shaffer, F., Venner, J.: Heart rate variability anatomy and physiology. Biofeedback (Online) **41**, 13 (2013)
21. Chen, Z., Chen, Z., Calhoun, V.: Blood oxygenation level-dependent functional MRI signal turbulence caused by ultrahigh spatial resolution: numerical simulation and theoretical explanation. NMR Biomed. **26**(3), 248–264 (2013)
22. Oladele, A.M., Tomomowo-Ayodele, S.O., Oluremi, O.Y., Olusola, A.M.: Health information needs and its sources among rural dwellers in Egbedore local government areas of state of Osun, Nigeria. Int. J. Humanit. Soc. Stud. **7**(7) (2019)
23. Wilder, J.: Basimetric approach (law of initial value) to biological rhythms. Ann. New York Acad. Sci. **98**(4), 1211–1220 (1962)
24. Schwarz, J., Fuchs, S.: Validating a "Real-Time Assessment of Multidimensional User State" (RASMUS) for adaptive human-computer interaction (2018)
25. Hoover, A., Singh, A., Fishel-Brown, S., Muth, E.: Real-time detection of workload changes using heart rate variability. Biomed. Signal Proc. Control **7**, 333–341 (2012)
26. Bruder, A., Schwarz, J.: Evaluation of diagnostic rules for real-time assessment of mental workload within a dynamic adaptation framework. In: Sottilare, R., Schwarz, J. (eds.) Adaptive Instructional Systems. Lecture Notes in Computer Science(), vol. 11597. Springer, Cham (2019). https://doi.org/10.1007/978-3-030-22341-0_31
27. Pedregosa, F., et al.: Scikit-learn: machine learning in Python (2011). ArXiv:abs/1201.0490
28. Gupta, M.D., et al.: COVID 19-related burnout among healthcare workers in India and ECG based predictive machine learning model: insights from the BRUCEE- Li study. Indian Heart J. **73**(6), 674–681 (2021)
29. Sammer, G.: Heart period variability and respiratory changes associated with physical and mental load: non-linear analysis. Ergonomics **41**(5), 746–755 (1998)
30. Bashiri, B., Mann, D.: Heart rate variability in response to task automation in agricultural semi-autonomous vehicles (2014)
31. Chamchad, D., et al.: Using heart rate variability to stratify risk of obstetric patients undergoing spinal Anesthesia (2004)

Exploring the Relationship Between Stress, Coping Strategies, and Performance in an Adapting Training System

Mira E. Gruber[1], Yazmin V. Diaz[1], Bradford L. Schroeder[2(✉)],
Gabriella M. Hancock[3], Jason E. Hochreiter[2], Javier Rivera[2], Sean C. Thayer[2],
and Wendi L. Van Buskirk[2]

[1] University of Central Florida, Orlando, FL 32816, USA
{mira.gruber,yazmin.diaz}@ucf.edu
[2] Naval Air Warfare Center Training Systems Division, Orlando, FL 32826, USA
{bradford.l.schroeder.civ,jason.e.hochreiter.civ,
javier.a.rivera59.civ,sean.c.thayer.ctr,
wendi.l.vanbuskirk.civ}@us.navy.mil
[3] California State University, Long Beach, Long Beach, CA 90840, USA
gabriella.hancock@csulb.edu

Abstract. Adaptive training (AT) is a promising avenue for training across multiple domains pertinent to warfighters, as it can adjust instruction and task difficulty based on the learner's real-time performance. Despite these benefits, such adaptations may elevate learner stress, hindering performance and learning. Coping strategies may be used to mitigate learner stress, but there is limited extant research comparing adaptive and maladaptive coping strategies in an adaptive training environment. This study explores the relationship between stress and task performance in the context of an adaptive training system that dynamically adjusted task difficulty based on real-time performance. Participants completed a radio frequency signal detection task in which they had to classify and report signals quickly and accurately. Participants were also instructed on adaptive or maladaptive stress coping strategies that either focused on the task itself (problem-focused) or on their emotions (emotion-focused). Results suggest that task engagement is related to performance and changes in micro-adaptive difficulty. Distress and worry, however, did not consistently relate to task performance or micro-adaptive difficulty. Problem-focused coping helped maintain task performance regardless of whether the strategy was adaptive or maladaptive. Maladaptive, emotion-focused coping led to the worst performance outcomes. Limitations and future directions are discussed.

Keywords: Adaptive Training · Adaptive Instructional System · Micro-Adaptive Training · Stress

© The Author(s), under exclusive license to Springer Nature Switzerland AG 2024
R. A. Sottilare and J. Schwarz (Eds.): HCII 2024, LNCS 14727, pp. 147–165, 2024.
https://doi.org/10.1007/978-3-031-60609-0_11

1 Introduction

Enhancing warfighter performance remains a central focus within Naval science (Williams, 2017), and training is imperative for effective task performance (Proctor & Vu, 2006). One possible training avenue is through adaptive training (AT). Difficulty micro-adaptations involve adjusting task difficulty based on a learner's real-time task performance (Landsberg et al. 2011). A concern with this approach is that increasing task difficulty might also increase operator stress, which, in turn, can negatively impact task performance and operator well-being (Hancock & Warm, 1989). In this paper, we explore the relationship between stress coping strategies, self-reported stress, and task performance in the context of an adaptive training system that dynamically adjusted task difficulty based on real-time performance.

1.1 Adaptive Training

Adaptive training (AT) systems are computer-based instructional systems that surpass standardized protocols by tailoring training at an individual level (Hancock et al. 2009). This training format holds promise for warfighter training as it can expedite learning by excluding already mastered material and facilitate swift progress toward higher proficiency levels (Goodwin et al. 2017).

There are three primary approaches to AT: macro-adaptation, micro-adaptations, and Aptitude Treatment Interaction (ATI) adaptations (Landsberg et al. 2011; Park & Lee, 2003). Macro-adaptations tailor the training based on broad characteristics of the learner such as their general ability to meet learning objectives measured prior to instruction (Park & Lee, 2003). In contrast, micro-adaptations adjust the instruction based on the learner's real-time performance or emotional state (Park & Lee, 2013). Finally, ATI adaptations match the instructional techniques to the learner's abilities or attributes such as learning styles or motivation (Park & Lee, 2003). Although ATI adaptations provide an even more tailored approach than macro-adaptations, aptitudes are typically measured pre-task, and the importance of certain aptitudes may change as the task progresses (Park & Lee, 2003). A hybrid approach that combines elements of micro-adaptivity and ATIs is one proposed solution to these limitations. A hybrid approach bases initial instruction on learner aptitudes and later uses real-time performance data as the instruction progresses (Park & Lee, 2003; Tennyson & Rothen, 1977; Van Buskirk et al. 2014). Because micro-adaptations rely on on-task measurements, they may be more sensitive to the learner's needs throughout the instruction compared to macro-level and ATI adaptations (Park & Lee, 2003).

While micro-adaptations can be delivered through one-on-one instructions (Landsberg et al. 2011), they can also be implemented through computer-based instructional systems that use algorithms and models to deliver appropriate instruction based on real-time assessments of the learner's performance (Landsberg et al. 2011). In essence, the algorithms deliver the right instruction for the learner at the right time. This can be decisions like selecting appropriate feedback for the learner at an appropriate level of detail, selecting a new lesson for a learner, or adjusting other elements of the task (e.g., making a scenario more difficult or reducing event rates). However, these types of systems may be more costly to implement due to additional processing requirements (Landsberg et al.

2011). Additionally, prior research suggests that any change in workload can increase stress and have a negative effect on performance (Cox-Fuenzalida, 2007; Helton et al. 2008). Thus, it is plausible that micro-adaptations may serve as a source of stress that may impair performance.

1.2 Stress

Stress is an interactive property of the immediate environment that induces an internal response from the individual that can be externally manifested given the magnitude of the stress and the individual's appraisal of it (Hancock & Szalma, 2008). Research has shown that stress has detrimental effects on task performance and operator well-being (Hancock & Warm, 1989). High stress levels have been found to impair operator judgment and lead to behavioral regressions (Barthol & Ku, 1959; Staal, 2004). Because military training aims to properly equip personnel with behaviors that enhance safety and efficiency, regression to incorrect or inefficient behavior would be undesired and likely a risk to safety. Therefore, understanding the impacts of stress on operator performance is a primary interest for maintaining optimal performance within military operations.

Given the critical role stress plays in performance, researchers have developed various assessment methods. Physiological measures of stress (e.g., Heart Rate Variability, blood pressure) allow for continuous assessment and real-time monitoring of stress (Hughes et al. 2019). However, such methods cannot capture the subjective experience of stress, are highly subject to individual differences and task parameters, and may be more costly to implement (Hughes et al. 2019). On the other hand, subjective measures such as the The Dundee Stress State Questionnaire (DSSQ; Matthews et al. 1999) provides a low cost alternative to measure the subjective stress experience. The DSSQ assesses three dimensions of stress: task engagement, distress, and worry. Task Engagement (TE) is characterized by the level of motivation an individual possesses to accomplish a given task, along with the intensity of energy and concentration directed toward that task (Matthews et al. 1999). TE is positively associated with performance on vigilance tasks (Matthews et al. 1999; Matthews et al. 2013). Distress is characterized by both mood and cognitive factors: high tension, unhappiness, low confidence, and low control. Worry, however, is cognitive and can be measured by levels of cognitive interference, self-focus of attention, self-esteem, and concentration (Matthews et al. 1999).

1.3 Stress Regulation Strategies

In addition to training individuals on the task itself to improve performance, training individuals to better manage their emotions and stress can also improve performance and operator well-being. Research has identified two primary stress coping strategies: problem-focused and emotion-focused coping. Problem-focused strategies involve solving the problem at hand through strategies such as seeking advice, gathering information, and generating steps to solve the problem (Lazarus & Folkman, 1984). In contrast, emotion-focused strategies involve regulating one's thoughts and emotional response to the stressor (Lazarus & Folkman, 1984). Examples of emotion-focused coping strategies include avoidance, self-blame, and cognitive reappraisal. Cognitive reappraisal is the process of re-interpreting the situation to change one's emotions towards it (Marroquín et al. 2017), and previous research suggests that it effectively reduces negative emotions (Gross, 2002).

Both problem-focused and emotion-focused coping can be considered adaptive or maladaptive depending on the specific strategy employed. Adaptive coping strategies are associated with greater well-being and are known to promote healthier, safer behaviors (Marroquín et al. 2017). For example, problem-solving, seeking social support, and cognitive reappraisal are all adaptive strategies. On the other hand, maladaptive strategies are more likely to lead to unsafe behaviors and are generally detrimental to well-being and performance (Marroquín et al. 2017). Examples of maladaptive strategies include rumination and emotional suppression. Emotional suppression involves attempting to retrain one's expression of negative emotions (Gross, 2002). Suppression can reduce the physiological arousal associated with stress, but it may be ineffective at reducing negative emotions (Gross & Levenson, 1993). Maladaptive strategies may be effective at managing stress in short-term, uncontrollable, and stressful situations, but they are detrimental in chronically stressful situations (Suls & Fletcher, 1985).

Despite the breadth of research on adaptive and maladaptive coping strategies, there is little research directly comparing these strategies in domains relevant to warfighters. Additionally, it remains unclear whether emotion or problem-focused coping have different effects on stress and in complex cognitive tasks. Matthews and Falconer (2000) compared adaptive, task-focused strategies (e.g., planning) with maladaptive emotion-focused strategies (e.g., self-criticism) and avoidance strategies (distracting oneself). The results suggested that task-focused strategies are the most optimal for performance. However, the study did not include maladaptive, task-focused and adaptive, emotion-focused strategies, making it challenging to draw conclusions about the best strategy for cognitive tasks.

1.4 The Present Study

The goal of the current research was to explore the relationship between stress, performance, and stress-coping strategies in a radio frequency (RF) detection adaptive training system that included micro-adaptive difficulty algorithms. We focused on stress in this paper, recognizing that the task itself can be considered the proximal and prime stressor (Hancock & Warm, 1989). Additionally, past research suggests that changes in workload, such as those produced by micro-adaptive difficulty, may be another source of stress (Cox-Fuenzalida, 2007; Helton et al. 2008). We hypothesized that participants using adaptive stress coping strategies would perform better on the experimental task and report less stress than those using maladaptive coping strategies. In addition to addressing these hypotheses, we were interested in comparing the effect of problem-focused and emotion-focused coping mechanisms on stress and performance given the limitations of past research (e.g., Matthews & Falconer, 2000). As the naming conventions of the constructs we are investigating may cause confusion, we reiterated the operational definitions we are using in this study in Table 1.

2 Method

2.1 Participants

Participants were recruited from an online research participation system at a large, southeastern university. Participants were 18 or older and had normal or corrected-to-normal vision and audition. A total of 75 participants were recruited through an online participant pool and paid $100 for their participation. Data was missing from one participant, resulting in a final sample size of $N = 74$.

Table 1. Operational definitions used in the present study.

Construct	Definition
Micro-adaptive difficulty	Adjustments in scenario difficulty based on the real-time measurement and assessment of performance
Adaptive stress coping strategy	Stress coping strategy that not only reduces stress but is also beneficial for task performance
Maladaptive stress coping strategy	Stress coping strategy that reduces stress but hinders task performance
Stress coping focal mechanism	The focus of a coping strategy that could be either focused on the task or problem at hand or focused on one's emotional or physical reaction to the stressor

2.2 Experimental Design

We used a $2 \times 2 \times 6$ mixed design. The between-subjects variables were coping strategy (adaptive or maladaptive coping) and the coping focal mechanism (problem-focused or emotion-focused). The within-subjects variable was scenario (pretest, scenarios 2, 3, and 4, posttest, and transfer). We randomly assigned participants to one of four coping and focal mechanism groups: adaptive, problem-focused coping (n = 18); adaptive, emotion-focused coping (n = 20); maladaptive, problem-focused coping (n = 19); and maladaptive, emotion-focused coping (n = 17). We instructed participants in the adaptive, problem-focused group to use the task interface to manage their stress (Chao et al. 2019). Specifically, we instructed them to use a "cheat sheet" that helped them complete the task. We trained participants in the adaptive, emotion-focused coping group on cognitive reappraisal. Specifically, we instructed them to re-interpret their stress as a beneficial learning experience that would help them perform better in the future (Marroquín et al. 2017). Participants in the maladaptive, problem-focused condition were instructed to other-blame the interface (Domaradzka & Fajkowska, 2018; Garnefski et al. 2001). Other-blame involves blaming another person or the environment for the negative circumstances (Garnefski et al. 2001). In the present study, participants criticized the system's interface by identifying aspects that impeded their performance. At the end of the study, participants had the opportunity to share their thoughts on the interface. In the maladaptive, emotion-focused condition participants were instructed on emotional suppression. They were told to ignore their stress and focus on the task instead (Marroquín et al. 2017).

2.3 Experimental Task

Participants. Completed a radio frequency (RF) detection task, in which they had to classify signals quickly and accurately. The signals differed in parameters related to their frequency, pulse type, and scan type. Participants used a two-screen setup to submit reports when signals began and ceased emitting along with classification reports based on their parameters. Participants analyzed these signals on the first screen, which displayed signal waveforms and other data, and submitted their reports on the second screen. The task was designed to be temporally demanding, as new signals entered the environment while participants were still attending to old ones. Participants completed five scenarios (micro-adaptive difficulty was only present for scenarios two (S2), three (S3), and four (S4)). Micro-adaptive difficulty algorithms determined the real-time adjustment to the number of signals in S2, S3, and S4 based on real-time assessment of performance. Each scenario started at a default number of signals, and this number increased, decreased, or stayed the same based on the participant's performance. Lastly, participants completed a transfer task in which they identified and reported on target signals amidst many distractor signals.

2.4 Materials

DSSQ. Stress was measured using the Dundee Stress State Questionnaire (DSSQ), a validated instrument designed to measure multiple dimensions of stress responses within various contexts (Matthews, et al. 1999). The DSSQ is a structured self-report questionnaire containing 30 items, organized into subscales that capture distress, worry, and task engagement. Scores for each subscale range from 0 to 30.

Performance. We used accuracy and timeliness of the reported signals to assess performance. Accuracy was the percentage of signal parameters correctly inputted in the report, and timeliness was the percentage of signals reported within the required timeframe. The accuracy and timeliness of reported signals dictated how many events were present and changed the task's difficulty. We computed a change in difficulty score to examine participants' micro-adaptive difficulty experience (in S2, S3, and S4 scenarios only). This calculation will be discussed in more detail in the results section.

2.5 Procedure

After providing informed consent, participants completed a comprehensive demographics questionnaire. Following this, they navigated a self-guided PowerPoint presentation explaining key terminology and concepts related to the experimental task and instructing them on their coping strategy. Participants then completed a guided instructional phase with a virtual instructor. An experimenter was present to answer questions and intervene when needed. The instructional phase introduced the signal detection and classification concepts as well as how to use the interface. After the instructional phase, participants completed five, 10-min scenarios. Participants watched a brief video prior to each scenario instructing them on their coping strategy (see Table 2). The content was excised from the initial, longer-form training video. The videos re-iterated key actions regarding how to cope (e.g., remember to disregard feelings, etc.) Participants completed the DSSQ after each scenario. Additionally, participants completed a 10-min transfer scenario consisting of an advanced signal detection task. They were required to identify and report only six specified target signals out of many distractor signals. After the transfer task and its corresponding DSSQ, we debriefed the participants and compensated them for their participation.

Table 2. Content of coping trainings.

Strategy	Video Script
Adaptive, problem-focused	"If you find yourself feeling stressed, overwhelmed, or experiencing any other emotions during the session, remember that you can always use the interface for support. Specifically, you can access a helpful guide about critical task information. You can use this feature of the interface for help at any time."
Adaptive, emotion-focused	"If you find yourself feeling stressed…remember that you are still learning, and your performance will improve with practice and time."
Maladaptive, problem-focused	"If you find yourself feeling stressed…remember any and all of these problematic design features of the interface as you will be asked to describe them later."
Maladaptive, emotion-focused	"If you find yourself feeling stressed…remember to disregard and not think about them so that you can successfully concentrate on your duties instead."

3   Results

3.1   Performance Metrics

Accuracy. A mixed-model Analysis of Variance (ANOVA) was conducted to examine the effect of coping strategy and focal mechanism on accuracy across the five scenarios and transfer scenario. There was a significant main effect of scenario on accuracy, $F(5, 350) = 2.74$, $p = .019$, $\eta_p^2 = .04$. Least Significant Difference (LSD) pairwise comparisons tests show that accuracy was lower at S1 ($M = 39.01$, $SD = 30.01$) compared to S3 ($M = 49.56$, $SD = 31.11$), S4 ($M = 49.16$, $SD = 28.86$), S5 ($M = 49.82$, $SD = 32.47$), and transfer ($M = 51.20$, $SD = 28.44$; $ps < .05$) (see Fig. 1).

There was also a significant two-way interaction between coping strategy and coping focal mechanism on accuracy, $F(1, 70) = 8.90$, $p = .004$, $\eta_p^2 = .11$. LSD pairwise comparisons show that, for the emotion-focused groups, the adaptive group ($M = 54.55$, $SD = 31.40$) had higher accuracy than the maladaptive group ($M = 30.91$, $SD = 27.20$; $p < .001$) as seen in Fig. 2.

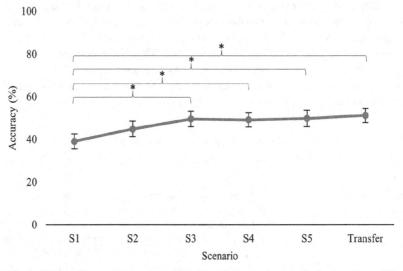

Fig. 1. Mean accuracy across scenarios. Error bars are ±1 standard error. *$p < .05$.

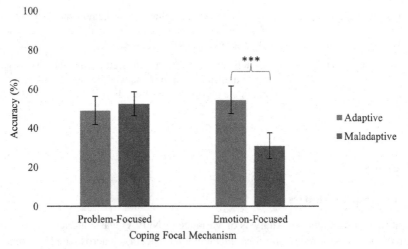

Fig. 2. Mean accuracy in the coping strategy and focal mechanism groups. Error bars are ± 1 standard error. ***$p < .001$.

Timeliness. A mixed-model ANOVA was conducted to examine the effect of coping strategy and focal mechanism on the percentage of signals reported on time across the five scenarios and transfer scenario. There was a significant main effect of scenario on timeliness, $F(5, 350) = 18.24$, $p < .001$, $\eta_p^2 = .21$. LSD pairwise comparisons show that the timeliness was significantly lower at S1 ($M = 12.43$, $SD = 22.13$) compared to S2 ($M = 19.96$, $SD = 24.30$), S3 ($M = 23.25$, $SD = 27.43$), S4 ($M = 21.51$, $SD = 22.91$), and S5 ($M = 20.17$, $SD = 24.83$; $ps < .01$). As can be seen in Fig. 3, timeliness significantly decreased at the transfer scenario ($M = 2.15$, $SD = 2.59$), and this value was lower compared to all other scenarios ($ps > .001$). Additionally, note that the percentage of signals reported on time was quite low across all scenarios (i.e., never surpassing 30%).

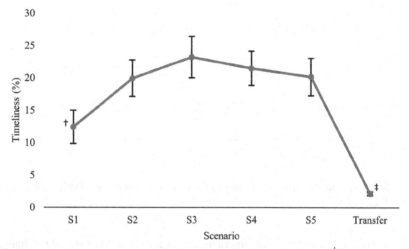

Fig. 3. Mean timeliness across scenarios. Error bars are ±1 standard error. †S1 is lower than S2, S3, S4, and S5 (**$ps < .01$). ‡Transfer is lower than all other scenarios (***$ps < .001$).

Change in Difficulty. We computed a change in difficulty score to examine participants' micro-adaptive difficulty experience. Change in difficulty was the percentage of presented signals to the maximum possible number of signals. Scenarios S2, S3, and S4 could present up to five additional signals for participants to analyze depending on their real-time task performance. The default for S2 was five, and the default for S3 and S4 was six. The means were centered around zero to account for when participants may have received less than the default number of signals. The value was also converted to a percentage so it could by compared across scenarios. Negative values indicate that fewer than the default was given (decreased difficulty) and positive values indicate that more than the default was given (increased difficulty). Zero indicates that there was no change in difficulty.

$$\Delta_{Difficulty} = 100\% * [(Number\ of\ Signals\ /(Default + 5)) - (Default/(Default + 5))]$$

A mixed-model ANOVA was conducted to examine the effect of coping strategy and focal mechanism on the average difficulty change across the micro-adaptivity scenarios (S2, S3, S4). There was a significant main effect of scenario, $F(2, 140) = 7.45$, $p < .001$, $\eta_p^2 = .10$. Change in difficulty was significantly lower at S2 ($M = +89\%$, $SD = $

17.3%) than at S4 ($M = + 16.7\%, SD = 20.5\%; p < .001$), as seen in Fig. 4. The change in difficulty was also significantly lower at S3 ($M = + 10.0\%, SD = 24.7\%$) than at S4 ($p = .001$).

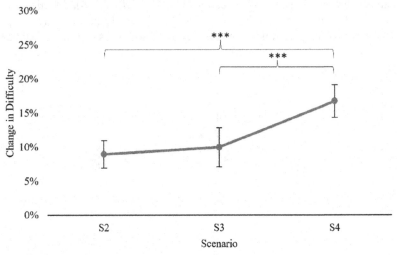

Fig. 4. Mean percentage change in difficulty in the scenarios with micro-adaptive difficulty, S2, S3, and S4. Error bars are ±1 standard error. ***$p < .001$.

There was a significant two-way interaction between scenario and focal mechanism, $F(2, 140) = 4.19, p = .017, \eta_p^2 = .06$. In the problem-focused coping condition, difficulty was significantly higher in S4 ($M = +22.9\%, SD = 18.1\%$) compared to S2 ($M = + 9.9\%, SD = 16.3\%$) and S3 ($M = + 99\%, SD = 23.5\%; ps < .001$) as seen in Fig. 5.

There was also a significant two-way interaction between focal mechanism and coping strategy on the average percentage change in difficulty, $F(1, 70) = 8.84, p = .004, \eta_p^2 = .11$. LSD pairwise comparisons show that, for emotion-focused coping, the change in difficulty was significantly higher in the adaptive group ($M = + 17.2\%, SD = 20.0\%$) than the maladaptive group ($M = 0.2\%, SD = 20.0\%; p = .003$) as seen in Fig. 6.

There was also a significant three-way interaction between scenario, coping strategy, and coping focal mechanism on the average percentage change in difficulty, $F(2, 140) = 3.91, p = .022, \eta_p^2 = .05$ (see Fig. 7). In S2, the change in difficulty was significantly higher in the maladaptive-problem focused group ($M = +14.2\%, SD = 15.7\%$) than the maladaptive-emotion focused group ($M = 0.5\%, SD = 15.2\%; p = .017$). In S3, the adaptive-emotion focused group ($M = +20.5\%, SD = 24.1\%$) had significantly higher difficulty than the adaptive-problem focused group ($M = + 3.5\%, SD = 26.3\%; p = .029$). Also, in S3, the maladaptive-problem focused group ($M = +16.3\%, SD = 20.7\%$) had a significantly higher change in difficulty than the maladaptive-emotion focused group ($M = -2.7\%, SD = 21.7\%; p = .018$). In S4, the maladaptive-problem focused group ($M = +21.1\%, SD = 18.7\%$) once again had a significantly higher change in difficulty than the maladaptive-emotion focused group ($M = +2.7\%, SD = 23.1\%$;

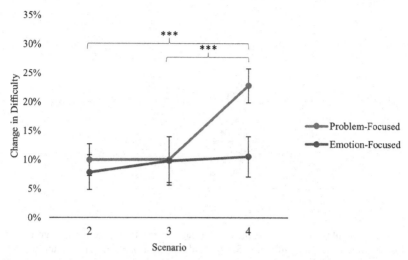

Fig. 5. Mean percentage change in difficulty for each coping strategy and focal mechanism groups. Error bars are ±1 standard error. **$p < .01$.

$p = .006$). Additionally, the change in difficulty for adaptive, problem-focused was significantly higher in S4 ($M = +24.7\%$, $SD = 17.6\%$) than S2 ($M = +5.6\%$, $SD = 16.9\%$) and S3 ($M = +3.5\%$, $SD = 26.3\%$; $ps < .001$).

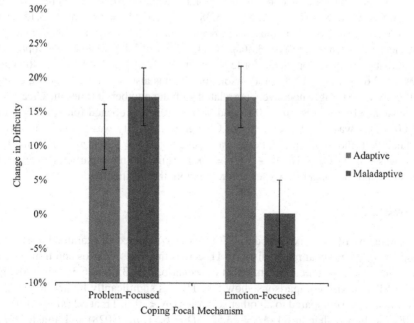

Fig. 6. Mean change in difficulty for focal mechanism groups across micro-adaptive difficulty scenarios, S2, S3, and S4. Error bars are ±1 standard error. **$p < .01$.

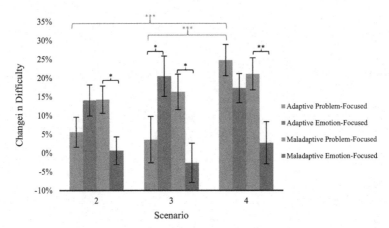

Fig. 7. Mean percentage change in difficulty for each of the four experimental groups in the scenarios with micro-adaptive difficulty, S2, S3, and S4. Error bars are ±1 standard error. *$p <$ .05. **$p <$ .01, ***$p <$ .001.

We followed-up the adaptive, problem-focused results with additional analysis to examine how participants' use of the cheat sheet may have contributed to the pattern of results. A repeated measures ANOVA was used to analyze how the number of times the cheat sheet was opened and the amount of time (in seconds) it was open changed over time. There were no significant differences in the number of time participants opened the cheat sheet. However, the data suggest a trend in which participants opened the cheat sheet fewer times in S4 ($M = 2.89$, SD = 3.38) than in S2 ($M = 3.67$, $SD = 3.12$) or S3 ($M = 4.67$, $SD = 5.14$). There was a significant main effect of scenario on seconds the cheat sheet was open, $F(2, 34) = 3.46, p = .043$, $\eta_p^2 = .17$. LSD pairwise comparisons suggest that participants opened it for less time in S4 ($M = 13.62$ s, $SD = 31.36$ s) than S4 ($M = 13.61$ s, $SD = 17.57$ s). Pearson correlations also reveal that, for S3, change in difficulty was strongly negatively correlated with the number of times the cheat sheet was opened, $r(16) = -.80, p < .001$, and how long it was opened for, $r(16) = -.65$, $p < .004$. This was also the case in S4. Change in difficulty was strongly negatively correlated with the number of times it was opened, $r(16) = -.69, p = .002$, and how long it was opened for, $r(16) = -.72, p = .001$. Thus, the more participants used the cheat sheet and the longer they used it for, the more their difficulty decreased.

3.2 DSSQ

Task Engagement. A mixed-model ANOVA was conducted to examine the effect of coping strategy and focal mechanism on TE across the five scenarios and transfer scenario. There was a significant main effect of scenario, $F(5, 355) = 6.31, p < .001$, $\eta_p^2 = .08$. LSD pairwise comparisons show that TE was significantly higher at S1 ($M = 25.71$, $SD = 4.74$) compared to all other scenarios ($ps < .05$). TE at S2 ($M = 24.65$, $SD = 5.48$) was higher than at S4 ($M = 23.52$, $SD = 6.51$; $p = .028$) and transfer ($M = 23.33$, $SD = 6.27$; $p = .028$). TE was also higher at S3 ($M = 24.43$, $SD = 6.29$) than at S4 ($p = .012$) and transfer ($p = .024$). See Fig. 8 below.

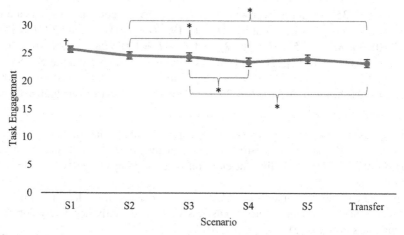

Fig. 8. Mean Task Engagement across scenarios. Error bars are ±1 standard error. *$p < .05$. †S1 is sig. Higher ($ps < .05$) than all other scenarios.

Pearson's correlations were run to examine the relationship between TE and the performance metrics. TE was significantly positively correlated with accuracy in the transfer scenario ($r(72) = .23, p < .05$). TE was also significantly positively correlated with timeliness during S2, $r(72) = .25, p = .032$. Finally, TE was significantly positively correlated with the change in difficulty in S2, $r(72) = .23, p = .046$; S3, $r(72) = .32, p = .006$; and S4 $r(72) = .29, p = .013$.

Distress. A mixed-model ANOVA was conducted to examine the effect of coping strategy and focal mechanism on the Distress DSSQ subscale across the five scenarios and transfer scenario. There was a significant main effect of scenario, $F(5, 355) = 38.02, p$

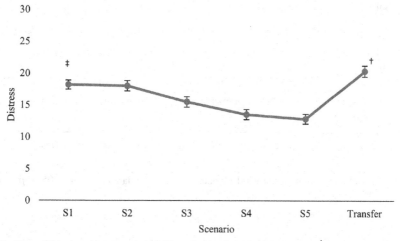

Fig. 9. Mean Distress across scenarios. Error bars are ±1 standard error. ‡S1 is sig. Higher than S2, S3, S4, and S4 ($p < .001$). †Transfer is sig. Higher ($ps < .01$) than all other scenarios.

< .001, $\eta_p{}^2$ = .35. As can be seen in Fig. 9, LSD pairwise comparisons indicate that Distress was significantly higher at S1 (M = 18.19, SD = 6.04) than at S3 (M = 15.49, SD = 7.10), S4 (M = 13.51, SD = 6.72), and S5 (M = 12.83, SD = 6.62; ps < .001). Distress at S2 (M = 18.01, SD = 7.00) was significantly higher than at S3, S4, and S5 (ps < .001). Distress at S3 was significantly higher than at S4 and S5 (ps < .001). Finally, Distress at the transfer (M = 20.33, SD = 7.56) scenario was higher than all other scenarios (ps < .01).

There was also a significant main effect of coping focal mechanism on distress, $F(1, 71)$ = 4.02, p = .049, $\eta_p{}^2$ = .05. Distress was significantly higher in the problem-focused group (M = 17.68, SD = 6.18) than the emotion-focused group (M = 15.14, SD = 7.28), as seen in Fig. 10.

Next, we examined Pearson's correlations to investigate the relationship between Distress and the performance metrics. Distress was significantly negatively correlated with timeliness in S2, $r(72)$ = −.29, p = .012.

Worry. A mixed-model ANOVA was conducted to examine the effect of coping strategy and focal mechanism on the Worry DSSQ subscale across the five scenarios and transfer scenario. There was a significant main effect of scenario, $F(5, 355)$ = 7.79, p < .001, $\eta_p{}^2$ = .10. Worry was significantly higher at S1 (M = 6.69, SD = 4.84) than S2 (M = 5.32, SD = 4.76; p < .003). Worry was significantly lower during the transfer scenario (M = 4.40, SD = 4.78), compared to S1, S2, and S3 (M = 5.09, SD = 4.74; ps < .05), see Fig. 10 below. Worry was not significantly correlated with accuracy, timeliness, or change in difficulty.

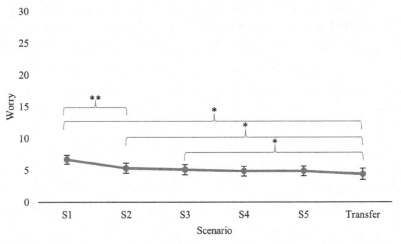

Fig. 10. Mean Worry across scenarios. Error bars are ± 1 standard error. *p < .05, **p < .01.

4 Discussion

4.1 Coping Strategies and Stress

We predicted that adaptive coping strategies would be associated with less stress than maladaptive strategies. Contrary to our prediction, there was no difference in TE dependence on coping strategy and focal mechanism. TE was highest during the first scenario, but decreased as time went on. TE was consistently related to the change in difficulty metric, as well as accuracy and timeliness in certain scenarios. Distress also tended to decrease over time; however, there was an increase in Distress during the transfer scenario. Distress was only correlated with timeliness in the second scenario and was not correlated with changes in difficulty. Also contradicting our prediction, participants in maladaptive groups did not experience more Distress than those in adaptive groups. However, we did find that participants in the problem-focused group were overall more distressed than those in the emotion-focused group. Worry was generally low throughout all scenarios, although it also decreased over time. Contrary to our predictions, there was no difference in Worry between any of the coping strategy groups. Worry was also not correlated with any performance metric or change in difficulty. Our findings for stress and performance are somewhat consistent with prior research. As expected, Task Engagement was correlated with performance (Matthews et al. 1999). It is possible that TE improved performance in the task and therefore, increased difficulty adaptations accordingly. Alternatively, performance may have improved due to other factors, and the resulting increase in difficulty may have made the task more engaging. Further research would be necessary to investigate the causal relationship between these variables.

Contrary to prior research (Matthews et al. 1999) and with the exception of scenario 2, Distress was not correlated with performance. Notably, this was the first scenario in which participants were introduced to micro-adaptive difficulty where nearly half of all participants experienced a change in workload (40.0% experienced an increase in signals and 9.3% a decrease). Although participants' level of distress did not change with changes to micro-adaptive difficulty, their distress may have had an influence on performance due to the task change. These results are interesting in light of prior research suggesting that changes in workload may increase stress and have a negative effect on performance (Cox-Fuenzalida, 2007; Helton et al. 2008).

4.2 Impact of Coping Strategies on Performance

We predicted that adaptive coping strategies would lead to better performance than maladaptive coping strategies. There was an interaction between coping strategy and coping focal mechanism on accuracy. Contrary to our prediction and to prior research (e.g., Brown et al. 2005; Matthews & Falconer, 2000), the maladaptive, problem-focused strategy did not impair performance. However, we did find the anticipated pattern for the emotion-focused coping groups such that the maladaptive, emotion-focused group performed worse than the adaptive, emotion-focused group. Potentially, the specific maladaptive, problem-focused strategy we used did not have the intended maladaptive effect. Participants in this group were instructed to blame the interface by identifying features that impeded performance. This strategy, known as other-blame, is typically used in the context of blaming another person (e.g., Garnefski et al. 2001) rather than a non-social agent such as the interface used in the present study. Some research has examined other-blame in interaction with social robots (Groom et al. 2010), but instead examines the *attribution* of blame, rather than its potential as a stress coping strategy. Thus, other-blame may only be a maladaptive strategy when used in response to social agents.

Additionally, we observed increases in accuracy between the first and third scenario. This may reflect the commonly observed trend in which learning improves rapidly early on, with decreases in improvement over time (Newell & Rosenbloom, 1981). Although Distress was generally not related to performance in the present study, it is unclear whether it might have had a negative impact on performance in a longer study or if it may impact well-being in the long run (Matthews et al. 1999).

4.3 Relationship Between Coping Strategies and Micro-adaptive Difficulty

The results suggest that coping strategies may impact participants' experience with the micro-adaptive difficulty. Overall, the change in difficulty increased in the fourth scenario, suggesting that performance improved over time. Analysis of the interaction between scenario and coping focal mechanism suggests that this increase in difficulty may be attributed to the adaptive, problem-focused group. It is particularly interesting that participants in this group experienced relatively low difficulty until the last adaptive scenario, where their change in difficulty increased considerably. Participants in this group may have taken longer to effectively use their adaptive strategy of utilizing the "cheat sheet" in the interface, but, when they did use it, performance improved. Perhaps participants relied more on the cheat sheet in earlier scenarios. Because the cheat sheet prevents them from viewing the interface, using it could have caused them to miss more incoming signals or direct more attentional resources to utilizing the tool, rather than reporting signals. The observed trend that participants used the cheat sheet fewer times and for less time in S4 supports this interpretation, as well as the strong negative correlations between cheat sheet use and change in difficulty.

Additionally, there was an interaction between coping strategy and coping focal mechanism. On average, the maladaptive, emotion-focused group experienced a lower change in difficulty compared to the maladaptive, problem-focused group. Specifically, this group did not receive any more signals than the default number. This suggests that the effect of the coping strategy may be moderated by the coping focal mechanism. Simply being a maladaptive strategy did not sufficiently impact a change in difficulty. Only the maladaptive, emotion-focused group was unable to "adapt upward." A similar pattern of data emerged for accuracy, where this group showed decreased accuracy compared to the other groups.

4.4 Limitations and Future Directions

One limitation of the present research is that the entire study took place within approximately two hours. Prior research suggests that while maladaptive coping strategies may be effective in abbreviated experiences, they may not be effective in prolonged periods of stress (Suls & Fletcher, 1985). Thus, a shorter time on task may not be long enough to elicit the negative effects of coping in a maladaptive, problem-focused group. Another limitation could be the operationalization of the maladaptive, problem-focused strategy. Future research could explore other types of maladaptive, problem-focused strategies that are more appropriate for this type of task. In this study we used one type of coping strategy for each experimental group. Alternatively, including multiple strategies within each condition might reveal different patterns or results. Future research could also examine the effects of coping strategies across multiple sessions to investigate their impact on long-term retention of task-relevant skills and knowledge.

4.5 Conclusions

Our findings provide new insight into the relationship between stress, and performance in a task that utilizes micro-adaptive difficulty. We also present novel findings on the impact of stress coping strategies in a micro-adaptive training environment. Task engagement was positively correlated with some performance measures and a change in difficulty, while Distress was not consistently related to these measures. Contrary to our predictions, the maladaptive, problem-focused coping strategy did not impede performance. These findings provide a foundation for future research investigating dimensions of stress and their predictive relationship with performance on military tasks using micro-adaptive training. These findings might also inform the development of stress regulation training. It would be advantageous to further analyze the role of stress in related contexts to determine whether the observed results extend to similar tasks.

Acknowledgments. This work was funded by the Office of Naval Research (N0001422WX01634) under Ms. Natalie Steinhauser, and the Office of Naval Research Sabbatical Leave Program. Presentation of this material does not constitute or imply its endorsement, recommendation, or favoring by the U.S. Navy or the Department of Defense (DoD). The opinions of the authors expressed herein do not necessarily state or reflect those of the U.S. Navy or DoD. NAWCTSD Public Release 24-ORL003 Distribution Statement A – Approved for public release; distribution is unlimited.

Disclosure of Interests. The authors have no competing interests to declare that are relevant to the content of this article.

References

Barthol, R.P., Ku, N.D.: Regression under stress to first learned behavior. Psychol. Sci. Public Interest **59**(1), 134 (1959)

Brown, S.P., Westbrook, R.A., Challagalla, G.: Good cope, bad cope: adaptive and maladaptive coping strategies following a critical negative work event. J. Appl. Psychol. **90**(4), 792–798 (2005). https://doi.org/10.1037/0021-9010.90.4.792

Chao, C.Y., et al.: Maladaptive coping, low self-efficacy and disease activity are associated with poorer patient-reported outcomes in inflammatory bowel disease. Saudi J. Gastroenterol. Official J. Saudi Gastroenterol. Assoc. **25**(3), 159–166 (2019). https://doi.org/10.4103/sjg.SJG_566_18

Domaradzka, E., Fajkowska, M.: Cognitive emotion regulation strategies in anxiety and depression understood as types of personality. Front. Psychol. **9**, 856 (2018). https://doi.org/10.3389/fpsyg.2018.00856

Goodwin, G.A., Kim, J.W., Niehaus, J.: Modeling training efficiency and return on investment for adaptive training. In: 5th Annual Generalized Intelligent Framework for Tutoring (GIFT) Users Symposium (GIFTSym5). US Army Research Laboratory (2017)

Hancock, G.M., Hancock, P.A., Janelle, C.M.: The impact of emotions and predominant emotion regulation technique on driving performance. Work (Reading, Mass.), 41 Suppl 1, 3608–3611 (2012). https://doi.org/10.3233/WOR-2012-0666-3608

Hancock, P.A., Warm, J.S.: A dynamic model of stress and sustained attention. Hum. Factors **31**(5), 519–537 (1989). https://doi.org/10.1177/001872088903100503

Hancock, P.A., Szalma, J.L.: Stress and Performance. In Performance Under Stress (pp. 17–34). CRC Press (2008)

Hughes, A.M., Hancock, G.M., Marlow, S.L., Stowers, K., Salas, E.: Cardiac measures of cognitive workload: a meta-analysis. Hum. Factors **61**, 393–414 (2019)

Landsberg, C.R., Van Buskirk, W.L., Astwood, R. S., Mercado, A.D., Aarke, A.J.: Adaptive training considerations for use in simulation-based systems (Special Report 2010–001). Defense Technical Information Center (DTIC) (2011). http://www.dtic.mil/cgi-bin/GetTRDoc?AD=ADA535421&Location=U2&doc=GetTRDoc.pdf

Marroquín, B., Tennen, H., Stanton, A.L.: Coping, emotion regulation, and well-being: intrapersonal and interpersonal processes. In: Robinson, M.D., Eid, M. (eds.) The happy mind: Cognitive contributions to well-being, pp. 253–274. Springer, Cham (2017). https://doi.org/10.1007/978-3-319-58763-9_14

Matthews, G., Campbell, S.E.: Task-induced stress and individual differences in coping. Proc. Hum. Factors Ergonomics Soc. Ann. Meeting **42**(11), 821–825 (1998)

Matthews, G., Joyner, L., Gilliland, K., Campbell, S.E., Falconer, S., Huggins, J.: Validation of a comprehensive stress state questionnaire: Towards a state "Big Three. " In: Mervielde, I., Dreary, I.J., DeFruyt, F., Ostendorf, F. (eds.), Personality psychology in Europe, vol. 7, pp. 335–350. Tilburg University Press (1999)

Matthews, G., Szalma, J., Panganiban, A.R., Neubauer, C., Warm, J.S.: Profiling task stress with the dundee stress state questionnaire. In: Cavalcanti, L., Azevedo, S. (eds.) Psychology of stress: New research, pp. 49–92. Nova Science (2013)

Newell, A., Rosenbloom, P.: Mechanisms of skill acquisition and the law of practice. In: Anderson, J.R. (ed.) Cognitive Skills and Their Acquisition, pp. 1–55. Erlbaum, Hillsdale (1981)

Park, O., Lee, J.: Adaptive instructional systems. In: *Handbook of research for educational communications and technology*, Edited by: Jonassen, D. 651 – 684. New York, NY : MacMillan Publishers (2003)

Proctor, R.W., Vu, K.-P.L.: Laboratory studies of training, skill acquisition, and retention of performance. In: Ericsson, K.A., Charness, N., Feltovich, P.J., Hoffman, R.R., (eds.), The Cambridge handbook of expertise and expert performance (pp. 265–286). Cambridge University Press (2006). https://doi.org/10.1017/CBO9780511816796.015

Staal, M. A. (2004). Stress, cognition, and human performance: A literature review and conceptual framework

Lazarus, R.S., Folkman, S.: Stress, Appraisal, and Coping. Springer (1984)

Suls, J., Fletcher, B.: The relative efficacy of avoidant and nonavoidant coping strategies: a meta-analysis. Health Psychol. **4**(3), 249–288 (1985). https://doi.org/10.1037/0278-6133.4.3.249

Gross, J.J., Levenson, R.W.: Emotional suppression: physiology, self-report, and expressive behavior. J. Pers. Soc. Psychol. **64**(6), 970–986 (1993). https://doi.org/10.1037/0022-3514.64.6.970

Gross, J.J.: Emotion regulation: affective, cognitive, and social consequences. Psychophysiology **39**(3), 281–291 (2002)

Garnefski, N., Kraaij, V., Spinhoven, P.: Negative life events, cognitive emotion regulation and emotional problems. Pers. Individ. Differ. **30**(8), 1311–1327 (2001). https://doi.org/10.1016/S0191-8869(00)00113-6

Groom, V., Chen, J., Johnson, T., Kara, F.A., Nass, C.: Critic, compatriot, or chump?: Responses to robot blame attribution. In: 2010 5th ACM/IEEE International Conference on Human-Robot Interaction (HRI), pp. 211–217 (2010).https://doi.org/10.1109/HRI.2010.5453192

Van Buskirk, W.L., Steinhauser, N. B., Mercado, A.D., Landsberg, C.R., Astwood, R.S.: A Comparison of the Micro-Adaptive and Hybrid Approaches to Adaptive Training. Proceedings of the Human Factors and Ergonomics Society Annual Meeting, vol. 58. no1, pp. 1159–1163 (2014).https://doi.org/10.1177/1541931214581242

Williams, R.: The Office of Naval Research-Science and Technology in Support of the US Navy and Marine Corps. In 2017 ERC (2017)

Directly Measuring Learning: A Community Engaged Learning Case Study with Implications for Human Centered Computing in Education

Emily Passera and Thomas Penniston[✉]

University of Maryland, Baltimore County, Baltimore, MD 21250, USA
{epassera,pthomas1}@umbc.edu

Abstract. This paper investigates the empirical measurement of student growth in relation to institutional learning outcomes, focusing on academic development supported through applied learning. The research specifically examines the impact of service-learning participation on practical, affective skills, and presents a method for directly measuring learning using rubric data, avoiding the use of proxies or indirect measures. It highlights the possibility of scaling course design to enhance learning outcomes, including self-paced remediation. The authors emphasize the importance of service-learning as a means to cultivate global citizenship and describe the potential for learning analytics to facilitate student success in other domains. This approach is tested in a public university setting and considers the unique challenges posed by its diverse student populations. The findings suggest that well-designed applied learning and service-learning programs can enhance student learning outcomes, aligning with institutional goals. The study concludes by advocating for the focused integration of ML and AI tools in education to tailor learning experiences to individual student needs in alignment with desired program outcomes.

Keywords: Learning Analytics · Applied Learning · Direct Measures

1 Introduction

How can education support student learning and growth that has direct applications for engaged participation in community and social life? Our paper operationalizes this question through an empirical lens, measuring student learning directly, rather than considering it as an abstract theoretical construct, or attempting to measure it imprecisely with proxies. We argue that through intentional, rigorous, course designs aligned with institutional objectives, educators can directly observe, and better facilitate student cognitive development. Furthermore, such an approach may be scaled using sophisticated machine learning (ML) and artificial intelligence (AI) platforms to improve student learning outcomes within a term (e.g., through self-paced remediation), and thereby simultaneously contribute to both learning as well as gains in institutional indicators of success, such as improved persistence and reduced time to graduate. The Shriver Center (the Center) at UMBC is a community engagement hub celebrating its 30th anniversary

R. A. Sottilare and J. Schwarz (Eds.): HCII 2024, LNCS 14727, pp. 166–177, 2024.
https://doi.org/10.1007/978-3-031-60609-0_12

of leading meaningful social change through transformational higher education and community partnerships. In advancement of this interconnected vision, the Center serves as a bridge joining the campus and broader regional community through engaged scholarship and applied learning designed to address community-identified needs. Specifically, the Service-Learning & Community Engagement (SLCE) team facilitates undergraduate and graduate student semester-long, zero credit practicum courses. Through this curriculum, students engage in supervised weekly service with a school district, non-profit, or government agency for an average of three hours per week. This framework provides students the opportunity to engage in three written reflections on their engagement, supporting metacognition to connect their service with their learning.

Measuring student learning ultimately measures student achievement across a span of time. The academic success of learners and their ability to define their metacognition is a priority of universities, especially as the enrollments of underrepresented students increases (Gregg, 2019). An essential component of this work is assessment of student learning. Utilizing Bloom's taxonomy of educational objectives, evaluation of the development of knowledge is a core component of our practicum experience (Bloom et al., 1956). Reflective practice facilitated through the Center is a central tool in connecting knowledge to experience through real-world application. The purpose of this case study is to demonstrate the value of direct measurement of student learning in a structured applied learning course.

2 Literature Review

This research investigates growth aligned with student learning outcomes in an applied learning course as a particular example of how institutions might support intentional course design as a means of scaling direct measures of learning across domains. Not only is this specific course relevant on its own as an example of service-learning programming supporting learner development, but it is also relevant to the broader community. This methodology may be used as a prototype describing how aligning and embedding direct outcome measurement within a course and across an institution might facilitate direct measurement of growth, which in turn can be used to ethically intervene in support of student success at scale.

2.1 Education Assessment

In 2020, the Center implemented a new rubric to standardize assessment criteria to measure achievement growth. This holistic rubric focused on gaining an overall impression of the students' thought processes and was used as the measurement tool for every example in our study. The use of rubrics across education keeps assessment of student growth uniform and standardized. Nine studies found students achieved higher cognitive achievement with the combination of a rubric and teacher feedback (English et al., 2022). Micro and macro level conditions affect the implementation of the assessment tool, including human condition. There is still room for academia to explore the use of ML and AI in grading interpersonal reflection.

2.2 Learning Analytics

EDUCAUSE's Director Learning Initiative Malcolm Brown (2011) surveyed and summarized the landscape of the burgeoning community of practice convened for the inaugural International Conference on Learning Analytics and Knowledge (LAK11). In defining this field, he states "At its core, learning analytics (LA) is the collection and analysis of usage data associated with student learning. The purpose of LA is to observe and understand learning behaviors in order to enable appropriate interventions" (Brown, 2011).

In other words, LA is data-informed action. What an administrator or practitioner knows is only as important as what they can do with the information to benefit learner outcomes. In turn, emphasis shifts from an academic interest in investigating relationships between variables to something much more substantial. Specifically, if we can accurately predict a student's academic pathway, what behavioral modification intervention might we ethically take in order to improve their chances of success? That's not a simple question to dig into, because not only does it require valid and reliable predictors, but also a means to positively influence an individual's unique learning trajectory. One potentially powerful option to improve learner outcomes is the use of behavioral nudging.

Thaler and Sunstein (2009) define *nudge* as "any aspect of the choice architecture that alters people's behavior in a predictable way without forbidding any options or significantly changing their economic incentives" (p. 6). Behavioral nudging presents an innocuous option to improve student outcomes. Thaler and Sunstein describe good "choice architecture" implemented through "libertarian paternalism." In an academic setting, we can intentionally design learning environments that help students make good, empirically grounded choices, and we establish means for monitoring and evaluating growth in these areas. Perhaps data indicate that it is difficult for a student to be successful unless they show up to class regularly and engage (as is typically the case in undergraduate settings)? In response, we might adjust our existing course policy to reward students for in-class participation. If they don't attend class, we could send them notifications highlighting what choices successful students make, or empathetically offer additional support. Learners can still choose to ignore the nudges, just as one can walk past healthy fruit and vegetable choices at the beginning of the line in the cafeteria. At a population level, however, these design adjustments can reduce negative outcomes, or increase the likelihood of socially preferable ones, like choosing to grab an apple instead of a bag of potato chips or reflecting on one's experiences through a service-learning. The key is in not making the decision for the individuals, but rather providing scaffolding for the learners in the form of external motivation, which will help them develop their own intrinsic motivation. As a learner progresses through a course and academic career, that scaffolding can be slowly scaled back and removed leaving intact the now fully developed, self-directed learner.

However, finding approaches to move the needle on institutionally valued measures is difficult under the best circumstances; at an exceptionally heterogeneous university, this proposition becomes even more complicated to realize. Students come to our university from all conceivable backgrounds, which is something we rightly pride ourselves on. Our diversity makes us stronger and helps break down the barriers that may be imposed at other institutions. At the same time, such equal representation also means we must be very intentional in the design of the scaffolding we employ. Simply nudging students to attend class, for example, may not be the best solution to discernibly increase persistence or reduce time to graduation. Perhaps such indirect indicators of success are not the best metrics to use in evaluating and supporting student success? After all, if our mission is *learning*, then shouldn't that be what we measure? Unfortunately, that's not always expedient, and institutions may nudge students in support of more easily measurable outcomes for which their funding is based (Penniston, 2023).

Through intentional design, we can gather multiple direct measures of growth, and thereby not only monitor student trajectories in relation to a given construct, but also intervene, or nudge, if there are deviations from successful patterns. For our current service-learning context, that intervention strategy might involve in-term adjustments at the service placement, while in a chemistry course, nudging may direct students to self-paced remediation of foundational concepts. The key, therefore, is in architecting courses that produce direct measures of learning that can be leveraged to inform nudging strategies early enough in a semester to help improve a student's chances of success. Although we may not be able to claim causality given such a design, we can inform an approach to intervene based on growth observed in the given construct, rather than relying solely on proxies (e.g., clicks on content).

2.3 Applied Learning

As service-learning practitioners and scholars, we are passionate about the transformational opportunity to change how students achieve learning goals. Students spend a significant amount of time practicing teamwork, problem solving, and communication skills with peers and community members with varied lived experiences. Applied learning is a general term that encompasses various types of applied learning, from research, to internships, to co-ops. Within that broad framework, this paper focuses on community engaged learning, and specifically service-learning, which is a particular pedagogy that balances, as implied by the term's hyphenation, student community service and learning as an outcome of these experiences. Unless otherwise specified (e.g. the word choice of a given author), the authors will use this terminology throughout the paper. Applied learning is a particular pedagogy intended to make metacognition more relevant and long-lasting to transferable environments such as class discussion, graduate research, and career exploration (Carnegie Classification of Institutions of Higher Education, 2023; Kolb, 1984).

Kolb's theory of integrating new information through the practice of reflection while practicing in concrete, experiential activities inform our curriculum (Kolb, 1984). The transfer of these new skills leads to student success across many domains, indicating deeper learning. Bloom et al.'s (1956) taxonomy summarizes how learning some skills is more difficult than others. To facilitate increasing complexity of cognition, processing knowledge is the lowest level of required skill. When students successfully repeat definitions or specific facts on a quiz, they are likely not demonstrating comprehension. Whereas when students are thinking critically about new information based on the prompts on social issues we provided, they are in a higher-level analysis category.

This category of the taxonomy is called synthesis (Bloom et al., 1956). While working with our community partners, undergraduate students are challenging prior beliefs and assumptions in real time while supporting a common mission driven by the experts within the site partnerships. The way they synthesize new information varies by student.

Traditionally, pedagogues, researchers, and evaluators have assessed learning through indirect measures, such as term grades. For example, Gregg (2019) found a statistically significant retention increase for college sophomores and juniors if they had participated in service-learning during their first year. In a similar vein, Penniston's (2014) research concluded a statistically significant four-year graduation rate for program participants. Applied learning is a practice of metacognition that is goal oriented. Previous theory on the best practices of facilitating reflection support adaptation after the introduction of new material (Gibbs, 1988). Highlighted in this cycle, more experimentation and review of an experience promotes engagement with learning; students are critically analyzing with their peers, alongside doing and thinking about what they are doing live in the moment. While they reconsider previous actions, they learn appropriate behaviors and cognitions for the future.

Previous empirical evidence supports the argument that service-learning is one of the high impact pedagogies that has a positive effect on graduation rates (Kuh, 2008; Gregg, 2019; Penniston, 2014; Lockeman & Pelco, 2013; Brownell & Swaner, 2009). Shea et al. (2023) and Hatcher et al. (2004) stressed the integration of classroom related academic content as being a core component of the experience. Applied learning experiences are the most successful when reflection is an integrated part of the curriculum. Regularly scheduled reflection assignments are more valuable than a single, summative reflection product (Hatcher, 2004). Throughout the practice of critically looking inward, written reflection creates a summary of learning, essentially "datalizing" changes in values, perspectives, and identities of students. Throughout our coursework, we are preparing students to engage in socially conscious reflective practice supporting internal adjustment preparedness to think, learn, and problem solve.

The environment where students learn in our case study includes community centers, neighborhood clinics, schools, places of worship, and campus buildings. Youth who have interactions with possible role models are more likely to have more positive mental and physical health outcomes (Ahea, 2016; Torres et al., 2021). Increased engagement in reflection oriented, hands-on learning improves active participation and social skills, positively benefiting the academic and personal well-being of college students (Hartman & Anderson, 2022; Torres et al., 2021). Mentorship is a significant asset when facilitating the cognitive learning process. While working with experienced peers, students begin assuming an active role in their internal and external processing of the world around them. With 40% of our participants identifying as first year students, there are university expectations to operate programs that retain students. Through reflection, program leaders help students consider the benefits and harms of their entry into communities, and to critically examine such things as their own implicit biases on the nearby local metropolitan city. With college educated students entering marginalized spaces, there are also power dynamics that need to be examined specifically to not reinforce harm. More stratification by income status hurts communities, especially if universities are becoming more unequal gatekeepers to social mobility. Notably, we have partners from many different backgrounds in our community, so more strata, which goes to the earlier discussion about intentional design of the experiences. Engaging in a process to improve social capital through emotional and attitudinal trial and error happens in real time while practicing teamwork and leadership skills in partnership with a community partner. Barriers to continued engagement in this work may include need for financial compensation, or perceived threat to financial stability, instead of participating in unpaid engagement opportunities in their field of choice. As described above, students will be transformed by this process of "learning by doing" because there is a natural and insepa-rable link between knowledge and experience (Gibbs, 1988). Our goal with this research is to assess their growth.

2.4 Direct Measures

For our current research, we look at direct measures of learning using rubric data to evaluate cognitive growth. Positive learning outcomes have been found for programs that incorporate feedback loops between graded assessments (Ahea et al., 2016). In our methodology, the loop is time based: students receive feedback before they have written another reflection, so they may make adjustments and gain clarification of expectations. In our assessment of learning, we did not use multiple choice or quiz form examinations. Rather, the use of long-form essay questions allowed for an open-ended response unique to the student's beginning, middle, and end of semester motivations and transference.

3 Methods

For this research, we drew from the undergraduate student population at an R1 STEM-focused institution, and the population is restricted to students who enrolled in a 10 to 30-h, 13-week practicum experience across four semesters post COVID-19 vaccine availability. Throughout 2020, most students enrolled in the practicum were not required to participate in in-person site placements, so data from that period was excluded. Many students, however, supported their hyper local communities through food distribution, elder wellness checks, and social media engagement in current events. Beginning fall 2021, our institution resumed in-person activities following masking protocols that grad-ually decreased in alignment with approved rules of quarantine in event of virus expo-sure. Our students worked with organizations prioritizing early childhood development, healthcare, policy, law, climate action, technical literacy, food justice, and animal advo-cacy. Deliverables varied, but expectations of engagement included teaching, mentor-ing, scholarship, and the delivery of creative products like digital stories. Program staff guided all students through orientation and training, goal setting, action, reflection, and evaluation.

The data set used for this analysis included 1536 total records, consisting of enroll-ments in a zero-credit undergraduate practicum course for the four successive primary terms beginning fall 2021 (i.e., fall 2021, spring 2022, fall 2022, and spring 2023). The date range corresponds with when the adoption of course rubrics aligned with insti-tutional functional competencies. Of these cases, two-thirds (1,032) represent students enrolling in more than one of these sections, meaning that there are 504 unique students enrolled in one or more practicum during this four-semester time frame.

To analyze our data, we reviewed the scores across three written reflections, distributed within the same academic term, where students had practiced and received feedback on their metacognitive growth. Reflections were assessed with a rubric by three instructors, within several days or weeks of the assignment's submission. Each row on the rubric included measures aligned with the university's functional competencies, which are the key affective skill measures at the top of our institutional hierarchy used for evaluating student success. Noting the four categories of our reflection rubric the assessment crosswalk was developed in alignment with the course's learning outcomes. The first of these outcomes to "increase awareness of community assets and needs," which directly challenges students to prioritize the knowledge, skills, and abilities of other community members. Positioned fifteen minutes from a major mid-Atlantic city with significant racial and economic stratification, the ability to thrive in contexts requiring cross cultural collaboration with people from different lived experiences is a highly valuable, employable skill set. To support this growth, students were paired with and worked for partner organizations within the community. They then observed, reflected upon, and conceptualized historic and current socio-political tensions. It's critical through this process for leadership to promote a growth rather than deficit mindset among families from minoritized backgrounds. Simultaneously, leadership must avoid inculcation centered on charity, salvation, or doing for, but rather to enhance social responsibility through grassroots democratic engagement by doing *with*. This is at the heart of guided reflection: making sure students take away the right lessons. The backgrounds and differing opinions at each placement require perspective-taking and intercultural awareness skills. Experiential learning deepens relationships with peers and community. We do not ask students to excel in these outcomes from the onset of the course, but we do expect growth across the term in these four constructs of global citizenship, internal reflection, leadership, and professionalism.

Two data sets were combined to visualize our student data in an accurate and comprehensive way. Looking to collect data during the activity, we collected the reflection scores from our learning management system (LMS) and integrated the unique student demographic data from our institutional data metrics. While there are rich qualitative opportunities for this data set, our focus was on direct measurement of student growth aligned with four primary constructs:

- **Construct 1**, *Global Citizenship*: demonstrate awareness of community partnership and perspective-taking.
- **Construct 2**, *Internal Reflection*: self-awareness of growth in placement, even amidst challenging value tensions and flexible career pathway examination.
- **Construct 3**, *Leadership*: skill development and organizational collaboration.
- **Construct 4**, *Professionalism*: student's ability to follow time management deadlines and writing prompt instructions.

We calculated traditional descriptive statistics while reviewing the direct measures of performance, the reflection grades, which were based on ordinal measures: Exceeds Expectations, Meets Expectations, or Didn't Meet Expectations. After analysis, our measurement of our grading rubric demonstrated 98% of students scored proficient on written reflection metrics.

Nearly 80% of total enrollments in this sample are students who entered the university as first-time freshman, with only 15% transfer student enrollment. Given that approximately half of the undergraduate population is composed of transfer students, we see that there's an underrepresentation of these students. The "first year experience" in American universities is designed to provide "traditional" students with opportunities to explore campus, develop communities of practice, build peer networks, and engage in campus life. However, this structure often means that students who transfer to the university after their first year may face challenges in accessing these established resources and integrating into pre-formed social networks (Brownell & Swaner, 2009). Notably, around 40% of all of these applied learning experiences take place during students' freshman year. In turn, transfer students are one of the groups our institution, as well as many others, focus on to intentionally foster these connections that are associated with academic success, including by connecting them to applied and service-learning experiences within the greater community, such as those experiences represented by the data in this analysis.

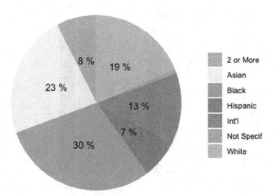

Fig. 1. Illustrates student enrollment in the course by ethnicity. Students of color tend to have greater representation in applied learning courses than they do in the general student population. For example, in 2023, African American students represented 20% of the population and White students 28% (University of Maryland, Baltimore County, n.d.)

Higher Education has a duty to study and support programming that increases students' sense of belonging and proof of economic return on investment (Brownell & Swaner, 2009). At our institution, 29% of students received Federal Pell grants, and over 60% identify with a non-white ethnicity (University of Maryland – Baltimore County, 2024). Operating as a minority serving institution, we believe in the need for systemic changes that promote retention and graduation while improving social capital for historically marginalized communities. From our total number of students, including the 66% of students who engaged in multiple semesters, we observed the highest racial participation group to be students of color. We provide an example in Fig. 1. Programs that are affirming to identity are positive drivers that create safe spaces to learn and practice skills. Our partner organizations also have high representations of black and

African American families. Extensive research in youth development supports models that include volunteers with a shared background to the community they are partnering with (Augustine et al., 2022).

4   Analysis

The data for this analysis provide some limitations, as well as opportunities. In regard to the former, we did not have a readily available comparison group, and quasi-experimental design is difficult to operationalize. In this case study, we did not observe the same learning outcomes for a random sample of students who have not participated in our service-learning course. Applied learning students are self-selected rather than randomly assigned, which inherently creates bias in the data. Figure 2, below, shows an overall increase in student proficiency across the semester for each individually measured construct, as indicated by the rubric values (i.e., R1, R2, and R3).

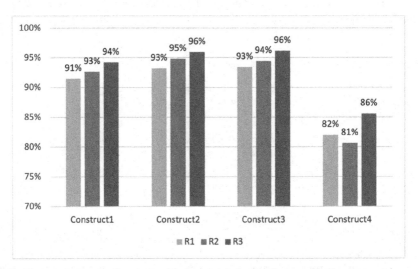

Fig. 2. Student measures of metacognition associated with Community Engagement increases across the semester.

Students demonstrated similar patterns across Constructs 1–3 "Global Citizenship," "Internal Reflection," and "Leadership," with 1–2% growth during the term. Construct 4, "Professionalism," shows the lowest overall scores, but also the highest gains (5%).

An observed semester-long increase in student proficiency across various constructs, per rubric evaluations, suggests adaptability of a similar approach to other diverse qualitative data types. Despite current reliance on manual coding, integration of ML and AI tools shows potential for advancement. These tools are crucial in course design, helping align learning objectives, assessments, and content, thus enhancing student outcomes. Importantly, they focus on direct learning measures, providing actionable intelligence and preventing institutional biases that might overlook student interests. In the future, leveraging ML and AI with similar course and data gathering design might enable institutions to adapt effectively to learners' unique needs and ensure educational strategies align with student learning outcomes as primary indicators of success.

5 Discussion

There's potential for this project to serve as a template for larger data-gathering initiatives designed to leverage direct measurement of learning. There are also some limitations. To begin, since we didn't have a traditional control group, we weren't able to fully account for social maturation or the endogeneity associated with student self-selection to enroll in these courses.

We're also not accounting for variation in grading. Maybe instructors tended to rate each successive reflection higher not out of merit, but because that's standard practice as the term progresses? Although these rubrics were developed in collaboration with our institution's Faculty Development Center, we can't say for certain that human bias didn't impact the scoring. However, rubric use is intended to reduce the impacts of arbitrary grading, and even if caprice was impacting our estimates of lift, there is no reason safeguards could not be incorporated in other contexts to address this concern moving forward, including training on rubric scoring, anonymous submissions, and blind scoring.

In the sample, the academic level of students represented most frequently was first-time freshman. We found no significant difference between academic level and overall performance after an ANOVA comparison. The Center partners with predominantly humanities faculty to integrate applied learning into course design, and reflection is a graded activity. This motivation could drive participation in direct measurement and skew toward first-year students. This suggests the program should engage in more recruitment of sophomore, junior, and senior students so the benefits of applied learning can be repeated. Barriers to engagement could include internships and research, which are more likely to be paid experiences. The highest number of reflections included in our sample came from fall 2022, followed by spring 2023. The t value of the evaluations was quite high, indicating with statistical significance that the number of submissions varied greatly. Data collection during staff transitions could have impacted these results. Another issue, as one may infer from reading this narrative, or in harkening back to their own experiences, is collecting direct measures of learning is exceptionally time consuming and requires a great deal of academic rigor. Good rubrics, unfortunately, do not grow on trees. They take time to develop and align, must be customized to the context and assignment, and can be arduous to deploy in the wild. But rigorously designing courses to gather accurate direct measures of growth can ultimately help us support student success.

Moving forward ML and AI might be integrated and leveraged through the full life of the experience, from conception, to deployment, to iterative monitoring and evaluation. Such a holistic approach would represent a win for both institutions of higher education and the students they serve. For example, one might imagine a scenario whereby an AI platform might ingest certain LMS data to autonomously grade assignments based on rubric parameters, and then curate a list of students for targeted outreach, or perhaps the AI itself might support content remediation through self-paced learning?

To achieve these ends, it's been argued that, in higher education, there's a need to emphasize course design to help align teaching with broader educational goals (Fritz, Hawken & Shin, 2021), from unit-level objectives to institutional competencies (Penniston, 2023). This approach, considered a best practice, not only helps in scaling educational interventions, but also in enhancing the quality of insights for instructors and administrators. Kuh's research on impactful practices to retain students includes many opportunities for collaboration (Kuh & Schneider, 2008). Service-learning is similarly designed to foster problem solving skills and self-reflection after exposure to novel ideas and values.

Systematically identifying behavioral inflection points allows us the opportunity to intervene – nudge – to influence the given outcome, in this case *learning*. Such an approach embracing top-down design with bottom-up alignment moves beyond superficial engagement metrics to focus on empirically measurable learning. Such a learner-centric paradigm in course design also allows for more precise identification of learning gaps, as may be illustrated in a chemistry course, for example, by enabling targeted interventions like self-paced study modules for specific topics like ionic bonds. Students in the practicum may be co-enrolled in courses in our College of Arts, Humanities, and Social Sciences. A future direction we want to explore is to integrate with more academic courses in the two other Science, Technology, Engineering, and Math Colleges.

6 Conclusion

The current project measures student growth aligned with institutional goals. With our university's 2026 accreditation fast approaching, prioritizing community engaged learning in strategic planning is essential, as illustrative stories directly from students learning while partnering with non-profit, school, and government agencies could lead to deeper understandings of career preparation phenomenon. A self-aware student is a successful student along their academic trajectory. At the same time, this research also represents a proof of concept for how we might generate and extract meaningful metrics of student learning at scale to support academic success across an institution.

References

Ahea, M.M.-A.-B., Ahea, M.R.K., Rahman, I.: The value and effectiveness of feedback in improving students' learning and professionalizing teaching in higher education. J. Educ. Pract. 7(16), 38–41 (2016). https://files.eric.ed.gov/fulltext/EJ1105282.pdf

Augustine, D., Smith, E., Witherspoon, D.: After-school connectedness, racial-ethnic identity, affirmation, and problem behaviors. J. Youth Dev. 17(4) (2022). https://doi.org/10.5195/jyd.2022.1137

Bloom, B.S., Englehart, M.D., Furst, E.J., Hill, W.H., Krathwohl, D.R.: Taxonomy of Educational Objectives. Handbook 1: Cognitive Domain. McCay, Inc. (1956)

Brownell, J.E., Swaner, L.E.: High-impact practices: applying the learning outcomes literature to the development of successful campus programs. Peer Rev. 11(2), 26–30 (2009)

Brown, M.: Learning analytics: the coming third wave [ELI Brief]. Educause Learn. Initiat. (2011). https://library.educause.edu/-/media/files/library/2011/4/elib1101-pdf.pdf

Carnegie Classification of Institutions of Higher Education. What is community engagement? (2023). https://carnegieclassifications.acenet.edu/elective-classifications/community-engagement/

English, N., Robertson, P., Gillis, S., Graham, L.: Rubrics and formative assessment in K-12 education: a scoping review of literature. Int. J. Educ. Res. **113**, 101964 (2022)

Fritz, J., Hawken, M., Shin, S.: Using learning analytics and instructional design to inform, find, and scale quality online learning. In Online Learning Analytics (1st ed., pp. 95–114). Auerbach Publications (2021)

Gibbs, G.: Learning by doing: a guide to teaching and learning methods. Further Education Unit. Oxford Polytechnic: Oxford. Accessed and Adapted from University of Edinburgh, Reflection Toolkit (1988). https://www.ed.ac.uk/reflection/reflectors-toolkit/reflecting-on-experience/gibbs-reflective-cycle

Gregg, D.: Effects of "high-impact practices" on first-generation college students' academic success (Doctoral dissertation, University of Maryland, Baltimore County) (2019)

Hartman, C.L., Anderson, D.M.: Psychosocial identity development and perception of free time among college-attending emerging adults. Leis. Sci. **44**(1), 77–95 (2022). https://doi.org/10.1080/01490400.2018.1483850

Hatcher, J.A., Bringle, R.G., Muthiah, R.: Designing effective reflection: what matters to service-learning? Mich. J. Community Serv. Learn. **11**(1), 38–46 (2004)

Kolb, D.: Experiential Learning: Experience as the Source of Learning and Development. Prentice-Hall Inc., Hoboken (1984)

Kuh, G.D., Schneider, C.G.: High-impact educational practices: what they are, who has access to them, and why they matter. Assoc. Am. Coll. Univ. (2008)

Lockeman, K., Pelco, L.: The relationship between service-learning and degree completion. Mich. J. Commun. Serv. Learn. **20**(1), 18–30 (2013)

Penniston, T.: The impacts of service-learning participation upon postsecondary students' academic and social development. (Doctoral dissertation, University of Maryland, Baltimore County) (2014)

Penniston, T.: Toward a new paradigm: learning analytics 2.0. In: Sottilare, R.A., Schwarz, J. (eds.) Adaptive Instructional Systems. Lecture Notes in Computer Science, vol. 14044, pp. 148–161. Springer, Cham (2023). https://doi.org/10.1007/978-3-031-34735-1_11

Shea, L.-M., Harkins, D., Ray, S., Grenier, L.I.: How critical is service-learning implementation? J. Experiential Edu. **46**(2), 197–214 (2023). https://doi.org/10.1177/10538259221122738

Thaler, R.H., Sunstein, C.R.: Nudge. Penguin (2009)

Torres, Y., Walsh, N., Tahvildary, N.: Impact of mediator mentors service-learning on college student social-emotional expertise and cultural competence. J. Pract. Stud. Educ. 3(1), 3–13 (2021). https://doi.org/10.46809/jpse.v3i1.38

University of Maryland – Baltimore County - Profile, rankings and data (2024).https://www.usnews.com/best-colleges/umbc-2105. Accessed 23 Jan 2024

Exploring the Impact of Automation Transparency on User Perceptions and Learning Outcomes in Adaptive Instructional Systems

Rebecca L. Pharmer[✉]

Colorado State University, Fort Collins, CO 80523, USA
rebecca.pharmer@colostate.edu

Abstract. In this study, an adaptive instructional system was developed to explore the impact of automation transparency on learning outcomes in an assembly task. Participants were instructed on an assembly process for 8 unique shapes. The system assigned either adaptive instruction, tailoring restudy based on learner performance, or static instruction, providing a fixed amount of restudy. Transparency was implemented in the form of text-based feedback provided to the learner. Participants either received transparent feedback with explanations regarding why the system assigned restudy of concepts, or general feedback without explanations. While adaptive instruction and automation transparency did not show a consistent performance improvement in the assembly task, they appeared to influence participants' perceptions of the training. The findings and implications will be discussed.

Keywords: Adaptive Instructional Systems · Automation Transparency · Learning · Adaptive Instruction

1 Introduction

The development of computerized training has had a profound impact on education and skill development. It offers unparalleled accessibility, enabling learners to access educational content anytime, anywhere, and at their own pace. The scalability of computerized training allows for reaching diverse and large audiences, and, additionally, provides valuable insights into learner progress, enabling continuous improvement in instructional strategies. Moreover, it facilitates interactive and engaging learning through multimedia elements and real-time feedback, enhancing comprehension and retention. The adaptive nature of computerized training tailors learning experiences to individual needs, promoting personalized and efficient skill acquisition. These benefits of computerized training are made possible through applying empirically supported learning strategies known to promote retention and skill acquisition, as well as implementing system design techniques that promote compliance with the automated training system. The impact of computerized training seamlessly extends to Adaptive Instructional Systems (AIS), automated training systems recognized for their capacity to tailor instruction to learners' proficiency levels and deliver a similar caliber of training to individualized human tutoring (Metzler-Baddeley & Baddeley, 2009; Ma, et al., 2014).

© The Author(s), under exclusive license to Springer Nature Switzerland AG 2024
R. A. Sottilare and J. Schwarz (Eds.): HCII 2024, LNCS 14727, pp. 178–188, 2024.
https://doi.org/10.1007/978-3-031-60609-0_13

1.1 Adaptive Instructional Systems

Adaptive Instructional Systems (AIS) have been characterized in the literature as automated training systems capable of adapting to a learner's proficiency level through altering instruction to provide an optimal level of support and challenge (Durlach & Lesgold, 2012; Landsberg et al., 2012). This adaptation is achieved by automatically measuring learner behavior, analyzing it to create a model of the learner's competency, and then using this model to customize learning content for individualized and optimal learning experiences (Durlach, 2019).

This type of tailored training reduces the imposed cognitive load on learners, eliminating the need for them to accurately assess their own performance while adjusting their study methods accordingly (Mayer, 2014; Sweller & Chandler, 1991). Modern AIS can implement various instructional strategies similarly to how a human tutor adapts their training approaches for skilled versus novice learners. As outlined in a review by Durlach & Ray (2011), these methods include adjusting content difficulty based on learner responses, modifying the spacing of content presentation, implementing "mastery criteria" to allow learners to focus on yet-to-be-mastered concepts, and incorporating metacognitive prompts for self-correction.

The present study aims to focus solely on error-sensitive feedback, which allows the system to inform users not only about their current learning performance but also the rationale behind the system's decision to reassign content presentations. However, the term "feedback", in this case, can range from general performance information to detailed corrective information. When implementing error-sensitive feedback into a training system, developers must consider the most appropriate information to include in feedback to ensure that it is helpful to the learning process. Literature in this area emphasizes the importance of task-focused feedback rather than feedback directed solely at learners' abilities (Kluger & DeNisi, 1996). In other words, performance benefits are more likely to be seen when corrective feedback offers insight into how to improve one's performance rather than simply stating an applicable score or other metric. This idea is especially relevant when considering the current domain-knowledge of the learner, as the Feedback Principle of Multimedia Learning posits that for novice participants, explanatory feedback recommending a change in task strategy over performance-focused feedback should produce better learning outcomes whereas the opposite is true of expert participants (Johnson & Priest, 2014).

1.2 Automation Transparency

Adaptive Instructional Systems (AIS) fall within the framework of Parasuraman et al.'s (2000) Levels of Automation, characterized by their ability to gather information from learners, assess the learners aptitudes, and then manipulate the instruction based on that assessment. When designing automated systems, careful consideration must be given to how users will interact with the system and, specifically, calibrate their compliance with the system (Parasuraman & Riley, 1997). The amount of transparency provided by such systems, defined as the amount of information provided to users about the decision-making process, significantly influences user compliance with automation (Sargent et al., 2023). Transparency is vital in allowing users to develop accurate mental models of the agent (Bhaskara et al., 2020). Recent research, including a meta-analysis by Sargent et al. (2023), indicates a strong positive effect of transparency on performance, with the understanding that it is implemented in a task-appropriate way.

Frameworks for implementing transparency exist, such as Chen's model (2014), which conceptualizes transparency based on Endsley's (1995) levels of situational awareness. This model includes implementing transparency by providing the user with: the purpose of the agent, the process of the agent, and the performance of the agent. While typically applied to automated systems supporting operators with shared goals, this model has yet to be applied to AIS in the education domain. In the context of AIS, Chen's model can be adapted to provide transparency to learners, offering insights into the system's goal, process, and expected performance outcomes. This application of transparency ensures that learners understand the system's objectives, the rationale behind recommendations for restudy, and the expected impact on learning outcomes.

Despite the importance of automation transparency, research on the role of automation transparency in educational technologies is limited. Putnam and Conati (2019) conducted a pilot study exploring user attitudes toward transparency in an AIS designed to teach constraint satisfaction problems. Participants interacting with the AIS expressed preferences for receiving explanations along with their performance feedback. This pilot study addressed the Kirkpatrick and Kirkpatrick's (2016) first level of training effectiveness, Reaction, indicating positive reactions from participants. The current study builds upon this previous research by focusing on Kirkpatrick and Kirkpatrick's (2016) Level 2 of training effectiveness – Learning. This level assesses the degree to which participants acquire intended skills and knowledge. The study sets the stage for a more comprehensive evaluation of transparency effects on learning outcomes in AIS, extending beyond user reactions to exploring the impact on skill acquisition.

1.3 The Current Study

The goal of the present study is to examine the impact that implementing transparency into an adaptive training system has on learning outcomes. Additionally, this study aims to assess how system transparency shapes users' perceptions of their knowledge and their attitudes toward the system itself. To empirically investigate these research goals, a simple adaptive training system for a shape assembly task was developed. This task was adapted from an assembly task used in previous research (Clegg et al., 2022). Adaptivity was introduced by incorporating error-sensitive feedback and assigning restudy for missed concepts. Transparency was manipulated through the delivery of text-based reasoning for assigning participants to restudy concepts they had missed. This form of transparency, chosen for its alignment with all three components of transparency in Chen et al.'s (2014) model, was considered easily interpretable for the learners, making it well-suited for the task at hand. With these research goals in mind, three hypotheses are tested:

1. Participants in the adaptive instruction conditions are expected to demonstrate superior learning performance compared to those in static instruction conditions (Metzler-Baddeley & Baddeley, 2009).

2. Learning performance is anticipated to be higher in transparent conditions than in non-transparent conditions, aligning with the observed performance benefits of automation transparency in other task domains (Bhaskara et al., 2020; Sargent et al., 2023).
3. Transparency is predicted to result in more positive perceptions of training, as indicated by previous research (Putnam & Conati, 2019).

2   Methods

2.1   Participants

A total of 60 (32 Male, 27 Female, 1 Non-Binary) Undergraduate Psychology students participated in this study for partial, optional course credit. Of the sample, 97% were between ages 18 to 24, 2% between 25–34, and 2% were over age fifty-five.

2.2   Study Design

The current study utilized a 2 (Instruction type: Adaptive or Static) X 2 (Transparency level: Transparent or Non-transparent) between-subjects yoked design to assess differences in learned content. Participants were randomly assigned to either an adaptive or static, preset, instruction with either transparent feedback related to the system's assessment of the user or standard performance feedback with no transparency.

Yoking Conditions. To reduce the confound of restudy, participants in the static condition were assigned to the same number of presentations as another participant in the adaptive condition. The first 30 participants were run in adaptive conditions, because their sequences (the amount of content restudied) provided the yoking input for the remaining 30 participants in the static conditions.

2.3   Materials

Stimuli. A set of 8 shapes adapted from Clegg et al., (2022) consisting of 4 to 9 components that attach as 3 to 8 possible positions to other parts. Figure 1 shows an example of one shape.

Participant Perceptions Questionnaire. Participants filled out a brief questionnaire gauging their perceptions of the training. This questionnaire also captured participants' self-assessments of their learning experience through their rated confidence in remembering the assembly process for each shape.

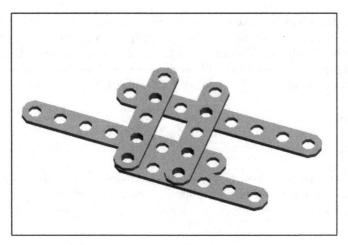

Fig. 1. Example of a shape that participants were instructed to build.

2.4 Procedure

Participants accessed the experiment via a Qualtrics survey on a computer, granting consent before receiving instructions. Participants engaged in an assembly task involving eight distinct shapes. The order in which the shapes were presented was randomized for each participant. For each shape, their learning process began with passive exposure through a 30-s assembly video, followed by a training phase focused on three key learning objectives: understanding the sequence of bars to construct each shape, identifying the attachment points for each piece, and mastering their proper placement. In the adaptive conditions, restudy was tailored to each participant's individual performance, while in static conditions, restudy was determined by another participant's study schedule (yoked participant). Following the completion of the training phase, participants were tested on all three learning objectives for each of the eight shapes. In adaptive instruction conditions, participants restudied items they answered incorrectly during training, while in static instruction conditions, restudy was based on an adaptive participant's schedule. For transparent conditions, participants received feedback explaining the system's decision to assign restudy, whereas non-transparent conditions received generic feedback. Table 1 provides examples of feedback in each condition. Following the completion of the training phase, participants underwent testing on the three learning objectives without feedback. Figure 2 illustrates examples of each learning objective.

Table 1. Examples of feedback given to each condition.

Adaptive & Transparent Feedback	Static & Transparent Feedback
"You responded to 2 of 3 questions incorrectly. Your areas for improvement are Attachment Points and Order of Assembly of each shape. The system has assigned restudy of these concepts based on your performance"	*"The system has assigned restudy for generally difficult concepts to improve your performance on the test"*
Adaptive & Non-Transparent Feedback	Static & Non-Transparent Feedback
"You will now restudy areas that need improvement"	*"You will now restudy areas that are generally difficult"*

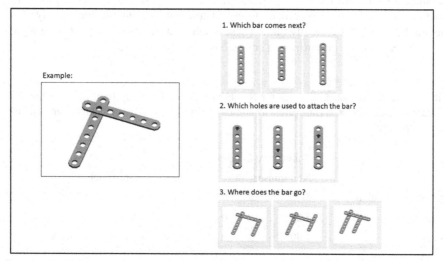

Fig. 2. Example of an incomplete shape and the items for each learning objective that participants were shown. Question 1 captures the learning objective of order of assembly, question 2 captures attachment points, and question 3 captures placement.

3   Results

3.1   Restudying

Participants in the adaptive conditions were assigned an average of 12.60 (SE = 1.23) items to restudy out of 48 possible items. In examining whether transparent feedback could lead to fewer cases of restudy, an Independent Samples t-test revealed no significant difference in the amount of items studies between the adaptive conditions (t(29) = −1.20, p = 0.24, d = -0.43). Participants in the Adaptive-Transparent condition restudied an average of 14 (SE = 1.62). Items, while those in the Adaptive Non-Transparent group restudied an average of 11 (SE = 1.62) items. Due to the yoked design, participants assigned to static instruction did not differ from the adaptive conditions in the number of items restudied.

3.2 Overall Learning Performance

Number of Shapes Correctly Built. Participants were scored on the number of shapes they were able to correctly build on the test. The test consisted of 6 items examining the steps in the build process they had received training on. Performance was assessed by whether participants responded to all 6 items for each shape correctly or had one or more errors. The average amounts of shapes correctly built at test for each condition are reported in Table 2. No main effect of adaptative condition ($F(1,56) < 1$), nor transparent condition ($F(1,56) < 1$) was observed. There was also no significant interaction between adaptive condition and transparency on the number of shapes correctly built ($F(1,56) < 1$). Planned contrasts revealed a non-significant difference between instruction types in non-transparent conditions ($t(28) = 0.66$, $p = 0.52$, $d = 0.21$) as well as in the transparent conditions ($t(28) = 0.33$, $p = .75$, $d = 0.14$), indicating that regardless of whether transparency was present or absent, instruction type did not significantly affect the number of shapes correctly built. These results indicate that H1, predicting higher learning performance in transparent conditions was not supported.

Table 2. Means and standard errors for Number of Shapes Correctly Built as a function of a 2(Adaptive Condition) X 2(Transparent Condition) design. M and SE represent mean and standard error, respectively.

	Transparent Condition			
	Non-transparent		Transparent	
Adaptive Condition	*M*	*SE*	*M*	*SE*
Adaptive	5.27	0.48	4.93	0.60
Static	4.80	0.52	4.67	0.55

Improvement from Pre-test to Post-test. To analyze the change in performance from initial training performance to test performance, participant gain scores were calculated using the following formula: ([Post-Test Score – Pre-Test Score]/[Total Score – Pre-Test Score]). Analyzing gain scores normalizes learning performance to account for the participant's scores at pre-test and room for improvement at test. Average gain scores between conditions are reported in Table 3. in A 2 (Adaptive Condition) by 2 (Transparency condition) ANOVA was used to determine if there were any significant differences in learning gains. There was no main effect of adaptive condition ($F(1,56) < 1$), nor was there a main effect of transparent condition ($F(1,56) < 1$). There was also no significant interaction between adaptive and transparent interventions ($F(1,56) = 2.62$, $p = 0.11$, $\eta2 = 0.04$). These results indicate that H2, predicting higher learning performance in transparent conditions was not supported.

However, when limiting the analyses to include only participants assigned to conditions that received transparent feedback, a one-tailed independent samples t-test revealed that the impact of instruction type on learning gains was significant $t(28) = -1.75$, $p =$

.05, $d = -0.64$). This trend was not observed in the non-transparent conditions $t(28 = 0.51, p = .61, d = 0.19)$, this shows a positive influence of adaptivity on learning gains, only when participants are given some indication that the system is in fact adapting to their performance.

Table 3. Means and standard errors for Gain Score as a function of a 2(Adaptive Condition) X 2(Transparent Condition) design. *M* and *SE* represent mean and standard error, respectively.

	Transparency Condition			
	Non-transparent		Transparent	
Adaptive Condition	*M*	*SE*	*M*	*SE*
Adaptive	36.96	9.91	56.60	8.91
Static	43.33	7.61	33.63	9.65

3.3 Performance in Learning Objectives

Order. Isolating test performance by the 16 test items related to the learning objective of Order (selecting the bar that came next in the sequence), participants in the Adaptive-Transparent condition correctly responded to an average of ($SE = 1.02$) items, the Adaptive-Non-transparent averaged 13.7 ($SE = 0.70$), the Static-Transparent condition averaged 13.5 ($SE = 0.77$), and Static Non-transparent averaged 13.70 ($SE = 0.49$). A between-subjects ANOVA showed no significant main effects of transparency ($F(1,56) < 1$), nor instruction type ($F(1,56) < 1$) and no significant interaction ($F(1,56) < 1$).

Attachment Point. Isolating test performance by the 16 test items related to the learning objective of Attachment Point (selecting the correct hole the bar uses to attach), participants in the Adaptive-Transparent condition correctly responded to an average of 13.9 ($SE = 0.73$) items, the Adaptive-Non-transparent averaged 13.9 ($SE = 0.61$), Static-Transparent averaged 14.2 ($SE = 0.47$), and Static Non-transparent averaged 14.30 ($SE = 0.37$). A between-subjects ANOVA showed no significant main effects of transparency ($F(1,56) < 1$), nor instruction type ($F(1,56) < 1$) and no significant interaction ($F(1,56) < 1$).

Location. Isolating test performance by the 16 test items related to the learning objective of Location (where the bar is placed on the shape), participants in the Adaptive-Transparent condition correctly responded to an average of 14 ($SE = 0.78$) items, the Adaptive-Non-transparent condition averaged 15.10 ($SE = 0.41$), Static-Transparent condition averaged 14.3 ($SE = 0.55$), and Static Non-transparent averaged 14.50 ($SE = 0.41$). A between-subjects ANOVA showed no significant main effects of transparency ($F(1,56) < 1$), nor instruction type ($F(1,56) < 1$) and no significant interaction ($F(1,56) < 1$).

3.4 Perceptions of Training

Enjoyment. Participants were asked to rate their enjoyment of the training on a 7-point Likert scale, with 1 being "Strongly Disagree" and 7 being "Strongly Agree". Overall, participants rated their enjoyment of the training fairly negatively, with the highest ratings in the Adaptive-Transparent condition ($M = 2.80$, $SD = 1.15$), followed by the Static-Transparent condition ($M = 2.60$, $SD = 0.91$), then the Adaptive-non-transparent condition ($M = 2.40$, $SD = 0.83$) and the Static-Non-transparent conditions ($M = 2.33$, $SD = 0.72$). There was no main effect of Adaptive condition ($F(1,56) < 1$) nor Transparency condition ($F(1,56) < 1$). There was also no significant interaction ($F(1,56) < 1$). This indicates no meaningful difference in participants enjoyment of the training, regardless of whether it was tailored to their performance or if they were given transparent feedback.

Perceived System Accuracy. Participants also rated how accurately the system was able to assess their performance. Surprisingly, participants rated the system is most accurate in the Static-Non-transparent condition ($M = 2.67$, $SD = 0.98$), followed by the Static-Transparent condition ($M = 2.53$, $SD = 0.64$), then Adaptive-Non-transparent condition ($M = 2.40$, $SD = 0.51$), and finally the least accurate in Adaptive-Transparent condition ($M = 1.80$, $SD = 0.68$). There was a significant effect of both adaptive condition ($F(1,56) = 7.22$, $p < .01$, $\eta2 = 0.11$) and transparency condition ($F(1.56) = 3.89$, $p < .05$, $\eta2 = 0.11$). No significant interaction was found ($F(1,56) < 1$). The system was actually rated as the least accurate in the adaptive-transparent condition, perhaps indicating that increased transparency in adaptive instruction could cause an unwanted decrement to perceptions of accuracy.

4 Discussion

Overall, there was not sufficient evidence to support the hypothesis that participants would show greater learning performance in the adaptive instruction conditions than in the static instruction conditions. Specifically, there was a lack of effects found between adaptive and static instruction in the overall test performance as well as minimal differences in learning gains from pre-test to post-test. Interestingly, though, when limiting analyses to include only conditions with transparency, adaptive instruction lead to greater learning gains than static instructions. This could be interpreted to mean that perhaps automated systems benefit from the inclusion of transparency, but the effects do not hold in traditional instructional systems. In other words, transparency only provides performance benefits when it is providing increased insight into an actual process the system is engaging in. Moreover, if this effect of adaptivity is only present in transparent systems, it may indicate that adaptive instruction is most beneficial to learning when learners understand how the system is responding to their performance.

The findings from this study did fail to replicate the robust effects of adaptive training that is seen in the literature. This could be due to mismatched instructional techniques implemented by the system to the optimal instructional techniques for this assembly task. Reviews of adaptive training system recommend careful consideration when determining the adaptive interventions implemented and their appropriateness for the task domain

(Durlach & Lesgold, 2012). Perhaps assigning a single restudy attempt was not a sufficient adaptation for training in the shape building task. Also, the feedback presented to the participants in this experiment may have been too simplistic and could instead have offered more insight into helpful strategies for encoding each step. A similar explanation can be given to the weak effects of transparency. In reviews of transparency interventions (see Bhaskara et al., 2020) there are several manipulations of system transparency. The present study only examined only text-based reasoning for assigning restudy during the training itself. It is possible that providing an pre-task description of how the system adapts and displays feedback would have produced different results.

When assessing participants' perceptions of training, there was no difference in the reported enjoyment of the training between conditions. However, some evidence was shown to support that adaptive instruction and transparency influence perceptions of how accurate the system is at understanding performance. Surprisingly, the adaptive conditions were rated as significantly less accurate in assessing participants' learning performance than the static condition. When transparency was added to these conditions, perception of accuracy also lowered by a marginally non-significant difference, suggesting that using transparency to explain the reasoning of adaptive interventions can sometimes be detrimental to a learner's perceptions of the system. Of course, this is in direct contrast to the findings reported in Putnam and Conati (2019). This discrepancy in findings highlights a need for further examination into how to best implement transparency into adaptive instructional systems to elicit positive perceptions and, in turn, promote more positive learning outcomes.

5 Conclusion

While the addition of transparency to adaptive instructional systems may not consistently impact learning performance, it has the potential to impact user perceptions of the system. Despite empirical reviews indicating performance benefits from transparent systems, more research is required to discern parameters influencing whether transparency aids or hinders task performance in adaptive instructional systems. These insights offer initial considerations for designing and implementing transparency in adaptive instructional systems, underscoring the need for further exploration of parameters influencing its impact on task performance.

Acknowledgments. The author would like to acknowledge Dr. Ben Clegg and Dr. Chris Wickens for their mentorship through this work.

Disclosure of Interests.   The author has no competing interests to declare that are relevant to the content of this article.

References

Bhaskara, A., Skinner, M., Loft, S.: Agent transparency: a review of current theory and evidence. IEEE Trans. Hum.-Mach. Syst. **50**(3), 215–224 (2020)

Chen, J.Y., Procci, K., Boyce, M., Wright, J., Garcia, A., Barnes, M.: Situation awareness-based agent transparency. Army research lab Aberdeen proving ground md human research and engineering directorate (2014)

Clegg, B.A., Karduna, A., Holen, E., Garcia, J., Rhodes, M.G., Ortega, F.R.: Multimedia and immersive training materials influence impressions of learning but not learning outcomes. In Interservice/Industry Training, Simulation, and Education Conference (I/ITSEC) (2022)

Durlach, P.J.: Fundamentals, flavors, and foibles of adaptive instructional systems. In: Adaptive Instructional Systems: First International Conference Proceedings, vol. 2, pp. 76-95 (2019)

Durlach, P.J., Lesgold, A.M. (eds.): Adaptive Technologies for Training and Education. Cambridge University Press, Cambridge (2012)

Durlach, P.J., Ray, J.M.: Designing adaptive instructional environments: Insights from empirical evidence. US Army Research Institute for the Behavioral and Social Sciences (2011)

Endsley, M.R.: Measurement of situation awareness in dynamic systems. Hum. Factors 37(1), 65–84 (1995)

Johnson, C.I., Priest, H.A.: 19 The Feedback Principle in Multimedia Learning. The Cambridge handbook of multimedia learning, vol. 449 (2014)

Kirkpatrick, J.D., Kirkpatrick, W.K.: Kirkpatrick's Four Levels of Training Evaluation. Association for Talent Development (2016)

Kluger, A.N., DeNisi, A.: The effects of feedback interventions on performance: a historical review, a meta-analysis, and a preliminary feedback intervention theory. Psychol. Bull. 119(2), 254 (1996)

Landsberg, C.R., Astwood, R.S., Jr., Van Buskirk, W.L., Townsend, L.N., Steinhauser, N.B., Mercado, A.D.: Review of adaptive training system techniques. Mil. Psychol. 24(2), 96–113 (2012)

Ma, W., Adesope, O.O., Nesbit, J.C., Liu, Q.: Intelligent tutoring systems and learning outcomes: a meta-analysis. J. Educ. Psychol. 106(4), 901–918 (2014). https://doi.org/10.1037/a0037123

Mayer, R.E.: Multimedia instruction. In: Spector, J., Merrill, M., Elen, J., Bishop, M. (eds.) Handbook of Research on Educational Communications and Technology, pp. 385–399. Springer, New York (2014). https://doi.org/10.1007/978-1-4614-3185-5_31

Metzler-Baddeley, C., Baddeley, R.J.: Does adaptive training work? Appl. Cogn. Psychol.: Official J. Soc. Appl. Res. Memory Cogn. 23(2), 254–266 (2009)

Parasuraman, R., Sheridan, T.B., Wickens, C.D.: A model for types and levels of human interaction with automation. IEEE Trans. Syst., Man Cybern.-Part A: Syst. Hum. 30(3), 286–297 (2000)

Parasuraman, R., Riley, V.: Humans and automation: use, misuse, disuse, abuse. Hum. Factors 39(2), 230–253 (1997)

Putnam, V., Conati, C.: Exploring the need for explainable artificial intelligence (XAI) in intelligent tutoring systems (ITS). In: IUI Workshops, vol. 19 (2019)

Sargent, R., Walters, B. Wickens, C.: Meta-analysis qualifying and quantifying the benefits of automation transparency to enhance models of human performance. In: Proceedings HCI-International. Copenhagen Denmark (2023)

Sweller, J., Chandler, P.: Evidence for cognitive load theory. Cogn. Instr. 8(4), 351–362 (1991)

Regulating Stress in Complex Tasks: Human Performance Implications of Adaptive and Maladaptive Coping Strategies

Bradford L. Schroeder[1]([⊠]), Wendi L. Van Buskirk[1], Jason E. Hochreiter[1], and Gabriella M. Hancock[2]

[1] Naval Air Warfare Center Training Systems Division, Orlando, FL 32826, USA
{bradford.l.schroeder.civ,wendi.l.vanbuskirk.civ,
jason.e.hochreiter.civ}@us.navy.mil
[2] California State University, Long Beach, Long Beach, CA 90840, USA
gabriella.hancock@csulb.edu

Abstract. Complex tasks impose varying demands on performers. When task demands change, stress increases, and performers must cope with this change in stress. Previous research indicates that people generally regulate this stress by coping in three ways: focusing on the task, focusing on their emotions, or disengaging from the task [1]. With all three approaches, performers' stress will decrease; however, performance decrements typically follow with emotion-focus or avoidance strategies. Previous research identifies task-focused coping (TFC) as an adaptive stress regulation technique, whereas emotion-focused coping (EFC) and avoidance coping (AC) are maladaptive techniques for performance [2]. This distinction has implications for a wide variety of human performance contexts, as training on the task and training stress regulation techniques are two interrelated means for improving performance [3, 4]. The present work examined how learners coped with a stressful, unfamiliar, complex task and how this affected their task performance. Results suggested that higher TFC was associated with higher accuracy and higher EFC was associated with lower accuracy and slower reaction times. Further analyses indicated that EFC indirectly harms accuracy and timeliness through its influence on slower reaction times. Training interventions that address learners' poor reaction times may serve as beneficial countermeasures when learners employ maladaptive stress coping techniques. Implications of these findings for adaptive training systems and future research directions will be discussed.

Keywords: Adaptive Training · Adaptive Instructional Systems · Stress Coping Strategies · Stress and Human Performance

1 Introduction

1.1 Stress and Human Performance

Performance under stress is a function of the individual and their environment, and these two elements interact to influence performance outcomes [5]. Lazarus and Folkman's transactional theory of stress [6] emphasizes the importance of the dynamic nature of

R. A. Sottilare and J. Schwarz (Eds.): HCII 2024, LNCS 14727, pp. 189–203, 2024.
https://doi.org/10.1007/978-3-031-60609-0_14

an individual's interactions with their environment or task. First, individuals execute a primary appraisal to determine how stressful the environment is. A secondary appraisal follows wherein they determine what resources they have available to cope with the stress identified in the primary appraisal. The outcome of these two appraisals is the selection and execution of a coping strategy to manage perceived stress while performing the task. This process is iterative, and re-appraisal occurs over time as the individual continues to interact with the environment.

In the context of training, this theory of stress considers the training material, the way the material is delivered to the trainee, and the means by which the trainee interacts with the training content. When learning something unfamiliar, the trainee will first appraise the training material to evaluate how stressful it will be to learn. Then, they will evaluate the resources available to them (e.g., prior knowledge, reference material, an instructor, etc.) to determine how they should cope with the training. Researchers have suggested that trainees may appraise the task as a challenge or a threat depending on the outcome of the primary and secondary appraisals [7]. Specifically, they argue if the task is perceived to be more stressful than the trainee has resources to cope with it, they will perceive the task as a threat. If the opposite is true, they will perceive the task as a challenge. As this is an iterative process, it is possible for training environments to shift from threat to challenge or challenge to threat over the duration of training. No two learners will have the same appraisal of their training environment, as individual differences (such as personality, cognitive ability, or emotional intelligence, to name a few) can influence the appraisal process and selection of coping strategies [8, 9].

1.2 Coping with Task Stress

When an environment or task has been appraised, there are a few coping strategies that trainees can select to manage the appraised stressors. These techniques can be adaptive (beneficial for task performance) or maladaptive (detrimental to task performance). Matthews and Campbell [1] identified three coping strategies based on clinical stress literature: task-focused coping (TFC), emotion-focused coping (EFC), and avoidance coping (AC). Importantly, these coping strategies are not mutually exclusive – individuals can engage in any of these coping strategies to varying extents simultaneously. TFC, also known as problem-focused or direct coping, is characterized by actions taken to succeed on the task, such as planning a strategy, remembering task procedures, or increasing one's concentration on the task. This technique is considered adaptive for performance [2] as it helps reduce perceived stress by making progress on the task. EFC serves to reduce stress by focusing on one's emotions and can include worrying about what to do, self-blame for poor performance, or trying to minimize one's emotional reactions to stressors. Although EFC reduces perceived stress, it is maladaptive for task performance [1]. AC manifests through task disengagement, daydreaming, or pretending the task is unimportant. Like EFC, AC is also maladaptive for task performance [2]. Though detrimental to performance, maladaptive strategies are effective at regulating emotional experiences and, as a result, prove very tempting for overwhelmed operators who often use them.

For human performance, Matthews and Campbell [1] argued the need for proper identification of coping strategies and developed the Coping Inventory for Task Stressors

(CITS). When properly identified, they asserted that countermeasures (i.e., some kind of aid or intervention) could be employed to mitigate the negative effects of maladaptive coping, particularly when learners are under high demand. Unfortunately, empirical research is lacking on potential immediate countermeasures for maladaptive coping in human performance contexts. We suspect there are a few reasons for this. Primarily, it is likely that countermeasures would be highly task specific. Someone who is coping maladaptively while driving may need a different countermeasure than someone who is performing a maintenance task. Second, much of the focus on coping interventions involves longer-term training, such as cognitive-behavioral workshops or consults with therapists (see Kent et al., 2018 for a systematic review [10]). In human performance contexts, it may not always be possible to commit the time investments necessary to execute the aforementioned interventions. We contend that immediate countermeasures may be better for improving human performance. For example, a countermeasure that effectively reduces one's maladaptive coping should benefit performance in real-time.

1.3 Implications for Adaptive Instructional Systems

Adaptive instructional systems (AIS) adapt training content to meet the specific training needs of each individual user, such as adjusting difficulty over time [11–13]. However, if an AIS is further able to detect when a learner is employing a maladaptive coping strategy, it could issue coping strategy countermeasures or other interventions to the learner as soon as it is detected and potentially mitigate any detrimental training effects due to this coping strategy. We posit that there may be two primary ways to improve training outcomes via such interventions. First, countermeasures that lead learners to switch to an adaptive coping strategy may mitigate the detrimental effects of maladaptive strategies on training. Second, it may be possible to provide interventions that specifically target the performance elements that suffered due to the employment of maladaptive coping strategies.

As trainees select coping strategies as part of an iterative process (according to Lazarus & Folkman's transactional model), these interventions could affect ongoing task appraisals to where perceived stress decreases. It follows that this could lead learners to re-appraise threats as challenges, which should improve performance on the task.

1.4 The Present Study

The present work examines coping with stress in an instructional system that trains radio frequency (RF) signal identification. The first goal was to examine whether expected relationships between performance and individual differences in coping were present in the experimental task. This yielded one hypothesis for each of the three stress coping strategies:

- H1: TFC will be positively associated with performance, such that higher TFC is associated with higher accuracy and faster Report Timeliness.
- H2: EFC will be negatively associated with performance, such that higher EFC is associated with lower accuracy and slower Report Timeliness.

- H3: AC will be negatively associated with performance, such that higher AC is associated with lower accuracy and slower Report Timeliness.

To test these hypotheses, we examined performance with RF signal Report Accuracy (how accurately all components of the RF signal were identified) and Report Timeliness (how quickly an RF signal report was submitted from the time the signal first appeared in the scenario). A more detailed description of the RF signal task is provided in the method section.

Next, we sought to explore behavioral variables associated with performance and individual differences in coping that could be potential targets for maladaptive coping countermeasures. Previous research suggests that different coping strategies employed by drivers can affect their behavior (e.g., longitudinal control [14]). Therefore, coping strategies may be associated with behavioral variables in the present task. We examined how long it took participants to click on an RF signal after it first appeared, and once they had selected that RF signal, how long it took them to analyze and classify the signal before submitting their report. We contend that these behavioral data account for attentional and processing elements of performance that complement the analysis of accuracy and timeliness.

For adaptive instructional systems, consideration for how learners are coping with the demands of their learning objectives could be informative for understanding how best to adapt or personalize the training experience. Understanding the implications of these coping strategies on performance as well as operators' cognitive and affective states can provide a useful basis for determining effective countermeasures to reduce the use of maladaptive approaches.

2 Method

2.1 Participants

We collected data from 76 participants (42 males, 34 female) who completed a military-style RF detection task. The average age of our participants was 21.05 (SD = 2.80). Participants were recruited from a large university in the southeastern United States and were paid $30 for 2 to 2.5 h of time to complete this study.

2.2 Testbed

Participants learned how to perform an RF signal identification and classification task using a simulated testbed comprising two side-by-side screens.

The first screen displayed RF signal information such as frequency and detection time. In addition, participants could tune the display to individual RF signals, allowing them to view live waveforms on a real-time display and take various signal measurements. The waveform audio was also played as an additional means of alerting participants to the signals currently present in the environment. These data were collected and analyzed by the participant to classify the type and threat priority of the signal.

The second screen was a reporting interface participants used to prepare and submit their classification reports for each signal using the signal parameter data collected from the first screen. Participants were instructed to identify and classify all signals. However, they were also instructed to prioritize certain types of signals and report them before attending to low-priority signals. Reports submitted via this interface were graded for the timeliness of the submission and the accuracy of the signal identification and classification information.

2.3 Procedure

Instructional Session. Participants began the research study with an informational presentation which acquainted them with the topic area (i.e., radar theory, radio frequency parameters, and signal characteristics) and instructed them how to perform the RF signal classification task on the testbed. Additionally, participants were trained to prioritize reports on signals having specific characteristics. Next, they practiced a scenario one-on-one with an experimenter, who first demonstrated how to use the interface and submit reports before passing control to the participant for them to complete the scenario on their own. During the entirety of this practice session, the participant was permitted to ask the experimenter questions, and the experimenter provided feedback to the participant. Experimenters had a "common questions and issues" document for their reference when providing feedback to participants and answering their questions. Afterward, participants completed a quiz on the material from the informational presentation and the procedures from the one-on-one practice scenario. The experimenter reviewed their responses with them to ensure they understood the terminology and task procedures before beginning the experimental task. Participants were instructed to consider the accuracy and timeliness components of their reports equally.

Radio Frequency Identification Task. Following the instructional session, participants completed five 10-min scenarios. In each training scenario, participants detected new signals as they came in and classified them based on their parameters. New signals continued to appear over time, and participants experienced time pressure relating to attending to multiple new signals, requiring them to perform an initial evaluation of each signal in order to prioritize them and submit reports accordingly. While signals presented with both auditory and visual components, participants were taught to use the former primarily as alerts to the onset of a new signal and the latter for actual parameter classification. Participants input their signal classification information into the reporting interface and submitted reports after completing classification.

The following steps reflect the general process participants followed during these RF classification scenarios:

1. Monitor the environment for newly appearing signals, using both auditory and visual cues.
2. When a new signal appears, click on it in the RF signals list to perform an initial visual inspection of its waveform and other parameters.
3. If no other higher-priority unclassified signals are present, assign an ID number to this signal and initiate a report in the reporting interface.
4. Using the waveform in the RF signals display, take measurements and classify signal parameters; enter these elements into the reporting interface.

5. Once the report is complete, submit it in the reporting interface.

During these steps, participants were also responsible for continuously monitoring the environment to determine if higher-priority signals appeared, which may require more immediate attention than their current task. Specific prioritization rules were provided during the instructional phase. On average, each of the 5 scenarios featured 15 new signals that participants were responsible for classifying.

Each report was assessed for performance through timeliness (time delta between signal onset and report submission; lower is better) and accuracy (correctness of each report element). In addition, we recorded behavioral data, such as how long participants took to click on a signal after onset (Click Time) and how much time they spent analyzing signals after assigning them ID numbers and before completing and submitting their reports (Analysis Time). It is important to note that participants were free to switch to another signal at any point after clicking on a given signal or even assigning it an ID number and beginning classification; as such, the total Analysis Time for a particular signal may include participant actions on other signals. Likewise, a participant's Click Time for a given signal may be inflated because they attended to other tasks between its onset and their click. These context switches have a variety of potential causes, including correct reprioritization of signal classifications or simply inspecting other signals in the interface. Additionally, Report Accuracy is measured entirely independently of Report Timeliness. Therefore, these behavioral data are necessary but insufficient components for proper Report Accuracy and Report Timeliness, but they are still useful potential indicators of attentiveness and timeliness.

Stress-Coping Questionnaire. At the end of each scenario, we assessed participants' stress-coping techniques with Matthews and Campbell's Coping Inventory for Task Stressors (CITS) [2] to determine to what extent they employed TFC, EFC, and AC during training. CITS is a 21-item questionnaire where participants self-report usage of items such as "Blamed myself for not knowing what to do" (EFC) and "Was careful to avoid mistakes" (TFC) using a 5-point Likert scale. Participants indicated the extent to which they used each strategy with a 0 for "not at all" to a 4 for "extremely." Each subscale (TFC, EFC, and AC) had 7 questions which yielded a range of values of 0–28 when all 7 items were summed.

Due to existing research linking these coping strategies with cognitive task performance, we examined how they impacted Report Accuracy and timeliness in this RF signal detection task. The following analyses examine learner performance on their final scenario.

3   Results

3.1   Planned Analyses

First, we examined the descriptive statistics and correlations among our variables in three conceptual blocks: performance (Report Accuracy, Report Timeliness), stress coping strategy usage (TFC, EFC, AC), and behavioral data (Click Time, Analysis Time; see Table 1 for descriptive statistics). There were significant correlations among all three groups of variables, such that performance was positively correlated with TFC ($r = .29$, $p = .011$ for Report Accuracy) and negatively correlated with EFC ($r = -.40, p < .001$ for Report Accuracy). Report Accuracy was negatively correlated with Click Time ($r = -.40, p < .001$) and Analysis Time ($r = -.41, p < .001$), but these relationships were positive for Report Timeliness ($r = .30$ for Click Time and $r = .48$ for Analysis Time, both $ps < .01$). However, the only significant relationship between coping measures and behavioral data was with EFC and Click Time ($r = .31, p = .006$; see Table 2). These analyses offer partial support for H1 and H2, such that TFC was positively associated with accuracy, and EFC was negatively associated with accuracy. However, we did not identify any significant associations between performance and AC.

Table 1. Descriptive Statistics for Analyzed Variables

Variable	*M*	*SD*	Min	Max
Report Accuracy (percent)	75.00	28.80	0.00	100.00
Report Timeliness (seconds)	87.10	42.47	40.00	251.00
TFC (0–28)	23.17	4.08	12.00	28.00
EFC (0–28)	6.63	6.06	0.00	24.00
AC (0–28)	4.32	4.74	0.00	19.00
Click Time (seconds)	33.32	29.29	10.17	189.90
Analysis Time (seconds)	81.35	43.02	27.51	243.92

Table 2. Correlation Coefficients for Analyzed Variables

Variable	1	2	3	4	5	6	7
1. Report Accuracy	–						
2. Report Timeliness	−.15	–					
3. TFC	−.29*	.01	–				
4. EFC	−.40**	.12	−.07	–			
5. AC	−.19	.15	−.32**	.34**	–		
6. Click Time	−.41**	.30**	−.07	.31**	.06	–	
7. Analysis Time	−.32**	.48**	−.02	.21	.01	.79**	–

Note. *$p < .05$, **$p < .01$; Pearson's r coefficients displayed.
Listwise N $= 76$

3.2 Exploratory Analyses

As our behavioral measures were correlated with EFC and performance, we considered exploring the nature of the interrelatedness among these variables. Report Accuracy and Report Timeliness are assessed upon submission of an RF signal report. Prior to submitting a report, learners must first click on the RF signal, and then analyze that signal. The time it takes to click on a signal and the time it takes to analyze that signal directly affect how timely that RF signal report will be. This is logical; however, it is unclear how these variables would be related to Report Accuracy. Nevertheless, theory suggests that EFC alters how one interacts with a task, which consequently affects task performance. To explore this idea, we examined whether EFC indirectly affected performance because of its influence on these behavioral variables through mediation analysis.

To conduct these analyses, we used Hayes's [15] PROCESS Macro version 4.1 for SPSS. We analyzed this as a serial mediation model, where the X variable was EFC, the first mediator was Click Time, the second mediator was Analysis Time, and the Y variables were our performance metrics (Report Accuracy and Report Timeliness, in two separate analyses).

The first analysis on performance accounted for 25.1% of the variance explained in Report Accuracy, such that EFC was both a direct and indirect predictor of Report Accuracy. Higher EFC predicted lower Report Accuracy, but higher EFC also increased Click Time, which partially explained the negative association with accuracy (regression model statistics are provided in Table 3, tests of total, direct, and indirect effects are provided in Table 4, and a path diagram with coefficients is displayed in Fig. 1).

Table 3. Regression Model Coefficients for EFC Predicting Report Accuracy through Click Time and Analysis Time

Variable	B (SE_B)	$95\%CI_B$	β	t
Constant (out of 100%)	94.74 (6.54)	81.71, 107.78		14.49**
EFC	−1.36 (0.50)	−2.34, -0.37	−0.29	−2.73**
Click Time (secs)	−0.32 (0.16)	−0.65, 0.01	−0.34	−1.95
Analysis Time (secs)	0.01 (0.10)	−0.19, 0.21	0.02	0.11

Note. ** $p < .01$, * $p < .05$. Model statistics: $F(3, 72) = 8.04$, $p = .0001$, $R^2 = 0.251$

The second analysis on performance accounted for 24.7% of the variance explained in Report Timeliness, such that EFC was an indirect predictor of Report Timeliness. Higher EFC indirectly predicted lower Report Timeliness because of its significant positive association with Click Time, which was significantly associated with Analysis Time, which led to slower report times (regression model statistics are provided in Table 5, tests of total, direct, and indirect effects are provided in Table 6, and a path diagram with coefficients is displayed in Fig. 2).

Table 4. Total, Direct, and Indirect Effects for Report Accuracy Model

Effect of EFC on Timeliness	$B (SE_B)$	$95\% CI_B$	β	t
Total Effect	−1.82 (0.49)	−2.81, −0.84	−0.39	−3.70**
Direct Effect	−1.36 (0.50)	−2.35, −0.37	−0.29	−2.73**
Indirect through Click Time	−0.49 (0.28)	−1.14, −0.06	−0.11	See note*
Indirect through Analysis Time	−0.004 (0.05)	−0.14, 0.08	−0.001	See note
Indirect through both Mediators	0.02 (0.18)	−0.34, 0.39	0.005	See note

Note. ** $p < .01$, * $p < .05$; estimated effect sizes of indirect effects were determined using 5,000 bootstrap samples to generate 95% confidence intervals. Bootstrapped confidence intervals were used to determine the significance of indirect effects as per Hayes' [15] recommendation.

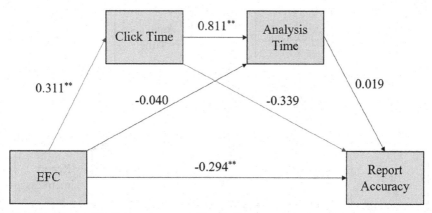

Fig. 1. Path Diagram for Report Accuracy. Standardized regression coefficients are indicated on each path. Significant indirect effect is indicated with red arrows.

Table 5. Regression Model Coefficients for EFC Predicting Report Timeliness through Click Time and Analysis Time

Variable	$B (SE_B)$	$95\% CI_B$	β	t
Constant (secs)	47.48 (9.75)	28.05, 66.91		4.87**
EFC	0.46 (0.74)	−1.02, 1.93	0.07	0.62
Click Time (secs)	−0.30 (0.25)	−0.79, 0.19	−0.21	−1.23
Analysis Time (secs)	0.56 (0.15)	0.26, 0.86	0.64	3.75**

Note. ** $p < .01$, * $p < .05$. Model statistics: $F(3, 72) = 7.90, p = .0001, R^2 = 0.248$

Table 6. Total, Direct, and Indirect Effects for Report Timeliness Model

Effect of EFC on Timeliness	$B\ (SE_B)$	$95\%CI_B$	β	t
Total Effect	0.93 (0.79)	−0.65, 2.50	0.13	1.17
Direct Effect	0.46 (0.74)	−1.02, 1.93	0.07	0.62
Indirect through Click Time	−0.46 (0.56)	−0.35, 1.88	−0.07	See note
Indirect through Analysis Time	−0.18 (0.30)	−0.79, 0.47	−0.03	See note
Indirect through both Mediators	1.10 (0.67)	0.07, 2.66	0.16	See note*

Note. [**] $p < .01,$ [*] $p < .05;$ estimated effect sizes of indirect effects were determined using 5,000 bootstrap samples to generate 95% confidence intervals. Bootstrapped confidence intervals were used to determine the significance of indirect effects as per Hayes' [15] recommendation.

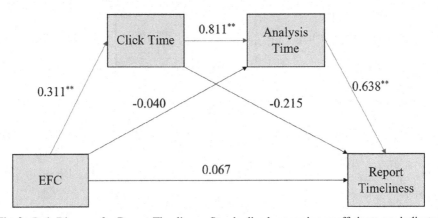

Fig. 2. Path Diagram for Report Timeliness. Standardized regression coefficients are indicated on each path. Significant indirect effect is indicated with red arrows.

4   Discussion

4.1   Summary of Results

We found partial support for our hypotheses with TFC and EFC, where TFC and Report Accuracy were positively associated (H1) and EFC and Report Accuracy were negatively associated (H2). We did not observe any significant relationships between AC and any performance or behavioral measures. Surprisingly, we did not observe any significant relationships between Report Timeliness and any of the coping measures. This is unexpected, as this task poses high temporal demand, and timeliness is a major component of successful performance (participants were instructed that timeliness and accuracy were of equal weight). It is possible that participants focused more heavily on accuracy than timeliness, but without self-report data assessing their priorities, we cannot determine why these relationships failed to replicate what has been observed in previous research.

Interestingly, EFC was the only coping measure associated with any of the behavioral variables. As theory suggests that coping strategy selection affects individuals' behaviors and consequently their performance, we examined the extent to which EFC directly and indirectly affected Report Accuracy and Report Timeliness on the RF identification task. In essence, our analysis supports that EFC's effects on task performance manifest through its effect on Click Time, which itself affects accuracy, or through its secondary effect on Analysis Time, which itself affects timeliness. Our interpretation of these exploratory analyses is provided in the next section.

4.2 What Do Click Time and Analysis Time Represent?

Click Time and Analysis Time may seem obvious to be so highly correlated; however, these two time measurements are completely independent. As a reminder, Click Time was calculated by measuring the time that a participant clicked on a signal minus the onset time of that signal. If they determined that signal was a priority, they would assign it an ID number. Analysis Time was calculated from the time they assigned the signal an ID number to the time that they submitted a report for it. The time a participant takes to click on a signal has no specific bearing on the time it takes them to subsequently analyze it and complete their report; they are two separate intervals of time that factor into the overall timeliness of a report submission.

As such, we posit that Click Time perhaps serves as an indication of the participant's attentional capabilities and procedural understanding of the task. For example, it might be that a participant is slow both to attend to new signals and to analyze them due to overall unfamiliarity with the task or the interface, or it might reflect a general inability to prioritize and task-switch properly throughout the task.

We suspect that Analysis Time may represent a performance efficiency. Its zero-order correlation was negative for accuracy and positive for timeliness, which suggests that participants who spent less time analyzing signals had higher overall performance. Participants who have learned the details of the task (e.g., how to submit accurate reports and when to submit them) may require less time to identify different signal characteristics to complete their reports. Conversely, participants who do not know what to do may languish in the analysis phase which increases their Analysis Time. This could further contribute to delays in Click Time on subsequent signals. We suspect that individual differences in cognitive load, training material encoding quality, or even working memory capacity may relate to this variable.

Taken together, there may be an external factor that could explain why these variables are so highly correlated. Some measure of perceptual speed may partially explain the high correlation between these two variables. In this task, someone who is slow to recognize new signals may also be slow to complete a subsequent analysis of those signals. Ultimately, improved data collection and data resolution capabilities are necessary to comprehensively understand how many behavioral components contribute to timeliness and accuracy.

4.3 Other Behavioral Variables to Consider

From these interpretations, we suggest that there are additional behavioral measurements that explain the variance in Report Timeliness. Report Timeliness is not simply the combination of Analysis Time and Click Time, and the timeliness of a report for a given signal may be impacted by the participant's ability to appropriately prioritize and attend to other concurrent signals. While TFC reflects increased focus on succeeding on the task, this might not specifically manifest as submitting reports earlier. High TFC may be accompanied by behaviors such as mentally planning a strategy or recalling task procedures, which may distract from or otherwise slow the completion of the task. We do note that TFC was positively correlated with Report Accuracy and that EFC was negatively correlated; in previous studies employing similar RF detection tasks, we have observed accuracy/timeliness tradeoffs such that participants who complete more reports on time tend to do so less successfully and vice versa [16]. During the instructional phase, participants were instructed to assign equal weight to the accuracy and timeliness components of their reports. Certainly, many of these findings are impacted by the specific task participants learned to perform, and further study is needed to determine if and how these results generalize to other tasks and domains.

4.4 Limitations and Future Research

Additionally, there are a few limitations of our data collection methods that are important to note. The specific RF signal simulation we used in our testbed is limited in the exact data it provides, and we were unable to obtain timing information about participant interactions in this interface beyond signal clicks and ID number assignments. In particular, the simulator exposes no timing details of signal measurements made in the real-time waveform displays. Furthermore, the complex and dynamic nature of this task, with multiple signals appearing that participants were responsible for classifying and prioritizing and with two displays with which they interacted, poses challenges for appropriately tracking time spent on a specific signal. A large Click Time for a given signal, which indicates that a participant spent a long time between its onset and their click in the RF signals list, could reflect that the participant overlooked this new signal due to focusing on another more important signal (suggesting TFC) or that they intentionally ignored it (suggesting AC). More fine-grained timing measurements relating to a participant's classification of signals, especially considering the multitasking nature of this task, may improve our understanding of how their employed coping strategy affected their Report Timeliness. Improvements in our ability to assess behavioral data from participants are under way and will continue in future research with this testbed.

Ultimately, future research with this task should investigate a targeted intervention to reduce Click Time, which could serve as a useful countermeasure for the maladaptive effects of EFC. For example, an AIS could determine an appropriate threshold for Click Time, and if that threshold is exceeded, deliver targeted feedback to train a participant how to attend to new signals more quickly. According to our analyses, such interventions may even further reduce Analysis Time, which may have further indirect benefits for performance. Interventions that train learners how to notice new signals coming in, or identify how they can reduce their time spent in other areas to prioritize addressing new signals could be beneficial.

4.5 Conclusion

We examined how learner stress-coping strategies in a complicated RF signal detection task related to their task performance, and affected their task behaviors. We saw that higher TFC was associated with higher accuracy, and higher EFC was associated with lower accuracy and slower report times. Stress-coping, particularly EFC, could be a valuable individual difference variable to integrate into the research and design of AISs.

When considering that individual difference measures may affect performance, AISs may benefit from the implementation of a hybrid adaptation approach (e.g., aptitude-treatment interaction (ATI) approach combined with micro-adaptation) as discussed by Park and Lee [17]. As EFC can change over time as part of the transactional process, a micro-adaptive ATI approach could be the best use of this individual difference measure to offset potential performance decrements [18]. For instance, if you can detect a learner is using an EFC or AC strategy, an AIS could intervene by providing tips or feedback to help the learner adjust to a TFC strategy, or reduce the usage of EFC or AC. At present, such an approach would be limited to the use of post-scenario subjective questionnaire data, but in the future, real-time measurements of these variables or physiological correlates may be feasible to trigger ATI adaptations.

In the context of AISs in general, we return to the potential of identifying stress-coping strategies for the purposes of both reducing the usage of maladaptive strategies and for reducing negative performance outcomes associated with maladaptive strategies. It may not always be possible to assess stress-coping techniques in real time through traditional questionnaires, particularly in highly time-sensitive, dynamic tasks. Indeed, as shown in this RF signal identification task, the impacts of maladaptive coping strategies on performance may not always be immediately apparent. When they are, such as the observed zero-order correlations between coping strategy and Report Accuracy, targeted countermeasures to guide participants toward an adaptive coping strategy can be effective at improving training. When they are not, such as the observed indirect correlations between EFC and Report Timeliness via Click Time and Analysis Time, interventions that aim to address these specific behavioral correlates may have a greater impact. As such, understanding the real-time behavioral correlates of these coping techniques for a given task may still inform preventive, tailored training interventions that can achieve these goals, and such correlates may be relatively simple to collect. In future work, we plan to investigate countermeasures to promote adaptive coping strategies in such training contexts as well as address participant behaviors that indirectly reflect maladaptive strategies to prevent detrimental training outcomes.

The present work represents a foray into the investigation of live countermeasures for maladaptive stress responses in human performance contexts. We believe that the challenge for understanding how to implement live countermeasures and how to understand their effects on performance lies in understanding a variety of factors. Researchers must first understand their impact on performance, subjective perceptions of stress, and individual differences in coping, as well as low-level behavioral data that links performance outcomes to these measures. It is highly likely that it may not be as easy to reduce maladaptive coping as it is to change behaviors like click time or analysis time. It may also be the case that causing a change in those behaviors induces no effect on performance outcomes when maladaptive coping stays unchanged. Understanding the nature of these underpinnings of human performance and their comingling influences will be critical to effectively implementing countermeasures in AISs.

Acknowledgments. This work was funded by the Office of Naval Research (N0001422WX01634) under Ms. Natalie Steinhauser, and the Office of Naval Research Sabbatical Leave Program. Presentation of this material does not constitute or imply its endorsement, recommendation, or favoring by the U.S. Navy or the Department of Defense (DoD). The opinions of the authors expressed herein do not necessarily state or reflect those of the U.S. Navy or DoD. NAWCTSD Public Release 24-ORL004 Distribution Statement A – Approved for public release; distribution is unlimited.

Disclosure of Interests. The authors have no competing interests to declare that are relevant to the content of this article.

References

1. Matthews, G., Campbell, S.E.: Task-induced stress and individual differences in coping. In: 42nd Proceedings of the Human Factors and Ergonomics Society Annual Meeting, pp. 821–825. SAGE Publications, Chicago, Illinois (1998)
2. Matthews, G.: Towards a transactional ergonomics for driver stress and fatigue. Theor. Issues Ergon. Sci. **3**, 195–211 (2002)
3. Mackworth, N.H.: Effects of heat on wireless operators. Br. J. Ind. Med. **3**(3), 143–158 (1946)
4. Morris, C.S., Hancock, P.A., Shirkey, E.C.: Motivational effects of adding context relevant stress in PC-based game training. Mil. Psychol. **16**(2), 135–147 (2004)
5. Hancock, P.A., Warm, J.S.: A dynamic model of stress and sustained attention. Hum. Factors **31**(5), 519–537 (1989)
6. Lazarus, R.S., Folkman, S.: Stress, Appraisal, and Coping. Springer, New York (1984)
7. Blascovich, J.: Challenge and threat appraisal. In: Elliot, A.J. (ed.) Handbook of Approach and Avoidance Motivation, pp. 431–445. Psychology Press, New York (2013)
8. Kilby, C.J., Sherman, K.A., Wuthrich, V.: Towards understanding interindividual differences in stressor appraisals: a systematic review. Pers. Individ. Differ. **135**, 92–100 (2018)
9. Parkes, K.R.: Coping in stressful episodes: the role of individual differences, environmental factors, and situational characteristics. J. Pers. Soc. Psychol. **51**(6), 1277–1292 (1986)
10. Kent, S., Devonport, T.J., Lane, A.M., Nicholls, W., Friesen, A.P.: The effects of coping interventions on ability to perform under pressure. J. Sports Sci. Med. **17**(1), 40–55 (2018)
11. Kelley, C.R.: What is adaptive training? Hum. Factors **11**(6), 547–556 (1969)
12. Marraffino, M.D., Schroeder, B.L., Fraulini, N.W., Van Buskirk, W.L., Johnson, C.I.: Adapting training in real time: an empirical test of adaptive difficulty schedules. Mil. Psychol. **33**(3), 136–151 (2021)
13. Van Buskirk, W.L., Fraulini, N.W., Schroeder, B.L., Johnson, C.I., Marraffino, M.D.: Application of theory to the development of an adaptive training system for a submarine electronic warfare task. In: Sottilare, R., Schwarz, J. (eds.) Adaptive Instructional Systems HCII 2019, LNCS, vol. 11597, pp. 352–362. Springer, Cham (2019)
14. Hancock, G.M., Hancock, P.A., Janelle, C.M.: The impact of emotions and predominant emotion regulation technique on driving performance. Work **41**(1), 3608–3611 (2012)
15. Hayes, A.F.: Introduction to mediation, moderation, and conditional process analysis: a regression-based approach. Guilford, New York, NY (2022)

16. Schroeder, B.L., Van Buskirk, W.L., Aros, M., Hochreiter, J.E., Fraulini, N.W.: Which is better individualized training for a novel, complex task? Learner control vs. feedback algorithms. In: Sottilare R., Schwarz J. (eds.) Adaptive Instructional Systems HCII 2019, LNCS, vol. 11597, pp. 236–252. Springer, Cham (2023). https://doi.org/10.1007/978-3-031-34735-1_17

17. Park, O.C., Lee, J.: Adaptive instructional systems. In: Spector, J.M., Merrill, M.D., Elen, J., Bishop, M.J. (eds.) Handbook of Research on Educational Communications and Technology, 4th edn., pp. 647–680. Springer, New York (2013)

18. Van Buskirk, W.L., Schroeder, B.L., Aros, M., Hochreiter, J.E.: Stress and coping with task difficulty. In: Sottilare, R., Schwarz, J. (eds.) Adaptive Instructional Systems HCII 2019, LNCS, vol. 11597, pp. 253–264. Springer, Cham (2023). https://doi.org/10.1007/978-3-031-34735-1_18

The Power of Performance and Physiological State: Approaches and Considerations in Adaptive Game-Based Simulation

William Stalker[1]([✉]) [iD], Summer Rebensky[1] [iD], Ramisha Knight[1] [iD],
Samantha K. B. Perry[1] [iD], and Winston Bennett[2]

[1] Aptima, Inc., Woburn, MA 01801, USA
lstalker@aptima.com
[2] 711th HPW AFRL, WPAFB., Woburn, OH, USA

Abstract. Adaptive training provides the capability to enable individualized learning and keep users in the Zone of Proximal Development. However, *when* an individual should adapt is still an open area of research. This paper discusses findings from a recent driving-based simulation in which trainee state was assessed by multiple means. Modalities of measurement included real-time physiological information (heart rate and fNIR data) used to assess mental workload, subjective ratings of workload from participants after each trial (NASA TLX), and performance data on key components related to the execution of the scenario tasks. These methods were used in tandem to adapt the difficulty of the simulation environment. Results show that measures with high sample rates, such as average miles per hour and reported relative oxygenated hemoglobin (HbO2) readings, outperformed less temporally sensitive measures, such as task response correct percentages or NASA TLX ratings. These findings emphasize that the granularity of the measurement method should align with the system's desired sensitivity in adapting to mental workload. More granular measures should be used when needing to adapt more frequently and/to smaller changes in workload. Empirically supported methods that lack high temporal or spatial resolution are unlikely to suffice for applications that require rapid adaptation.

Keywords: Adaptive Training · Performance Measures · State Measures · Game-based Simulation

1 Introduction

Adaptation is the hallmark of a high-quality educator. Skilled instructors frequently assess their students, formally and informally, to gauge their level of understanding and current emotional and mental state. These skilled instructors that adapt their own strategies, providing extra emphasis on key points or may introduce a discussion exercise to recapture attention or address student skill deficiencies. Not all educators adapt well to the dynamic of the audience, instructional tutoring software even less so. Software-based

R. A. Sottilare and J. Schwarz (Eds.): HCII 2024, LNCS 14727, pp. 204–215, 2024.
https://doi.org/10.1007/978-3-031-60609-0_15

systems such as Adaptive Instructional Systems (AIS) are uniquely able to tailor instruction to its user, improving the efficiency and effectiveness of the knowledge acquisition process. AIS, and similar software-based solutions, can balance the mental workload demand by dynamically adjusting the task difficulty, keeping trainees within the Zone of Proximal Development [1]. The capability of developing highly adaptive systems is more feasible than ever—particularly within game engines [2]. An AIS can adapt between trial sessions to accommodate rapid skill change and ensure that the instruction difficulty is matched to student skill level. These features are highly desirable but designing a sophisticated AIS that is capable of responding appropriately to an individual's complex skill state is no simple task. A common struggle during the development of an AIS is discerning what data to use and how different datapoints should be factorially weighted, or if weighted is warranted at all. Included factors will vary, but most sophisticated AISs will incorporate estimations about the user's mental workload in their algorithm. Considering the growth of automation across job fields, more training and learning systems will have to train skills such as monitoring, sense-making, and human-agent teaming. These skills rely heavily on a user's mental state and the ability to juggle many sources of information at once, making the measurement of mental workload more important than ever before.

Mental workload, though conceptualized as a cohesive concept, can be measured by a multitude of means such as subjective assessments, objective performance data, and physiological indicators. Research suggests advantages associated with using a combination of these measures [3], but how these related ratings compare in different task scenarios requires further study (e.g., [4]). Implementations of adaptive systems with multiple measures of mental workload are rare due to the inherit difficulty in development. This paper discusses the considerations of utilizing a diverse arsenal of mental workload measures in AIS design, within the context of an adaptive training simulator designed in collaboration with AFRL's Gaming Research Integration for Learning Laboratory (GRILL®). This paper also reflects on converging and diverging measures of mental workload and closes with empirically supported considerations when implementing similar measures within different adaptive instructional systems.

2 Methods

2.1 Participants

A total of 28 participants were recruited from the Dayton Metropolitan Area. All participants were 18 years or older, a U.S. citizen, and had normal or corrected hearing. Participants were compensated $40 for participating. Demographic information, experience with driving, hobbies, and whether they had recently consumed any stimulants such as nicotine or caffeine was collected following consent to participate in the study and prior to beginning the study. The overall median age was 25.5 years old (18 – 68). The sample consisted of 19 males and 9 females. A total of 75% ($N = 21$) of participants identified as Caucasian, followed by Asian 14.29% ($N = 4$), African American 7.14% ($N = 2$), and other 3.57% ($N = 1$).

Participants' weekly experience and comfort with driving was sampled prior to the study. Participants' weekly driving experience was distributed between 35.71% (N = 10) having less than 5 h but more than none, and 64.29% (N = 18) with equal to or more than 5 h of weekly driving experience. The amount of anxiety that participants experience while driving varied between 35.71% (N = 10) experience no anxiety, 35.71% (N = 10) experience only a little anxiety, 25% (N = 7) experience some anxiety, and 3.57% (N = 1) experience quite a lot of anxiety while driving.

Participants' exercise and video game habits were also sampled prior to the study. The majority of participants indicated they exercised a few times per week (57.14%, N = 16), while the others indicated exercising either a few times per month (17.86%, N = 6), a few times per year (14.29%, N = 4), or never / rarely (10.71%, N = 3). Most of the participants currently played video games (71.43%, N = 20). The top three most popular gaming platforms that participants regularly used was PC 60.71% (N = 17), followed by Nintendo 25% (N = 7), and Mobile 21.43% (N = 6). The most popular genres of games that participants most regularly played included: Shooters 42.86% (N = 12), Casual 35.71% (N = 10), Action 32.14% (N = 9), Role-playing 32.14% (N = 9), Strategy 32.14% (N = 9), and Adventure 28.57% (N = 8).

Half of the participants consumed some form of a normal amount of caffeine during the day prior to participating in the study and it did not impact their physiological metrics or task performance during the study. None of the participants consumed nicotine prior to participating in the study.

2.2 Driving-Based Adaptive Research Testbed (DART)

The data was collected from the dual task experimental testbed, the Driving-based Adaptive Research Testbed (DART). The DART was developed at and is hosted by the GRILL®. The DART acts similarly to traditional AISs; it adapts the difficulty of the task based on the user's state and their previous performance. The DART was initially designed with two goals in mind. First, an engineering goal intended to demonstrate that Unreal Engine can receive live-streamed physiological data related to mental workload and that it can dynamically update the environment and tasks based on the user state. Second, a scientific goal to determine what physiological responses are most effective to adapt a simulation environment to based on performer state. This testbed dually challenges its users, requiring them to drive through a simulated hostile terrain using a mock-HMMWV while simultaneously engaging in various cue discrimination tasks. The DART uses multiple measures of mental workload, including the NASA-TLX [5], a functional near-infrared (fNIR) sensor developed by BionicaLabs called NIRSense, and a Polar H10 Heart Rate Monitor sensor for tracking blood flow and heart rate (HR), and performance metrics from the simulated tasks themselves.

2.3 Experimental Design and Procedure

The experiment consisted of a short Qualtrics survey (5 min) followed by training and a longer behavioral task (45 min) that consisted of three phases. Participants completed all steps on site. After consenting to participate, participants were outfitted with a fNIR sensor developed by BionicaLabs called NIRSense, and a Polar H10 Heart Rate Monitor sensor for tracking blood flow and heart rate. Participants were then asked to complete a short demographic Qualtrics survey. After completing the Qualtrics survey, participants reviewed an instructional slide deck for the behavioral task for controlling the vehicle. Participants were then informed that their driving performance score during the experimental task would be based on how well they managed to stay on the road and the speed they maintained while doing so. Participants were reassured that they would be returned to the road automatically should they drive off after 5 s or they could also choose to reset themselves at any given time by selecting a specific button. Participants were then asked to drive along a 3-min straight road driving baseline for physiological metrics, and then given more instructions for the secondary cognitive task that would be performed while driving.

2.4 Auditory n-Back

The cognitive task was a form of the classic n-back task [6]. Participants in this case were required to perform a 2-back auditory memory tasks. Letters were played aloud through the participant's headset every 2.5 s in a random order. Participants were instructed to press either of the paddles on the steering wheel when a letter matched the same letter played 2 letters back. Participants were shown a video demonstrating the task and given the chance to do a practice run. If the participants had no questions, they began the n-back task.

Participants were tasked with driving along a series of different levels of road curviness while completing an auditory 2-back task. Participants completed one 2-min segment with 46 n-back letters played with the first trial on a "medium" level difficulty or somewhat curvy road. Then, participants completed five additional 2-min trials (12-min total). After each trial, the adaptation algorithm would run and determine whether to make the roads easier (straighter) or harder (more curved) to drive. Before beginning the next trial, participants would fill out mental workload prompts from the NASA-TLX [5]. The total duration of this phase was 14 min.

2.5 Measures of Performance and State

Multiple measures of workload were collected in the DART simulation environment as well as several additional physiological sensors. Measures were chosen based on their past success in previous experimental studies and their ability to distinguish between good and bad performance. All the selected measures are sensitive to increasing workload, but to different degrees. Seemingly redundant measures were included to account for these differences in sensitivity.

Two types of physiological sensors were used to measure workload. Mental workload was captured using an fNIR sensor, a lightweight optical brain monitoring device. fNIRs have been used in previous research with n-back related tasks and have been found to reliably discriminate different workload levels [7, 8]. The fNIRS sensor reported relative oxygenated hemoglobin (HbO2) and derived hemoglobin difference (HBD) levels. In pilot testing, users were exposed to both high workload and low workload scenarios to determine the average change in the fNIR sensor levels. This allowed us to determine a sufficient threshold for adaptation for the selected experimental task. The other physiological sensor used was a Polar H10, a chest strap heart rate sensor. While fNIRs do derive heart rate information, the Polar H10 was also utilized as a more validated and standard sensor for heart rate information. Heart rate sensor data also has been used in previous experimental studies and demonstrated sensitivity to increased workload levels in n-back tasks [9]. In addition, the NASA-TLX collected a subjective measure of workload, serving as a non-sensor based/non-performance dependent comparison metric.

Driving performance measures included the average miles per hour (Avg MPH) driven and times off road (TOR). During pilot testing, many users attempted to regulate their own workload by reducing the difficulty of driving by slowing down. As a result, users were instructed to maintain a minimum driving speed of 45 miles an hour. The road within the DART was designed to be like a one lane road. Participants were instructed to stay on the road and not deviate. The program tracked the amount of time that at least one of the wheels was off the road. The wheel physically rumbled when the participant deviated from the road to remind them to correct their course. These two driving performance metrics have been used in previous experimental studies [9, 10]. Users were transferred back to the road they if they lost control of the vehicle or if they themselves manually inputted that they wanted to reset on the road. Resets was also tracked as a driving performance metric.

N-back performance is commonly calculated as the number of letters correctly responded to or incorrectly responded to [8, 11]. The number of letters that resulted in correct rejections, false alarms, misses, and hits were tracked across all the 46 n-back letters presented in the experiment. Responses were sorted by type and then combined into to create a percent-correct metric that represented how many letters were hits and correct rejections.

2.6 DART Adaptation

The initial version of DART utilized for this study included a rule-based algorithm. DART assessed user's state and performance to adapt the difficulty of the task appropriately. DART included two kinds of potential adaptations including providing scaffolded levels of multimodal cueing or changing the physical environment road to become more or less difficult via the curviness of the road. Heart rate-based, rule adaptations were adapted based on findings form [9], with the remaining thresholds determined by pilot testing due to the unique design of the DART testbed. This made it difficult to utilize previous driving performance metrics and relative nature of fNIR data. The matrices and logic for the rule-based adaptation are described below in Table 1.

If the HR was 4bpm higher than baseline or HbO2 was greater than the sum of the current HbO2 and absolute value of the baseline HbO2, the state was marked as stressed. 2If the number of times the vehicle went off road remained below 15 and the Avg MPH was greater than 45, performance was marked as good. 3If the percentage of n-back letters correctly responded to was greater than or equal to 75%, the performance was marked as good. Any other conditions than listed in each category resulted in a low or poor marking.

Table 1. Adaptation Logic.

Adaptation	Stress Level[1]	Driving Performance[2]	N-back Performance[3]
Harder	High	Good	Good
Easier	High	Poor	Good
Harder	Low	Good	Good
Harder	Low	Poor	Good
Easier	High	Good	Poor
Easier	High	Poor	Poor
Harder	Low	Good	Poor
Easier	Low	Poor	Poor

3   Results

3.1   Common Indicators of Performance

A correlation matrix was constructed to better understand the relationship between the selected measures (see Fig. 1). All driving performance measures were statistically significantly correlated with one another. Time off road (TOR) was significantly correlated with the Avg MPH ($p < .01$) and the number of Resets ($p < .05$). The Avg MPH and the number of Resets was also significantly correlated ($p < .05$). These findings suggest that these performance indices are all appropriate measures of the driving task and have some degree of convergent validity.

Multiple of the physiological measures were statistically significantly correlated with one another. As one would expect, the reciprocal fNIR measures, HBD and HbO2, were negatively correlated with the reciprocal Polar measures, HR and RRInt, respectively ($p < .01$). The fNIR sensor's derived HR measure significantly correlated with the Polar's ECG HR ($p < .05$), adding merit to the derived approach of measuring of heart rate.

TOR was significantly positively related to fNIR HbO2 ($p < .05$), and significantly negatively related to its reciprocal, HBD ($p < .05$). fNIR HbO2 was also significantly positively correlated with the two other driving performance measures, number of Resets ($p < .05$) and Avg MPH ($p < .05$). Measures from the Polar H10 sensor were not significantly correlated with any of the performance measures, yet fNIR HR was statistically significantly positively correlated with TOR ($p < .05$). The NASA TLX and n-back performance did not significantly correlate with any measure.

3.2   MANOVA

A one-way multivariate ANOVA was conducted to determine whether there was a difference in the task performance measures (Avg MPH, n-back accuracy, Resets, TOR), physiological measures (fNIR HbD, fNIR HbO2, HR, HRV), and the subjective NASA-TLX measure based on n-back task difficulty (Easy vs. Hard).

The multivariate result was statistically significant for n-back task difficulty (Pillai's Trace $= .60$, $F(10,77) = 11.57$, $p < .001$), indicating a difference in the measures between

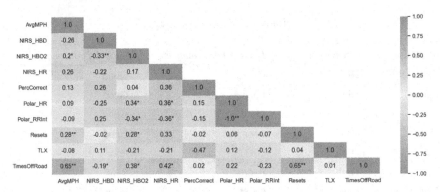

Fig. 1. Measure Correlation Matrix – *N*-Back Easy and Hard Trials

the Easy and Hard *n*-back task difficulty levels. The univariate F-tests showed that there was a significant difference between the Easy and Hard *n*-back task difficulty for Avg MPH (F(1,86) = 39.88, $p < .001$), fNIRS HbD (F(1,86) = 9.49, $p = .002$), fNIRS HbO2 (F(1,86) = 16.30, $p < .001$), and TOR (F(1,86) = 11.63, $p < .001$). However, the remaining variables were not statistically significant (see Table 2 and 3).

Table 2. MANOVA Results.

	Value	F	df	P
Wilks' lambda	0.3396	11.57	10, 77	0.000
Pillai's trace	0.6004	11.57	10, 77	0.000
Hotelling's trace	1.5028	11.57	10, 77	0.000

Table 3. Univariate Analyses.

	Easy		Hard		F	df	p
	M	SD	M	SD			
Average MPH	67.34	13.58	50.29	11.72	39.88	1, 86	0.000
Time Off Road	19.07	8.80	12.28	9.77	11.64	1, 86	0.000
Road Resets	1.07	1.39	0.93	2.38	0.11	1, 86	0.746
N-back % Correct	87.61	12.57	91.02	6.24	2.65	1, 86	0.107
Polar HR	77.83	12.21	79.84	14.10	0.51	1, 86	0.477

(continued)

Table 3. (*continued*)

	Easy		Hard		F	df	p
	M	SD	M	SD			
Polar RR Int	793.75	140.65	773.20	131.40	0.50	1, 86	0.481
fNIR HR	65.54	10.51	68.57	10.64	1.80	1, 86	0.183
fNIR HBD	−0.0003	0.0008	0.00002	0.0002	9.49	1, 86	0.002
fNIR HBO2	0.001	0.001	−0.00007	0.0006	16.31	1, 86	0.000
TLX	12.66	4.74	12.39	3.02	0.11	1, 86	0.744

As can be seen in Table 3, participants drove faster during the Easy *n*-back difficulty level than in the Hard level. Participants also experienced a greater change in HbD during the Easy level than compared to the Hard level. Avg MPH, TOR, HbO2 and HBD were all statistically significant as can be seen in the univariate analysis. The *n*-back and NASA TLX results were non-significant.

4 Discussion

Adaptive instruction is a highly desired component of training and learning, and has been, ever since the empirical data about its benefits first emerged (e.g., [11]). Most of the publications following this trend highlight the benefit of an adaptive system over a non-adaptive system. Few, however, make comparisons between one adaptive system and another. Even fewer provide recommendations for how to build and evaluate one's own system [12]. This is, in part, because of the notion that adaptive frameworks should be designed and tailored specifically to the specific system in use for the particular effort. While this concept is an effective best practice from a practical execution standpoint, the practice does not provide scientists much insight into which measures to choose to positively impact performance or mental workload across studies. This conundrum is only exaggerated when measures, that are expected be related, correlate in only one scenario, and then do not correlate in subsequent or related experiments. This happens all too often in studies that utilize measures of workload (e.g., [4]). The DART was designed to alleviate some of this confusion by acting as an all-in-one system that could be modified to test different adaptive conditions. While we do not offer a panacea, this experiment has provided further evidence that the relationship of common workload measures can vary across types of tasks.

This paper articulates the findings of a simulation-based effort in which mental workload was assessed by multiple means. These methods include real-time measurement of physiological information related to mental workload, subjective ratings of perceived workload from the participant after each trial, and performance data on key components related to the performance tasks. These methods were used in tandem to adapt the difficulty of the simulation environment.

The following discussion focuses on the relationships that are important to consider when developing adaptive systems that use measures related to mental workload as the source for modification. The meaningful trends in the data that are the most prominent are presented below.

4.1 fNIR

fNIR sensors provide a relatively streamless alternative for collecting neurophysiological information. fNIR sensors also derive heart rate information and significantly correlate with standard chest strap sensors like the Polar H10. This sensor's relative simplicity has been met with skepticism but the results from this study support the notion that fNIR data can be used as a mental workload measure. The fNIR HbO2 data collected significantly correlated with all the driving task performance measures. These findings lead to the conclusion that the fNIR HbO2 is the prime candidate for measuring workload in a driving task. The fNIR HR data also significantly correlated with TOR when the Polar H10 did not. The authors believe that this is in part due to the higher sensitivity of the fNIR sensor. These findings lead to the recommendation others interested in adaptive driving systems should consider using fNIR because of its high temporal resolution, low invasiveness, and its relative ability to capture the mental demands during this type of task.

However, the fNIR sensor did not correlate with our chosen n-back performance metric, percent correct. While this was initially surprising, as other studies have found relations between physiological data and n-back performance [8, 13], this difference might be related to how different researchers quantified n-back performance, using only "missed targets" instead of percent correct (e.g., [7]). One might suspect that the finding of this study is the result of a temporal miss-match, but that does not explain why the n-back did not correlate with any other measure. Further research is needed to understand what n-back metrics might best be representative of high workload such as reaction time, d-prime, or a sub-measure of performance such as missed targets.

4.2 Adaptation and Measure Relationship

Participants varied in how they balanced increasing difficulty of task demands. This is expected when no primary task is given; participants were tasked with balancing both tasks. Some, however, choose to focus primarily on the driving task while others prioritized the arguably more cognitively intense n-back. This finding raises the question: which performance measure should the adaptive algorithm use—and subsequently, what to adapt. If a participant is struggling with the audio-response task, they could choose to ignore their performance on that task and instead continue to keep their driving performance excellent, allowing their physiological data to suggest that they are unbothered. Based upon the current rule-based adaptation logic, this would lead the adaptive algorithm to incorrectly suggest increasing the difficulty and make the roads more difficult. However, driving for that individual is not the skill that needs attention. Considering the MANOVA uncovered that driving performance metrics and physiological state metrics were both significant, each of those metrics actually reflect different aspects of the trainee state. One might suggest that because of this, the adaptive logic should be split and adapt driving with driving performance, while the level of n-back would adapt the difficulty of the n-back task, and the pacing of the tasks with the stress level. Future research aims to adapt the environment based upon the state measures to determine if this approach provides more tailored adaptive experiences.

However, future research should be cautioned that the measures described here may not be representative enough to enable a task split adaptive system. Some of the measures, including heart rate variability and oxygenation data from fNIRs, were used in this study as more general measures of workload. These measures do not allow for clearly parsing apart which task is causing the strain. Further research will continue to identify measures that provide the greatest insight into the trainee state and which measures could enable such a complex task split adaptive design.

4.3 Limitations

One limitation of the study was a potential lack of granularity in difficulty levels. Most participants ($n = 14$) ended up in the Easy or the Hard level, with little variability after the second trial. This is the primary reason that "Medium" difficulty was not represented in the analyses. Assuming that these participants were assigned the appropriate difficulty, this may suggest that more difficulty options should be included. This can be done either by adding more levels of difficulty but at the cost of development complexity. Alternatively, the system could switch to a continuous difficulty spectrum step-based adaptation strategy instead (see [14] for review). Step-based adaptation may be too volatile in a multi-tasking environment such as the DART. The authors choose to stay within the trial-based adaptation out of concern of over-correcting and based on our pilot data; however, the pilot participants did not represent as wide of a range of ability as later sampled. Future researchers should consider airing on the side of caution and include more levels of difficulty, especially when beginning with a limited pilot sample to prevent floor and ceiling effects.

Another limitation was the weighting of workload measures. There is broad consensus about the benefits of adapting based on workload; however, there are far fewer examples of instantiations or formulas that proclaim the appropriate balance of factors. The equation used in this experiment came from a handful of pilot-tests and general recommendations from across the literature. An alternative explanation to the lack of variability in trial difficulty could be caused by an imbalanced weighting of workload. Further fine tuning of the formula may still be required after collecting more data or a switch to more data-driven methods in which a model is trained to adapt.

5 Conclusion

Designers of adaptive instructional systems have long sought clarity, both in the context of the tutoring or traditional education-sense, and in the training simulators realm. AISs, and similar systems, offer desirable benefits, but are rarely implemented due to their complexity. While different from other, more typical AISs, the insights from the DART's development and data collection efforts may assist future implementations of adaptive instruction by providing findings on the various factors impacting AIS effectiveness.

This paper has covered the findings from the actualized implementation of an adaptive system, DART, and share considerations for researchers designing systems with similar interest areas. The conclusions provide insights into the structure of metrics that enable adaptive training, and present considerations for future AIS design. This work contributes to other, relatable adaptive instruction topics, including when a system should adapt, the appropriate number of difficulty options, considerations when using multiple tasks, and the relationship between different measures of mental workload.

Acknowledgments. We would like to thank our engineers who created the DART testbed, Shawn Turk and Jonathan Reynolds.

Disclosure of Interests. This work was supported by the Air Force Research Laboratory (AFRL) 711th Human Performance Wing (HPW/RHW) Gaming Research Integration for Learning Laboratory (GRILL) Contract Number: FA8650-21-C-6273. The views, opinions and/or findings are those of the authors and should not be construed as an official Department of the Air Force position, policy, or decision unless so designated by other documentation.

References

1. Vygotsky, L.S.: Mind in Society: The Development of Higher Psychological Processes. Harvard University Press, Cambridge, MA (1978)
2. Rebensky, S., Perry, S., Bennett, W.: How, when, and what to adapt: effective adaptive training through game-based development technology. In: 2022 Interservice/Industry Training, Simulation, and Education Conference. I/ITSEC, Orlando, FL (2022)
3. Vidulich, M.A., Tsang, P.S.: Mental workload and situation awareness. Handbook of human factors and ergonomics, pp. 243–273 (2012)
4. Matthews, G., Reinerman-Jones, L.E., Barber, D.J., Abich, J., IV.: The psychometrics of mental workload: multiple measures are sensitive but divergent. Hum. Factors **57**(1), 125–143 (2015)
5. Hart, S.G., Staveland, L.E.: Development of NASA-TLX (Task Load Index): results of empirical and theoretical research. Adv. Psychol. **52**, 139–183 (1988)
6. Kirchner, W.K.: Age differences in short-term retention of rapidly changing information. J. Exp. Psychol. **55**(4), 352 (1958)
7. Heff, C., Heger, D., Fortmann, O., Hennrich, J., Putze, F., Schultz, T.: Mental workload during n-back task—quantified in the prefrontal cortex using fNIRS. Front. Hum. Neurosci. **7**, 935 (2014)
8. Unni, A., et al.: Brain activity measured with fNIRS for the prediction of cognitive workload. In: 6th IEEE International Conference on Cognitive Infocommunications (CogInfoCom), pp. 349–354. IEEE. Gyor, Hungary (2015)
9. Mehler, B., Reimer, B., Coughlin, J.F., Dusek, J.A.: Impact of incremental increases in cognitive workload on physiological arousal and performance in young adult drivers. Transp. Res. Rec. **2138**(1), 6–12 (2009)
10. Lenné, M.G., Hoggan, B.L., Fidock, J., Stuart, G., Aidman, E.: The impact of auditory task complexity on primary task performance in military land vehicle crew. In: 58th Proceedings of the Human Factors and Ergonomics Society Annual Meeting, pp. 2185–2189. HFES. Chicago, IL (2014)
11. Kelley, C.R.: What is adaptive training? Hum. Factors **11**(6), 547–556 (1969)

12. Landsberg, C.R., Astwood, R.S., Jr., Van Buskirk, W.L., Townsend, L.N., Steinhauser, N.B., Mercado, A.D.: Review of adaptive training system techniques. Mil. Psychol. **24**(2), 96–113 (2012)
13. Vine, C., Coakley, S., Myers, S.D., Blacker, S.D., Runswick, O.R.: The reliability of a military specific auditory n-back task and shoot/don't-shoot task. Exp. Results **3**, 1–14 (2022)
14. Aleven, V., McLaughlin, E.A., Glenn, R.A., Koedinger, K.R.: Instruction based on adaptive learning technologies. Handb. Res. Learn. Instr. **2**, 522–560 (2016)

AI in Adaptive Learning

Opportunities and Challenges in Developing Educational AI-Assistants for the Metaverse

Christopher Krauss[1]([✉]), Louay Bassbouss[1], Max Upravitelev[1], Truong-Sinh An[1],
Daniela Altun[2], Lisa Reray[2], Emil Balitzki[3], Tarek El Tamimi[3],
and Mehmet Karagülle[3]

[1] Fraunhofer Institute for Open Communication Systems FOKUS, Kaiserin-Augusta-Allee 31,
10589 Berlin, Germany
`{christopher.krauss,louay.bassbouss,max.upravitelev,`
`truong-sinh.an}@fokus.fraunhofer.de`
[2] Fraunhofer Institute for Communication, Information Processing and Ergonomics FKIE,
Fraunhoferstraße 20, 53343 Wachtberg, Germany
`{daniela.altun,lisa.reray}@fkie.fraunhofer.de`
[3] Technical University Berlin, Advanced Web Technologies, Straße des 17. Juni 135,
10623 Berlin, Germany
`{balitzki,eltamimi,mehmet.karaguelle}@campus.tu-berlin.de`

Abstract. The paper explores the opportunities and challenges for metaverse learning environments with AI-Assistants based on Large Language Models. A proof of concept based on popular but proprietary technologies is presented that enables a natural language exchange between the user and an AI-based medical expert in a highly immersive environment based on the Unreal Engine. The answers generated by ChatGPT are not only played back lip-synchronously, but also visualized in the VR environment using a 3D model of a skeleton. Usability and user experience play a particularly important role in the development of the highly immersive AI-Assistant. The proof of concept serves to illustrate the opportunities and challenges that lie in the merging of large language models, metaverse applications and educational ecosystems, which are self-contained research areas. Development strategies, tools and interoperability standards will be presented to facilitate future developments in this triangle of tension.

Keywords: Large Language Models · LLM · AI-Assistants · Virtual Assistants · Virtual Reality · Metaverse · Education · Learning Technologies · Interoperability

1 Introduction

AI-based Assistants can already answer learners' technical questions, provide explanations, guide interactive exercises, and offer personalized recommendations based on user interactions. They also suggest customized content and exercises, act as virtual instructors, and assist with questions. Moreover, they provide feedback on given answers, identify knowledge gaps, and track learning progress. To ensure the quality of the learning experience, these bots should generate appropriate content such as learning media, quizzes and exercises tailored to the learners' needs. The enormous potential of processing large amounts of data through generative artificial intelligence has become widely recognized in education, particularly with the advances of technologies from OpenAI (such as ChatGPT[1], GPT-4[2], DALL-E[3] and Sora[4]). The challenges lie in making technology transparent and user-friendly. From a technical perspective, this requires data sovereignty, vendor openness and diversity. Proprietary cloud solutions have limitations in terms of data control and access. Open-source alternatives and federated learning offer promising approaches to address these challenges.

At the same time, there are many areas of application in which the explanation of relationships in a classic learning platform "in 2D" or text form appears more complicated than illustrating the same relationships in the real world. This is usually the case when complex objects, such as machines, buildings, or the human body, need to be presented and explained. As access to these objects can sometimes be very expensive (e.g., in the case of large machines) - they may even break in the process - or it is not even possible to gain an insight into the relationships between the various components, as is the case with living bodies, there are many 3D simulations in suitable virtual and augmented realities.

The metaverse is more than the sum of AR and VR. Thereby, metaverse refers to a multi-vendor environment and not necessarily to Meta's Metaverse. Three features of the metaverse clearly distinguish it from normal AR and VR applications: "shared", "persistent", and "decentralized". "Shared" means, that people can interact with others using a new identity; a social world is created with social connection and interaction. "Persistent" means that this world cannot simply be restarted; it continues to run and thus learns from the interactions that take place in it over time. It means that experience inside cannot be paused or restarted. Additionally, decentralized technologies (e.g., blockchains) are needed to ensure that economic activities can be safely conducted and that personal property and logs in the metaverse will not be modified by others. AR and VR are used to present the content of the metaverse, and AI enables this parallel world to follow the rules defined by the creator. In the metaverse, AI is mainly used for arbitration (solving conflicts), simulation (enabling non-human avatars characteristics to act like humans), and decision-making (following rules pre-defined by the creator) in the metaverse [1].

In this paper, we show the potential and challenges that arise when AI-Assistants are introduced directly in metaverse learning environments. The integration of standards from the educational domain as well as standards from the metaverse world will, among others, play a critical role to effectively integrate Large Language Models into virtual education ecosystems and enhance learning in the digital space.

From a non-technical view, many aspects should be considered to successfully implement such a future-oriented technology and gain user acceptance in a variety of application scenarios. Studies have shown that considering the learners emotion in a virtual learning environment can be beneficial for the learning outcomes, users' motivation, and perceived usability [2]. In this paper, we not only introduce the technology and its

---

[1] ChatGPT: https://chat.openai.com/.

[2] GPT-4: https://openai.com/gpt-4/.

[3] DALL-E 3: https://openai.com/dall-e-3/.

[4] Sora - Creating video from text: https://openai.com/sora/.

functionality but also discuss the implications and possibilities it offers for training. For instance, we explore additional application scenarios, such as how it can be combined with user state analysis to support individual needs adaptively, and how this technology can complement competency-based education in many ways. We also investigate human-centered design, methods to test its usability, and how to evaluate the system.

The paper is structured as follows: First, the scientific literature from the areas of metaverse, LLM-based AI-Assistants and virtual training is analyzed in the context of the fusion of these three topics. Then the possibilities and added values of merging the technologies are discussed and a first proof of concept of an AI assistant in a lecture hall of a medical study program is presented. The development of this still proprietary prototype is explained and followed by a detailed discussion of why the future of sustainable developments lies in a loose coupling of components, interoperability standards and the interchangeability of technologies instead of permanently relying on monolithic, non-transparent, and proprietary technology stacks. The paper concludes with a summary and an outlook.

2   Related Work

The work operates at the intersection of a triangle of tension between artificial intelligence, educational ecosystems and metaverse applications. While AI in educational ecosystems is a recent but heavily researched area, there are many new developments around Large Language Models (LLMs), and thus implicitly also their potentials and challenges for education. However, the use of LLM-based AI-Assistants for educational environments has hardly been researched yet.

2.1   MetaHumans

Over the past years, major technology companies have demonstrated significant interest in virtual technologies, particularly virtual character, and face capture technologies [3, p. 154–155]. Epic Games[5] being one of the leading companies and early investors in the virtual sector developed their own solution for the effortless creation of photo-realistic digital humans, called MetaHumans, using Unreal Engine[6]. These can be described as high-fidelity digital human models, allowing for a high level of immersion and photo-realism. MetaHuman models include a full model of a person, including the body and the head, with the latter being able to express many emotions and poses. Additionally, Epic Games provided a MetaHuman Creator framework enabling users to easily create, shape and fine-tune their avatars [3]. With the Unreal Engine itself, users can effortlessly acquire the asset in a fully rigged and prepared state, enabling seamless animation and motion capture. The asset offers users an array of performance capture tools to enhance the animation and realism of digital humans in whatever virtual environment [3].

---

[5] Epic Games: https://www.epicgames.com/site/en-US/home/.

[6] MetaHumans: https://www.epicgames.com/site/en-US/news/announcing-metahuman-creator-fast-high-fidelity-digital-humans-in-unreal-engine/.

2.2 Metaverse Applications for Training

Several papers have already discussed the benefits of using virtual reality for training certain professions, including [4–6], where VR seems to have a great potential to enhance the learning process. A similar development can be seen in the AI field, with large language models, such as ChatGPT, having a big impact on the future of learning [7–9]. Several online platforms offer an immersive virtual training experience, both for onboarding processes as well as for developing and acquiring new skills. Mursion[7] specialized in virtual reality simulations to foster interpersonal competencies, is offering an environment wherein users engage and interact to refine their skills through practice. Alternatively, STRIVR[8] provides a wide range of training modules covering domains such as customer service and safety protocols, employing virtual reality technology. Through this approach, STRIVR aims to enhance learning retention and encourage user engagement. Labster[9] primarily serves as the educational domain, with a focus of virtual laboratory simulations. These immersive virtual labs encourage and motivate students to conduct experiments and explore scientific concepts, within a secure environment.

The possibilities of learning in the metaverse go beyond these applications if we consider the characteristics of shared, persistent, and decentralized, as mentioned above. An illustrative example is language learning: In the metaverse, the goal of language learning is more than a course or learning activity; instead, it is to provide learners with a different life, a living environment in which English is used for working, learning, socializing and entertainment as if they were native English speakers. The two learning experiences (i.e., VR and the metaverse) are very different [1]. However, because research on the metaverse is still in its infancy, we need to use the results from research with VR and AR as a basis.

Studies comparing a variety of learning media often show no difference in the effectiveness of the training using VR as media or another learning media. Especially in the medical sector, where patients could be hurt and training resources may be rare or expensive, research focused on different learning scenarios where training can benefit by using VR technology. For example, Gurusamy et al. [10] summarized 23 studies in a review for the field of laparoscopy alone, which compared either VR with video, VR with a classic laparoscopy trainer or VR with no training. In this review, the authors concluded that training with VR achieves at least as good results as training with video trainers and recommend the use of VR as a supplement to classic laparoscopy trainers. Torkington et al. [11], who compared the learning outcomes of VR trainers with the classic laparoscopy trainer, found no differences in performance between the two learning media. Altun and Schulz [12] compared training with VR to two other learning media (paper instructions, video instructions) in the context of learning how to solve a Rubik's Cube and found no disadvantages of using VR regarding learning outcomes, sustainable learning outcomes, learning transfer, usability, or workload.

All of them adopt immersive technologies to drive training and skill development. However, they do not integrate AI-based assistants, which enables dynamic and personalized guidance and better tailoring to each user's needs.

---

[7] Mursion: https://www.mursion.com/team/.

[8] STRIVR: https://www.strivr.com/why-strivr/.

[9] Labster: https://www.labster.com/.

2.3 LLM-Based AI-Assistants

Artificial intelligence-based tutoring agents (bots) are playing a significant role in modern education and training. The introduction of AI-Assistants for 2D learning environments is a greater challenge, as there are many prerequisites to be met. According to Altun et al. [13] these include:

1. Methodological prerequisites: What is the task and how can the success of the AI be measured? The goal, the methodology and the evaluation framework must be tailored to each other.
2. Organizational prerequisites: Stakeholders must be identified, picked up, taken seriously, and trained. The necessary processes must be initiated or established for this.
3. Didactic prerequisites: An AI does not replace the teacher but should assist with certain tasks. AI functions must be very well embedded in the didactic setting.
4. Content requirements: Appealing and varied digital content must be available, well-described by metadata, and interoperable.
5. Technical requirements: Ideally, interoperable data and services should be integrated using common interfaces and format standards.

AI tools are transforming the approach to academics, as learning in virtual environments enhances students' performance and motivation, offering an interactive and engaging learning experience. Among these, one of the most popular chatting bots is the ChatGPT, a conversational AI tool based on the GPT-3 large-scale language model, developed by OpenAI[10]. This natural language model is a conversational agent that dialogues with a user. Using machine learning, it generates responses from user text inputs, mimicking human conversation. The GPT-3 language model is trained on diverse internet text data, including books and other publicly available data [14]. ChatGPT demonstrated the capacity to comprehend users' input, enabling almost any topic consultation between the bot and the user [7]. By asking a series of questions (prompts), the model can help identify potential problems and recommend appropriate action, including facilitating access to healthcare information [8]. Thereby, ChatGPT's huge potential as virtual learning assistant has already been shown for educational settings [15].

Several strategies have emerged to steer the generation of LLM outputs, which can be roughly divided into (1) fine-tuning LLMs and (2) prompt engineering. The fine-tuning of a model can be described as the continuation of an LLM training by providing a (possibly self-created) dataset on which the LLM is further trained on. Prompt engineering focuses on methods of how prompts can be strategically modified to guide LLMs towards an expected behavior. Currently, different approaches like [16] are discussed for harnessing prompt engineering within general chat-based learning content. Furthermore, these techniques are also explored within subject-specific approaches, e.g., within prompt engineering in medical education [17].

While these strategies focus on refining a prompt directly by users, another approach is to modify prompts programmatically, for example, by adding more context information to a prompt. One field that follows this approach is Retrieval Augmented Generation

---

[10] OpenAI: https://openai.com/.

(RAG), where semantically fitting chunks of information are retrieved from specified documents and are included within the context information of a user prompt to generate an answer. While this approach introduces new possibilities of adaptivity of learning content by, for example, including insights from learning analytics into the prompts to further adapt the content to learners needs, it is currently still in preliminary development in the education domain.

2.4 Usability and User Experience

Usability and user experience are decisive success criteria for high-immersive applications. Usability forms the basis for a user's interaction that is satisfactory. In the best-known definition of usability by Nielsen [18], terms such as learnability, efficiency, memorability, errors, and satisfaction are central. User experience (UX) reflects the experiences, sensations and feelings of a person while using a product [19] and therefore includes concepts such as the so-called "joy of use".

Usability focuses on the problems, frustrations and barriers that can occur when using a product and attempts to eliminate these. UX, on the other hand, deals with the (3) positive effects of using or owning a product. Positive emotions such as joy or pride play a role here. These differences also result in a difference in the research methodology. While usability mainly uses qualitative methods (e.g., interviews), quantitative methods (e.g., questionnaires) are often used to record user experience. In summary, it can be said that UX should be seen as a complement to usability. A high usability does not guarantee that a product will be perceived as good, although usability is often seen as a prerequisite for a positive user experience.

In the ongoing development of the metaverse, UX is crucial because it determines how people interact with and perceive virtual environments. A good user experience can make a virtual world feel immersive and engaging, while a poor user experience can make it feel confusing or frustrating to use [20]. When designing the metaverse, there are several factors to consider that can affect the user experience, including the physical and virtual environment, the user's goals and motivations, and the hardware and software used. Bisset Delgado names three big challenges for UX design in the metaverse: (1) The first challenge is the need to design for a new and rapidly evolving medium. The metaverse is still in its early stages of development, and there are many unknowns and uncertainties about how it will evolve and what users will expect from it. Additionally, (2), because the metaverse will be accessed through a variety of devices (including VR headsets, AR glasses, and smartphones), there is a need to design the metaverse for a wide range of platforms. As a third factor (3), it will be a challenge to design immersive and interactive environments that can be difficult to predict and control, because users will be able to move freely and interact with the environment in ways that are not possible in the physical world [21]. To meet these challenges appropriately, proven methods from VR and AR research must be combined to create new methods. AI-based avatars allow highly immersive dialogs in which users can have almost free and human-like conversations with the bots. First game technologies already show non-player characters that are based on large language models and practically play a role by sticking to a given script like a human actor[11]. While in principle attempts are made to create the highest possible user acceptance through more realistic virtual avatars, the uncanny valley effect can also occur, for example, in which more realistic-looking figures in virtual realities are sometimes less accepted than more abstract figures. We make some suggestions for dealing with challenges later in this article.

2.5 Opportunities of Virtual AI-Assistants for Education

By combining high-fidelity, hyper-realistic human models, together with the growing abilities of large language models, a highly realistic virtual training experience can be created, allowing for easy, cheaper, and more effective training of personnel for many job professions. In contrast to normal text-based learning through online research, we wanted our users to be able to walk around a digital world representing familiar locations depending on the scenario of the virtual training.

However, it is very important to mention that AI-Assistants should only be used in the metaverse where the metaverse itself benefits existing training. For example, a pure transfer of classroom learning, where the user sits at a desk in front of a computer or book or listens to a PowerPoint presentation in a lecture hall, to virtual reality should be avoided. Of course, familiar locations, such as offices or training rooms, can still be presented to strengthen the emotional connection and immersion based on familiar environments. However, it should always be investigated which interactions with the environment offer added value in the metaverse.

Virtual AI-Assistants can play to their strengths where VR environments can, except that the integration of AI-based avatars in VR means that practical knowledge can be presented more flexibly and interactively. Thereby, AI-Assistants are always available, so users don't have to wait for answers from human experts. Currently, they are not yet able to react emotionally to the user's input, which can be both a blessing and a curse. For this reason, they should not be seen as a replacement for human teachers, but as a useful addition that needs to be well integrated into an existing teaching setting - e.g., for specific exercises or as a temporary assistant for the human tutor.

Examples would include a fire department for ongoing firefighters, an operating theater for future surgeons, or maybe even more complicated or dangerous training grounds, such as a space station, military camp, or an airplane. For our project, we have chosen an example of medical training inside a university classroom. The environment is full of high-quality props and assets, allowing for a better understanding of complicated topics.

3 Realization of a Proof of Concept

With an initial proof of concept, we wanted to highlight the characteristics and benefits of virtual training with AI-bots and showcase its feasibility with a proprietary tech stack consisting of an Unreal Engine based Metaverse experience using MetaHuman models and a connection to ChatGPT.

---

[11] ConvAI - Conversational AI for Virtual Worlds: https://convai.com/.

We allow the user to join our virtual training environment, being represented as a MetaHuman model, and learn medical subjects by asking questions to a Virtual Assistant (VA), called Gavin (see Fig. 1). The VA is a realistic-looking, non-player character (NPC) at who the user can look at, speak to, and ask questions related to the scenario of the virtual training. The logic inside of the VA was implemented using a pipeline of different APIs, including the conversion of text-to-speech, speech-to-text, ChatGPT-based comprehension, and creation of text messages. Additionally, the VA supports lip-speech synchronization and additional animations, such as gestures or neck rotation.

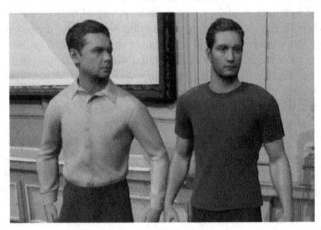

Fig. 1. MetaHuman models of the Player (right) and the Virtual Assistant Gavin (left).

Next to the virtual assistant stands a skeleton 3D model of the human body. When VA Gavin is asked about certain bones, he can interact with the environment and highlight bones on the skeleton. With the use of syntax and text analysis, a simple environmental awareness of the VA was implemented, enabling the VA to highlight specific bones of a skeleton located next to him. By doing so, we are making an example of the usage of teaching props for easier visualization of complex topics, such as human anatomy. This is intended to demonstrate how separate services (such as the Large Language Model in the cloud) can generate answers that can in turn be followed up in virtual reality. A short video of the proof of concept can be found online[12].

3.1 The Environment, MetaHuman Models and Player Controller

As the main game engine, we have used the Unreal Engine 5.1 created by Epic Games. It is a modern, flexible, and artists-friendly game engine, which allows for the development of next-generation, realistic video games and 3D projects. We have heavily used one of its prominent features, the blueprint programming, providing visual nodes for creating projects without the need for deep programming knowledge. While code can be seamlessly exchanged between blueprint and C++ programming, we opted for the

---

[12] See the video of the proof of concept: https://www.youtube.com/watch?v=exFRG7l8cjw.

latter only to implement the reading of custom environment variables from the operating system; and for all other components, blueprint programming has proven to be sufficient. For source control, we used Perforce's Helix Core[13] mostly due to its support for bigger binary files (often used in game development).

The virtual environment of our Metaverse experience was created solely using the above-mentioned Unreal Engine (UE), allowing for easy arrangement of in-game assets. The assets themselves have been taken from multiple sources. High-quality, realistic objects, including indoor walls, windows, doors, and some of the classroom props were added to increase immersion (but have no function beyond that) and primarily sourced from Quixel Bridge Megascans[14], which is a pre-installed extension to UE. The remaining assets were obtained from freely available third-party projects that were uploaded to the Unreal Engine Marketplace. With the above-mentioned assets, we were able to create a hyper-realistic scene (environment), including the inside of a university's medical classroom, and the main lobby, functioning as the joining room for the players and featuring instructions and in-game controls. The classroom was designed to look like a replica of a life-like medical classroom, including the blackboard, human skeleton, desks, and other classroom decorations.

MetaHuman Models. One of the most important assets was the use of the MetaHuman (MH) models. *We have* represented both the player and the virtual assistant with the MH models mentioned, available at Quixel Bridge Megascans. When uploaded to UE, the MH models contain several assets' layers, including skeletal meshes, textures, and meshes, and are possible to be scripted and/or animated. Overall, all human models in our project make use of MH technology, combined with high-quality, free animations for third-party services.

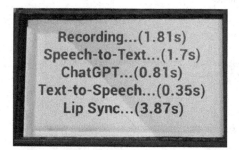

Fig. 2. The logging blackboard, located in the classroom.

Fig. 3. The skeleton teaching prop, with the bones of an arm being highlighted (humerus, ulna, and radius).

Logging Blackboard. The goal of the board is to display the latency for each of the API calls, giving feedback on the players' actions, and being vital for keeping user engagement and focus (see Fig. 2). The board is an object with a TextRender component, and a custom board tag attached. By default, it displays a note encouraging the user to

---

[13] Helix Core: https://www.perforce.com/products/helix-core/.

[14] Quixel Bridge Megascans: https://quixel.com/bridge/.

press the R key to start recording. When pressed it logs the future API calls, i.e., the required time (in seconds) it took for each call to finish. The logging is done by simply appending more text to the currently displayed text. Time difference calculation has been done with the GetRealTimeSeconds node, subtracting the current time and one before the call. Finally, when finished and after a fixed time interval, the board is cleared back to the default note.

Skeleton Teaching Prop. The replica of a human skeleton serves as an example of a training prop; thus, how virtual assets can be used for immersive and effective learning. The skeleton is built from several individually named bones (such as an atlas, patella, skull, or sternum), with an option to highlight each set of bones individually (see Fig. 3). Each set of bones has a custom tag attached, making it easy to search them within the scene. When a highlight is turned on, a custom material is applied to the bone, and the Render CustomDepth Pass option is set to true. Finally, the script part of the highlighting is done using a special syntax, therefore by analyzing the assistant's replies.

Player Controller. We have used the default player blueprint from a default Third Person project demo in the UE. It was further extended by the addition of the MetaHuman model and an option to switch to the first-person camera by pressing the V key, overall increasing the immersive feel of the experience. The player movement supports only keyboard and mouse input, with key binds set in the Input section in the UE's Projects settings screen. The mentioned key binds include simple movement (A, S, D and W) and camera rotation (mouse movement and scroll wheel), and other player actions, including the above-mentioned camera toggle. Finally, the animations for the player (thus the MetaHuman model) have been created with the use of Live Retarget Mode inside of the player's blueprint, re-targeting the animation from the mesh bones of the default player model to the MetaHuman model.

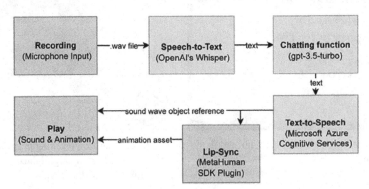

Fig. 4. The main blueprint pipeline of the virtual assistant. (Purple: processed within the Unreal Engine; green: services from OpenAI; blue: service from MS Azure). (Color figure online)

3.2 Virtual AI-Assistant

The Virtual Assistant Gavin consists of several functions, executing API calls in a fixed order. The main pipeline can be seen in Fig. 4. Additionally, two main animation blueprints are attached to the VA object - responsible for the MetaHuman body and face animations.

Player Voice Recording. Firstly, the player's voice from a microphone is recorded. The voice is saved into a.wav file and propagated further into the pipeline. Additionally, in this section, the logging board gets cleared and starts logging the times.

Speech-to-Text (OpenAI's Whisper). For the Speech-to-Text we have used the OpenAI API Plugin[15], implementing an Unreal Engine interface for calling OpenAI APIs. The.wav sound file is passed to OpenAI's Whisper[16] which transcribed the player's voice into a text. The text is then fed as input to the AI service.

Large Language Model (ChatGPT). We have selected the GPT-3.5- turbo model for the chat function (chat settings of temperature 1.0 and max tokens 250), once again implemented with the use of the OpenAI API Plugin. At the beginning, an array of chat messages is initialized, with a starting message explaining the function and requirements to the AI model. The array keeps track of the chat history and is fed in whole at each following question, making sure the AI model remembers the previous questions and information. The virtual assistant should have a basic understanding of the environment surrounding him, namely the skeleton prop located next to him. This is done by teaching the LLM the location of the skeleton with the use of simple text and the starting message.

> *"You are a medical teacher called Gavin. You are 35 years old and are an expert in medicine. The responses need to be short and no longer than 2 sentences. Behave to me as if I was a student and do not mention anything that suggests you are a language model or AI. There is a human skeleton to your left, whenever a student asks you anything related to human bones you mention that the bone can be seen on the skeleton. When mentioning any skeleton bone add at the very end of the message a single symbol #, following a list separated by a comma of the following bone names: atlas, axis, clavicle, feet, femur, fibula, hands, hipbone, humerus, patella, radius, ribs, scapula, skull, spine, sternum, tibia, ulna or thoracic vertebra. If the name is not on the list, choose one of the closest bones from the list. Do not put any more text after the syntax, do not put more than one # symbol, and remember to always update the list at the end when asked a new question."*

After calling the AI service, a response is received with a # symbol at the end of the message, followed by the names of the bones to highlight. The message is then cut into parts at the place of the # symbol, with the first part propagated to the text-to-speech component. The second part is a comma separated list of bones' tags, which is then parsed into an array and used to find bone objects in the scene. A new material is then attached to the selected bones for a fixed period, making them visible to the player.

---

[15] OpenAI API Plugin for Unreal: https://github.com/KellanM/OpenAI-Api-Unreal/.
[16] Whisper: https://openai.com/research/whisper/.

Text-to-Speech (Azure Cognitive Services). For the text-to-speech component, we have used Microsoft's Azure Cognitive Services API[17] and the free Unreal Engine plugin called AzSpeech - Voice and Text[18]. The response text from the Chat AI is propagated to the Azure API which outputs a reference to a generated sound wave object. For the voice, we have used the en-US-BrandonNeural voice option.

Lip-sync (MetahumanSDK). The lip-sync animation is created with the use of the MetahumanSKD plugin[19]. Additionally, we have selected the eye movement mode mocap option, generating additional eye movements in addition to mouth movement. As a result, an animation asset for the player's MetaHuman model is created.

Sound and Animations. The last step in the VA pipeline is to play the generated sound and animation. The sound is quite simple and uses the "Play Sound 2D" node, playing together with the animation. Playing the animations is more complex and used the Animation Montages from the UE. For each of the dynamic (playing at runtime) animations a designated slot is required inside the animation blueprint. During the runtime a reference to an animation asset is assigned to the slot, finally playing the requested animation. In addition to the discussed lip synchronization, we also include three other animations: neck rotation, idle animation, and hand gestures. The neck rotation makes the virtual assistant face the player when inside his proximity. Idle animation moves the body of the VA a little, making him look like he's breathing (instead of keeping a static T-pose). Lastly, we support a list of short animations featuring hand movement. The animation is picked randomly from a list and triggered whenever the VA is replying to the player's question (for our project, we use only a single animation inside the list). All the above-mentioned animations are then blended inside the VA's face and MetaHuman body animation blueprints, allowing the model to play multiple animations at once and making VA's behavior more natural and immersive.

4   Preliminary Findings of the Proof of Concept

The main goal of the proof of concept was to create an end-to-end high-immersive prototype in which the user can interact in a virtual environment with a realistic LLM-based avatar, which can respond autonomously to the user's questions and uses its environment to illustrate its answers. Technically, the end-to-end integration of the various components could be fully demonstrated. However, there are a few fundamental aspects that stand in the way of a fully immersive experience.

The correctness of the responses is again hard to assess, but generally positive. While the answers of the AI and the highlighting of the bones are correct most of the time, they can get odd and inaccurate with more questions asked. This is true, due to the new questions being appended to old ones, creating more noise in answers. This can

---

[17] Cognitive Services: https://azure.microsoft.com/en-gb/products/cognitive-services.

[18] AzSpeech: https://www.unrealengine.com/marketplace/en-US/product/azspeech-async-text-to-voice-and-voice-to-text/.

[19] Metahuman SDK for Unreal Engine: https://www.unrealengine.com/marketplace/en-US/item/66b869fa0d3748e78d422e59716597b6/.

also lead to highlighting the wrong bones on the skeleton or losing the correct syntax, not highlighting the bones at all. Finally, situations like the recording being not clear enough, and speech getting converted to the wrong text or different language, can create artifacts in the further pipeline components. In addition to the known limitations of LLMs (such as discriminatory biases, the hallucination of wrong information, the lack of explainability of the generated answers, or security concerns, e.g., in regard to prompt injection attacks [22]), which have already been demonstrated by various publications of large language models, the lack of expertise (which needs to be tackled by fine-tuning or retrieval-augmented strategies) due to the lack of sufficient information in the original training data [23]) and the still limited scope of action (this is only related to the skeleton next to Gavin) are particularly noticeable in our prototype.

In addition, Gavin takes a comparatively long time to respond to a user's question. The responsiveness of the VA's replies can be better evaluated, with the overall delays depending on the individual component of the API pipeline. The average response time for components (in order of recording, speech-to-text, chat AI, text-to-speech, and lip-sync) can be seen in Table 1. It can be noticed, that most time during the processing is lost during the lip synchronization (skipping the recording time), counting roughly as 51% of the overall delay. The fastest component is the text-to-speech from Azure Cognitive Services, approximating around 9% of the total delay. Additionally, we observe that ChatGPT's response times depend on the requested information, if they were already known during the initialization phase, answers tend to be quicker.

Our virtual assistant consists of several cooperating, decoupled components, executed one after another. From our evaluation, we can assess that the overall answer delay for the virtual assistant can get often quite long, with the lip-sync part taking the most time. In interpersonal communication, the answers are rather quick and engaging, thus the delays should be kept at the lowest possible levels, making it easier for the user to be entertained and stay focused. According to ITU-T Rec. P.1305 [24], a "mouth-to-ear" delay in speech application of less than 150 ms leads to transparent interactivity.

While we tried to achieve the lowest delays, the result varies and depends on multiple factors, including the internet connectivity and bandwidth, API's resources availability, time of the day, or the length and complexity of questions. While it is impossible to exactly predict the length of queries, more work could be done to further minimize the delay. Additionally, we did not compare the execution times of our components to their counterparts available on the market and open-source alternatives which allow for on-premise hosting. Therefore, possibly exchanging one or more components can speed up the answer times of the VA. A comprehensive analysis could be done, to find the most performant component for each of the tasks. Finally, each of the functions can be achieved by following multiple solutions for the implementation and deployment (either self-deployed locally or through cloud API calls). For example, it is possible to create a lip-sync animation from text or sound files. Moreover, one can prepare parts of it beforehand and stitch them together later or generate it from scratch during run-time, thus generally dealing with a compromise between performance and the visuals. Each of the components could be analyzed in a similar fashion, regarding its implementation and deployment strategies, and their best solutions could be chosen.

Table 1. Average response times (in seconds) for individual components, with each question asked 5 times.

Question	REC	S2T	GPT	T2S	LS
Hi, what's your name?	1,872s	1,494s	1,05s	0,604s	3,53s
Can you show me the bones of the human legs?	3,138s	1,508s	1,746s	0,712s	4,218s
Can you explain to me the function of the human pelvis?	4,126s	2,450s	1,914s	0,848s	5,360s
Hi, today is my first day in this medical school and I haven't met anyone here yet, so can I ask you what's your name?	8,532s	2,438s	1,092s	1,034s	4,102s
The last time I have seen a human skeleton must have been in high school, so I was wondering if you could show me which bones are part of the human leg.	9,976s	2,370s	1,518s	0,820s	4,580s
I will be writing an exam about human anatomy next week and I was wondering if you could tell me the function of the human pelvis, please?	9,684s	2,536s	1,628s	0,868s	5,648s

5   Challenges and Recommendations

Complex interconnected AI-systems are likely to exceed the maintenance capabilities of one organization. Relying on a one-fits-all solution with no or only proprietary interfaces would not be sustainable at a certain point, since the number of additional functions and modifications would make it very difficult to maintain and expand those monolithic systems. In service-oriented architectures (SOA), services can either be integrated as flexible Software-as-a-Service (SaaS) solutions (e.g., cloud-based), or hosted entirely or in part on premise, usually when deeper integration with existing systems is needed. This results in a shift towards a Best-of-Breed (BoB) strategy, especially in the education domain [25] which describes the strategy to utilize the best possible component to take care of a specific problem. Compared to monolithic approaches, BOB strategies tend to be more complex during the initial setup phase. A leading system, such as a middleware or the blueprint pipeline in our Proof of Concept, must be defined, which takes care of the orchestration of all other services [26]. This can also be supported by abstraction frameworks (e.g., LangChain[20] for LLMs).

   AI, educational ecosystems and metaverse are three very different topics, each with dedicated subject matter experts. The modular nature of the Best-of-Breed strategy

---
[20] LangChain: https://www.langchain.com/.

allows several focused developer teams from different organizations to work on different parts of the overall system in parallel. The entire system, as well as parts of it, can be better operated and maintained, or quickly replaced if necessary. This leads to parallelizable, scalable systems. However, for the corresponding services to offer users transitions that are as seamless and imperceptible as possible, the services must exchange information efficiently. To achieve this, all services involved must agree on a common understanding of the exchange mechanisms, data formats and semantics. For example, the proprietary syntax for marking bones from the proof of concept would hardly be reusable if the relevant services were replaced. For this reason, it is highly advisable to use established interoperability standards when opting for the Best-of-Breed strategy when developing educational AI-Assistants for VR environments.

5.1 Learning Technologies

The support of common standards and open specifications from education plays an essential role in the context of educational AI-Assistants, without which sustainable use in educational ecosystems is hardly conceivable. To date, hardly any educational standards have been supported by LLMs. On the one hand, different input data must be able to be processed and interpreted by the LLM, such as learning content (accessible via LTI or cmi5) from curated sources, their pedagogical and didactic metadata (e.g. according to the IEEE LOM standard) in terms of findability and interpretation, didactic contexts (as a common cartridge) as well as user interactions (as xAPI) for the description of user behavior and profiles. This is needed so that LLMs can also access the knowledge that has already been uploaded to the learning platforms.

On the other hand, the same AI-Assistants should be re-usable for learning management systems (e.g., as textual chatbots) as well as for VR environments for direct user interaction. Launch mechanisms such as LTI (Learning Tools Interoperability) and cmi5 (a protocol based on xAPI) are suitable for the integration into learning platforms. The support of these standards by LLMs must be realized via corresponding adapters.

5.2 Large Language Models

The use of sophisticated, but proprietary cloud services, such as the ones from OpenAI, in training scenarios faces significant challenges: First and foremost, these solutions are only offered as black-box services in the cloud, with training content and user data transferred via commercial interfaces on a pay-per-use basis. How copyright-protected content and personal data of the users is (further) processed by the content publishers and educational institutions is not transparent and is the sole responsibility of the provider. However, with ChatGPT having been available to the public for just over a year (as of January 2024), many promising free or even open-source alternatives have been released, enabling on-premise hosting, which enables the use of LLMs in domains with higher requirements for data security and privacy. This is particularly relevant for scenarios involving sensitive user prompts or documents processed within Retrieval Augmented Generation (RAG) strategies. OS Models also enable the use of LLMs within offline environments or networks with connection restrictions and eliminate the reliance on outside services. Benchmark results and surveys [27] indicate that models like LLama 2 by Meta[21] (not open-source, but free to use) or Mistral by MistralAI[22] are catching up to their closed-sourced counterparts or even surpass them in some cases. Their deployment is, however, accompanied by its own caveats, which can be partly attributed to their hardware requirements.

Since these models and the local server infrastructures of educational institutions are not designed for processing large and all-encompassing information, federated learning emerges as a viable solution for specific use cases. Here, the respective specialized knowledge bases are trained separately and can be used individually or linked for larger use cases. As this field of application is still very young and hardly researched, there is currently minimal experience in the context of specific educational areas.

5.3 Metaverse Enablers

The voice integration of LLM-based AI-Assistants in the VR environments can be realized via various speech-to-text (e.g., OpenAI Whisper, Mozilla DeepSpeech[23]) and text-to-speech (e.g., OpenTTS[24], Mozilla TTS[25], Mimic[26]) technologies on an open-source basis. Open standards such as WebXR[27], OpenXR[28] and glTF[29] enable the integration of generated interactive content and voicebots in VR applications. WebXR, a key initiative of the W3C Immersive Web Working Group, is an API designed to empower developers in seamlessly creating immersive experiences across various XR devices directly in the browser. This capability opens virtual learning assistant applications to a broader audience. Notably, Meta Quest devices and the new Apple Vision Pro support WebXR technology, further expanding the reach and accessibility of these immersive experiences. OpenXR offers a uniform interface for system-independent development and glTF optimizes the transmission of 3D content. They enable the seamless integration of virtual experiences, 3D models and interactive content. These standards make learning in the digital space open, effective, and more engaging.

5.4 User Acceptance as Key Driver

In the area of AI-supported learning in the metaverse the assessment of usability and user experience is particularly difficult. Morales et al. did not find any evaluation studies on metaverse user experience in their literature review [28]. There are still many open questions regarding the use of AI in learning environments, especially in the metaverse: Do additional factors such as explainability play a role in the UX-perception of AI-supporting learning? How does the initial user's trust or distrust in the AI-support influence the "joy of use" and the willingness to use the product in general? How do the new

---

[21] Llama: https://ai.meta.com/llama/.

[22] Mistral: https://mistral.ai/.

[23] Mozilla DeepSpeech: https://github.com/mozilla/DeepSpeech/.

[24] OpenTTS: https://github.com/synesthesiam/opentts/.

[25] MozillaTTS: https://github.com/mozilla/TTS/.

[26] Mimic: https://mycroft.ai/mimic-3/.

[27] WebXR: https://immersiveweb.dev/.

[28] OpenXR: https://www.khronos.org/openxr/.

[29] Graphic Language Transmission Format: https://www.khronos.org/gltf/.

characteristics of the metaverse ("shared", "persistent" and "decentralized"), that make the difference between existing VR and AR applications and the metaverse, change the user's experience of this medium? How can general Usability/ UX- guidelines be adjusted to these new products? Some existing guidelines, as for example recommendations for front size, may differ for applications in VR.

In addition, the rapid development of hardware and software leads to poor comparability of usability between products and is accompanied by high demands on the systems. Despite the relevance of AI-supported learning applications and learning applications in VR, no design guidelines for interaction in such a system have yet been defined. However, there are initial efforts to expand the guidelines for AI systems in general[30]. Due to the novelty of the system, there is no experience in evaluating its usability and user experience. Nevertheless, to some extent, usability guidelines can be transferred from other application areas. For example, many of the design guidelines in DIN EN ISO 9241 apply to AI-supported learning applications as well as to VR applications. However, these guidelines may need to be re-examined for this specific application purpose and tailored to the needs and context of the relevant application.

Therefore, the authors of this paper suggest a hands-on approach where existing guidelines are being used in the design-process in a meaningful way wherever possible. For the evaluation of such a system we recommend using a mixed-method approach [29] with quantitative and qualitative methods. Depending on the application context, the evaluation should not only focus on usability aspects, but also on other factors such as (sustainable) learning outcomes, organizational benefits, and user's motivation.

For the use case of learning in the metaverse, stakeholders in this area must be given special consideration when researching the UX. Stakeholders in the learning environment might be, for example, teachers, learners, authors, technical support staff, etc., each of whom has their own specific requirements for the learning environment. In future, individual roles or at least aspects of individual roles may also be replaced by intelligent agents, as in our use case. The potential of such applications is huge. For example, it enables new ways of reflection, an important component of learning according to the principle of complete action. According to the principle of complete action, a good learning situation consists of the phases of informing or analyzing, planning, deciding, implementing, monitoring, or evaluating and reflecting[31]. While interacting with an intelligent agent in the metaverse, learners could obtain feedback on their work in a protected space before submitting it to the teacher. Especially when there is a shortage of specialists in the field of education, safe learning environments can be created in which learners can try things out and make mistakes to learn from them.

Didactic principles, such as the principle of complete action, must be considered from the outset to provide users with a good UX in the learning environment and to make learning effective. The guiding principle of "didactics first" should lead to the meaning and purpose of a learning environment being seen as a key factor in the decision for a particular learning environment and its development.

---

[30] AI: First New UI Paradigm in 60 Years: https://www.nngroup.com/articles/ai-paradigm/.

[31] School curriculum for vocational schools (German): https://schucu-bbs.nline.nibis.de/nibis.php?menid=171.

In the context of combining these systems with user-state-assessment [30] additional factors should be considered. Does the system compliment the user-state in a beneficial way? Barriers to the use of user-state-assessment can also be the acquisition costs of equipment or the feeling of being monitored. The measurement methods should therefore be easy to use and as unobtrusive as possible. The use of body values to adapt the learning environment to the user state is particularly useful when a special user state is required to learn how to act in a particularly challenging situation, such as landing an aircraft under stress or the actions of emergency services under fire or during a flood. However, too much realism should not be used at all costs - especially when it comes to simulating stressful or dangerous situations - as this can also lead to a lower acceptance or even blunting of real situations. The metaverse offers many new possibilities for creating learning situations that are dangerous or costly in real life. The user experience is crucial for learning success, especially in the above-mentioned critical contexts.

6   Summary

While high-quality training in real life is very important, virtual training offers different advantages, such as higher immersion with experiences not possible in real life, lower costs and needed equipment, better availability, easier visualization, or no exposure to dangers. This can be crucial while training for certain professions or the training equipment is so expensive the training can happen only once. AI-based outputs, such as those recently often generated by large language models, can relieve the burden on teachers and, thus, that LLM-based AI-Assistants at the same time respond more quickly to individual questions from students, especially when no experts are available.

We presented an end-to-end proof of concept based on currently very popular technologies such as ChatGPT and the Unreal Engine and used this example to demonstrate added value and challenges. We particularly advocate the use of the best-of-breed strategy [25] to allow individual functions to be developed as independently as possible by experts from the domains of education technologies, LLMs and metaverse applications. To this end, interoperable technologies, interfaces, and standards from the respective areas should be reused as much as possible.

In summary, the combination of education and technology is having a transformative impact. Chatbots and the metaverse offer innovative ways to personalize learning content, foster collaboration and enable deeper understanding. The integration of technology standards optimizes this experience and reshapes the educational landscape in a sustainable way. However, it is important to emphasize that technology serves as a tool, while the pedagogical approach and human interaction remain essential for learning.

References

1. Hwang, G.J., Chien, S.Y.: Definition, roles, and potential research issues of the metaverse in education: an artificial intelligence perspective. Comput. Educ. Artif. Intell. **3**, 100082 (2022)
2. Malekzadeh, M., Mustafa, M.B., Lahsasna, A.: A review of emotion regulation in intelligent tutoring systems. J. Educ. Technol. Soc. **18**(4), 435–445

3. Fang, Z., Cai, L., Wang, G.: Metahuman creator the starting point of the metaverse. In: 2021 International Symposium on Computer Technology and Information Science (ISCTIS), 2021, pp. 154–157 (2015)
4. Fast, K., Gifford, T., Yancey, R.: Virtual training for welding. In: Third IEEE and ACM International Symposium on Mixed and Augmented Reality, pp. 298–299. IEEE (2004)
5. Pantelidis, V.S.: Reasons to use virtual reality in education. VR in the Schools (1995)
6. Salzman, M.C., Dede, C., Loftin, R.B., Chen, J.: A model for understanding how virtual reality aids complex conceptual learning. Presence: Teleoperators Virtual Environ. **8**(3), 293–316 (1999)
7. Deng, J., Lin, Y.: The benefits and challenges of chatgpt: an overview. Front. Comput. Intell. Syst. **2**(2), 81–83 (2023)
8. Srinivasa, K.G., Kurni, M., Saritha, K.: Harnessing the Power of AI to Education. In: Learning, Teaching, and Assessment Methods for Contemporary Learners, pp. 311–342. LNCS. Springer Texts in Education. Springer, Singapore (2022). https://doi.org/10.1007/978-981-19-6734-4_13
9. Sok, S., Heng, K.: Chatgpt for education and research: a review of benefits and risks (2023)
10. Gurusamy, K.S., Aggarwal, R., Palanivelu, L., Davidson, B.R.: Virtual reality training for surgical trainees in laparoscopic surgery. Cochrane database of systematic reviews, 2009
11. Torkington, J., Smith, S.G., Rees, B.I., Darzi, A.: Skill transfer from virtual reality to a real laparoscopic task. Surg. Endosc. **15**(10), 1076–1079 (2001)
12. Altun, D., Schulz, D.: Learning to take the right turn–which learning media is best suited to learn a sequence of actions to solve a Rubik's cube? In: International Conference on Human-Computer Interaction, pp. 191–205, July 2023
13. Altun, D., et al.: Lessons learned from creating, implementing and evaluating assisted e-learning incorporating adaptivity, recommendations and learning analytics. In: Sottilare, R.A., Schwarz, J. (eds.) Adaptive Instructional Systems. HCII 2022. LNCS, vol. 13332, pp. 257–270. Springer, Cham (2022). https://doi.org/10.1007/978-3-031-05887-5_18
14. Brown, T., et al.: Language models are few-shot learners. Adv. Neural. Inf. Process. Syst. **33**, 1877–1901 (2020)
15. Prodinger, M., Stampfl, R., Deissl-O'Meara, M.: ChatGPT as a learning assistant in distance learning. In: The European Research Consortium for Informatics and Mathematics (ERCIM) News 136 - Special Theme: Large Language Models. ISSN 0926–4981
16. William. C.: Prompting change: exploring prompt engineering in large language model AI and its potential to transform education. TechTrends **68**(1), 47–57 (2024)
17. Heston, T.F., Khun, C.: Prompt engineering in medical education. Int. Med. Educ. **2**(3), 198–205 (2023)
18. Nielsen, J.: Usability inspection methods. In: Conference Companion on Human Factors in Computing Systems, pp. 413–414, April 1994
19. Hassenzahl, M., Law, E.L.C., Hvannberg, E.T.: User Experience-Towards a unified view. Ux Ws Nordichi **6**, 1–3 (2006)
20. Choi, Y.: A study on factors affecting the user experience of metaverse service. Int. J. Inf. Syst. Serv. Sect. (IJISSS) **14**(1), 1–17 (2022)
21. Delgado, C.B.: User experience (UX) in metaverse: realities and challenges. Metaverse Basic Appl. Res. **1**, 9 (2022)
22. Hadi, M.U., Al-Tashi, Q., Qureshi, R., Shah, A., et al.: Large Language Models: A Comprehensive Survey of Its Applications, Challenges, Limitations, and Future Prospects, 2023
23. Kandpal, N., Deng, H., Roberts, A., Wallace, E., Raffel, C.: Large language models struggle to learn long-tail knowledge. In: Proceedings of the 40th International Conference on Machine Learning, pp. 15696–707. PMLR (2023)

24. ITU-T, "Telemeeting assessment - Effect of delays on telemeeting quality," ITU-T Recommendation P.1305, July 2016
25. Krauss, C., et al.: Best-of-breed: service-oriented integration of artificial intelligence in interoperable educational ecosystems. In: Uden, L., Liberona, D. (eds.) Learning Technology for Education Challenges. LTEC 2023. CCIS, vol. 1830, pp. 267–283. Springer, Cham (2023). https://doi.org/10.1007/978-3-031-34754-2_22
26. Krauss, C., Hauswirth, M.: Interoperable education infrastructures: a middleware that brings together adaptive, social and virtual learning technologies. In: The European Research Consortium for Informatics and Mathematics (ERCIM) News 120 - Special Theme: Educational Technology, pp. 9–10. ISSN 0926-4981
27. Chen, H., et al.: ChatGPT's One-Year Anniversary: Are Open-Source Large Language Models Catching Up? arXiv, 2023
28. Morales, J., Cornide-Reyes, H., Rossel, P.O., Sáez, P., Silva-Aravena, F.: Virtual reality, augmented reality and metaverse: customer experience approach and user experience evaluation methods. literature review. In: Coman, A., Vasilache, S. (eds.) Social Computing and Social Media. HCII 2023. LNCS, vol. 14025, pp. 554–566. Springer, Cham (2023). https://doi.org/10.1007/978-3-031-35915-6_40
29. Rerhaye, L., Altun, D., Krauss, C., Müller, C.: Evaluation methods for an AI-supported learning management system: quantifying and qualifying added values for teaching and learning. In: Sottilare, R.A., Schwarz, J. (eds.) Adaptive Instructional Systems. Design and Evaluation. HCII 2021. LNCS, vol. 12792, pp. 394–411. Springer, Cham (2021). https://doi.org/10.1007/978-3-030-77857-6_28
30. Schwarz, J., Fuchs, S., Flemisch, F.: Towards a more holistic view on user state assessment in adaptive human-computer interaction. In: 2014 IEEE International Conference on Systems, Man, and Cybernetics (SMC), San Diego, CA, USA, pp. 1228–1234 (2014). https://doi.org/10.1109/SMC.2014.6974082

Revamping the RAMPAGE Adaptive Intelligence Analysis Framework in the Age of Generative AI

Ashley F. McDermott[1] , Elizabeth Whitaker[2] , and Sarah J. Stager[3]($\boxtimes$)

[1] Hilltown Engineering, Chesterfield, MA 01012, USA
[2] Georgia Tech Research Institute, Atlanta, GA 30332, USA
[3] College of Information Science and Technology, Pennsylvania State University, University Park, PA 16801, USA
profstager@psu.edu

Abstract. Generative AI (GenAI) has the capability to revolutionize even modern processes and framework implementations. In 2021 we presented the Reasoning about Multiple Paths and Alternatives to Generate Effective Forecasts (RAMPAGE) process framework to support hypothesis generation for counterfactual forecasting and intelligence analysis [1]. This framework provided a structure to organize and order analysis methods to maximize the number and quality of hypotheses generated to improve forecasts. Different instantiations of GenAI could be used to improve the results of each stage in this framework. Adding GenAI to implementations of the RAMPAGE framework would greatly expand the number and quality of hypotheses than analysts can generate, which would lead to better forecasts. In this paper, we present examples of how GenAI tools could be used to enhance the RAMPAGE framework.

Keywords: Generative AI · Intelligence Analysis · learning tools · hypothesis generation

1 Introduction

The best use of technology is to extend human potential. Advances in artificial intelligence, machine learning, and natural language processing have led to the creation of Generative AI (GenAI) tools and the integration of these tools into many fields including teaching and learning. While the role of technology in teaching and learning is not new, the integration of GenAI holds promise [2].

In a systematic review of GenAI in education published in 2023, Wang, Wang, and Su [3] analyzed 27 academic articles published between 2020 and 2023, using an "inductive grounded approach the coding revealed four technological affordances: accessibility, personalization, automation, and interactivity; and five challenges: academic integrity risk, response errors and bias, over-dependence risk, the widening digital divide, and privacy and security." The technological affordances inform our recommendations to support intelligence analysis with these tools. The challenge of academic integrity risk

R. A. Sottilare and J. Schwarz (Eds.): HCII 2024, LNCS 14727, pp. 239–249, 2024.
https://doi.org/10.1007/978-3-031-60609-0_17

and widening the digital divide are nominal issues for this work as we are using the tool to generate alternate hypotheses outside of an academic setting. Response errors and bias, over-dependence risk, and privacy and security present significant issues that need to be considered because of the nature of the context, extending the hypothesis for intelligence analysts. To combat these challenges, we encourage users to incorporate the integration of GenAI as an extension of their capability, not as a replacement for their not a replacement of their labor.

When used to support teaching and learning, GenAI tools can become thought partners to extend the process of inquiry. To apply this integration of GenAI this paper will extend the role of Intelligence analysis and the process of developing the skill related to generating alternative hypotheses using a framework developed in part by two of the authors of this paper (McDermott and Whitaker).

2 RAMPAGE Framework

The Reasoning About Multiple Paths and Alternatives to Generate Effective Forecasts (RAMPAGE) process was originally developed as part of the Forecasting Counterfactuals in Uncontrolled Settings (FOCUS) program run by IARPA. The RAMPAGE process was designed to be a flexible framework to support intelligence analysts, while being somewhat agnostic to the particular methods used. The key to the RAMPAGE process is hypothesis generation. In general, hypothesis generation is the foundation of all analytical methods. However, hypothesis generation is also greatly affected by human biases and cognitive limitations, including framing bias and confirmation bias [4–7]. These biases and limitations lead to neglect of key regions of the possible hypothesis space. The RAMPAGE process created an order for when to use which types of methods to maximize multi-path reasoning (MPR). Multi-path reasoning uses iterative, convergent broadening and narrowing hypothesis assessment through four cognitive processes: framing and contextualization, down-collect, conflict and corroboration, and hypothesis exploration [8]. The goal of MPR is for the hypotheses generated to relate to independent parts of the hypothesis space and to differentiate potentially valid alternative explanations rather than dismiss alternate hypotheses too early. The RAMPAGE process brings analysts through each of these four cognitive processes in discrete stages. The RAMPAGE framework is designed to support independent analysis and differentiation through selecting iterative broadening and narrowing methods. When using the RAMPAGE process, it is expected that each analyst works independently with minimal, strategic idea sharing. In the original framework, there are five stages, but the fifth stage was an artificial add-on based on the needs of the FOCUS program, so we only consider the first four stages in this paper.

The first stage is the Information Gathering and Evaluation stage, which is when initial data is collected and distributed to analysts. The ways in which information is presented have profound effects on how analysts will explore the hypothesis space. Controlling how much information an analyst can access at a given time will influence the kinds of hypotheses they generate. Order of presentation can help to overcome problems with anchoring bias by ensuring different analysts will anchor to different parts of the hypothesis space. Similar effects can be achieved by presenting analysts with all of the information but with different initial questions to answer. In this stage, a key piece is also to evaluate the quality of different evidence sources to establish which evidence should be given the most weight when generating hypotheses.

The second stage, The Multi-Path Generation stage, is where analysts create their initial formal hypotheses. When done in conjunction with the data being distributed into different conceptual bins, the analyst would receive a set of data, review and check the quality of the evidence, and generate hypotheses based on the available data before adding more data. During this stage, the goal is to maintain independence and build and evaluate hypotheses working on the subsets of data or specific questions. This stage involves both an initial generation method to maximize the number of independent hypotheses under consideration and an initial narrowing method to focus on the most likely or most informative hypotheses (i.e., best differentiate between parts of the hypothesis space).

The third stage is Problem Visualization. The goal of problem visualization is to assist the analyst in visualizing the hypothesis space, helping them to identify gaps in their knowledge or possibilities they had not considered. This visualization can take on many forms from creating a "mind map" picture of how information is related to creating a comparative matrix of the leading hypotheses. This stage can be treated as it's own independent stage or a process used throughout the other stages to support broader analysis.

The fourth stage, the Multi-Path Reasoning stage, is where analysts broaden their outlook by considering a range of different perspectives on the analysis. These perspectives could include different roles within the scenario, such as a health official versus a head of state, or roles external to the scenario, such as how would a game developer imagine this scenario into a game. The change in perspective could also involve new ways of evaluating different hypotheses, such as trying to identify malleable factors and the effects of changing those factors or assuming that the best supported hypotheses are catastrophically wrong and identifying what alternative would offer answers that avoided this catastrophic interpretation.

3 Generative AI Enhancements

Generative AI programs, such as Chat GPT and Bing's Copilot, are powerful tools to supplement human knowledge. While GenAI programs have several limitations, including a lack of actual cognitive processing to check their responses, the tool can quickly consolidate and organize information, develop examples, and visualize information. Here we consider how existing GenAI tools could be used to support intelligence analysis using the RAMPAGE framework. We used both Chat GPT and Bing's Copilot as examples to illustrate how the tools can be used.

3.1 Stage 1: Information Evaluation Enhancements

The key to stage 1 is to evaluate the quality and reliability of available evidence. GenAI can be used in this stage to test assumptions, identify potential cognitive biases, and provide information about important considerations for evaluating evidence. GenAI can also optimize the order of presentation to address problems with anchoring bias. GenAI can also suggest different anchors to parts of the hypothesis space.

Below is an example of using Chat GPT to identify important considerations for evaluating evidence. The list could be used as what would essentially be a checklist of considerations for each piece of evidence. Given specific types of evidence, Chat GPT could even create a list more specifically relevant to that particular type of evidence. Analysts could also ask questions about existing evidence or to get a quick understanding of a topic or area of expertise that is unfamiliar to them.

Prompt: What are important factors to consider when evaluating the available information in the Undercover Infiltration scenario?

ChatGPT:

When evaluating available information in the Undercover Infiltration scenario, several important factors should be considered. These factors help ensure the accuracy, reliability, and relevance of the gathered intelligence. Here are key considerations:

Source Credibility:

Assess the credibility of the undercover agent as a source. Consider the agent's track record, past performance, and reliability in providing accurate information. Evaluate their ability to maintain cover and avoid detection.

Consistency and Corroboration:

Look for consistency and corroboration within the information provided by the undercover agent. Cross-reference details with other sources, both human and technical, to validate the accuracy of the intelligence. Consistency across different pieces of information enhances its reliability.

Contextual Understanding:

Gain a deep understanding of the context in which the information is collected. Consider the cultural, social, and organizational dynamics of the infiltrated group. Understanding the context helps interpret the significance of the gathered intelligence accurately.

Operational Security (OPSEC):

Evaluate the operational security measures implemented by the undercover agent. Assess the risk of exposure and potential counterintelligence threats. Adequate OPSEC ensures the safety of the agent and the integrity of the operation.

Motivations and Biases:
Consider the motivations and potential biases of the undercover agent. Assess how personal beliefs or incentives might influence the information provided. Understanding the agent's perspective helps in interpreting the data more effectively.

Timeliness of Information:
Evaluate the timeliness of the gathered intelligence. Assess how recent the information is and whether it aligns with the operational timeline. Timely information is often more valuable for making informed decisions.

Reliability of Communication Channels:
Assess the reliability and security of the communication channels used by the undercover agent. Consider potential risks of interception or compromise. Secure and reliable communication is essential for maintaining the confidentiality of the intelligence.

Potential Deception:
Be vigilant for signs of potential deception or misinformation. Consider the possibility that the infiltrated group may be aware of the undercover operation and intentionally providing false information. Critical analysis helps identify deceptive tactics.

Impact on Operations:
Evaluate the potential impact of the gathered intelligence on ongoing and future operations. Consider how the information aligns with strategic objectives and whether it contributes to the overall success of the mission.

Legal and Ethical Considerations:
Ensure that the methods used by the undercover agent adhere to legal and ethical standards. Evaluate whether the information was obtained through lawful means and whether the operation complies with relevant regulations.

By carefully considering these factors, intelligence analysts can assess the quality and reliability of information gathered in the Undercover Infiltration scenario. This comprehensive evaluation contributes to informed decision-making and effective intelligence analysis.

3.2 Stage 2: Multi-path Generation Enhancements

The goal of stage 2 is to identify as many hypotheses as possible. GenAI can help to develop a large number of hypotheses quickly. For instance, ChatGPT can be directly prompted to provide hypotheses for a given scenario or for a very specific aspect of a scenario or piece of evidence (see example below). While these hypotheses are likely to be the more common or obvious hypotheses, this easily collates the hypotheses the

GenAI can create from available knowledge. Analysts can build on these hypotheses to develop more creative hypotheses with the initial hypotheses already identified for them.

Prompt: Please give me 10 hypotheses about what could happen to the Undercover Agent if the target organization knows that someone has infiltrated their organization

ChatGPT:

Certainly, here are ten hypotheses about potential outcomes for the Undercover Agent if the target organization becomes aware of the infiltration:

Exposure and Capture:

The undercover agent may face the risk of exposure, leading to capture by the target organization. This could result in the agent being detained, interrogated, or subjected to harm.

Counterintelligence Measures:

The target organization might implement counterintelligence measures upon discovering the infiltration, making it more challenging for the agent to operate covertly. Increased scrutiny and surveillance could compromise the agent's cover.

Compromised Mission Objectives:

The organization, upon learning of the infiltration, may change its tactics, operations, or locations to counteract the agent's activities. This could lead to a compromised mission with diminished chances of achieving the original objectives.

Informant Turned:

The target organization might attempt to turn the undercover agent into an informant, coercing or convincing them to provide false or misleading information to their handlers.

Internal Investigations:

The organization may launch internal investigations to identify other potential infiltrators or sources of information leaks. This could expose other undercover agents or compromise the entire intelligence network.

Increased Security Measures:
The target organization may tighten security measures in response to the infiltration, making it more difficult for the agent to gather information or communicate with handlers.

Attempted Recruitment:
The organization might attempt to recruit the undercover agent, exploiting their insider knowledge for their own purposes. This could involve coercion, blackmail, or other methods to gain the agent as an asset.

Retaliation Against Associates:
The target organization may retaliate against individuals believed to be associated with the undercover agent. This could include harm or threats against family members, friends, or colleagues.

Coordinated Disinformation Campaign:
The organization may launch a disinformation campaign to mislead intelligence agencies and disrupt their ability to distinguish between accurate and false information provided by the undercover agent.

Attempted Extraction or Escape:
The undercover agent might need to execute an emergency extraction or escape plan to evade capture or harm once the infiltration is exposed. This could involve a carefully planned extraction operation by the intelligence agency.

These hypotheses highlight the various potential risks and consequences that an undercover agent may face if the target organization becomes aware of their presence. It emphasizes the dynamic and high-stakes nature of undercover operations and the need for thorough risk assessment and contingency planning.

GenAI tools can also prompt analysts with additional questions to ask. For instance, Bing's Copilot, which is built on ChatGPT, provides 3 possible next inquiries after each prompt response. This can be a double-edged sword, as the GenAI could lead an analyst down a particular way of thinking, preventing them from developing other hypotheses, but it can also guide them to questions they may not have otherwise asked. The power of using GenAI for this is that it is easy to go back and explore multiple paths, avoiding narrowing of the hypotheses under consideration too soon.

3.3 Stage 3: Problem Visualization Enhancements

The goal of stage 3 is to visualize the information. In previous instances of the RAMPAGE framework, this has meant either creating a mind map or creating comparison matrices. Ultimately, this stage is about changing how an analyst perceives the problem by representing it in a visual format. This could take a multitude of forms, including pictures or vignettes of particular events, mind maps, or Kanban-style boards grouping evidence with hypotheses. Tools like Bing's Copilot supply relevant images when requesting intelligence analysis scenarios, for example. Tools like Taskade can either use existing text or generate text based on prompts and then turn the text into a variety of visualizations, including a mind map, task board, or even a calendar. Below are examples of the mind map and task board for the steps to learn to ride a bike (Figs. 1 and 2).

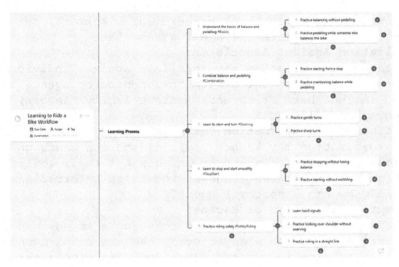

Fig. 1. Example of a mind map from Taskade

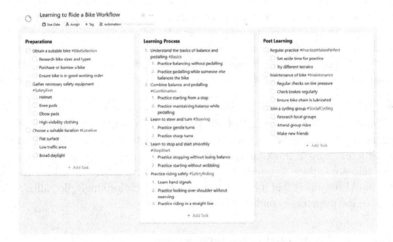

Fig. 2. Example of a task board from Taskade

3.4   Stage 4: Multi-Path Reasoning Enhancements

The goal of stage 4 is to again broaden the hypothesis space by incorporating differ-ent perspectives and considerations. As shown with the following prompt, ChatGPT can provide multiple relevant perspectives for a given scenario. It can also develop counterfactuals or assist in a pre-mortem analysis.

Prompt: Please elaborate on the Undercover Infiltration from 3 different perspectives

ChatGPT:

Certainly, let's elaborate on the Undercover Infiltration scenario from three different perspectives:

Undercover Agent's Perspective:

From the perspective of the undercover agent, successful infiltration involves building and maintaining trust within the target organization. The agent must navigate a delicate balance between blending in and gathering criti-cal information without compromising their true identity. They face challenges such as maintaining cover under scrutiny, adapting to the organization's culture, and es-tablishing relationships with key figures. The agent's perspective involves assessing risks, managing stress, and making quick decisions to ensure the success of the mission while safeguarding personal safety.

Operational Commander's Perspective:

The operational commander overseeing the undercover oper-ation has a strategic viewpoint. They must evaluate the progress of the undercover agent in gathering valuable intelligence, assess the risks posed to the agent's safety, and consider the overall impact on the organiza-tion's objectives. The commander must weigh the benefits of continued infiltration against potential risks, such as exposure, compromise of the operation, or harm to the undercover agent. Strategic decisions involve considering the long-term goals of the operation and balancing the need for information with the safety of the operative.

Intelligence Analyst's Perspective:

The intelligence analyst is responsible for assessing the information provided by the undercover agent and provid-ing ongoing analysis to support decision-making. From this perspective, the analyst must evaluate the credibil-ity of the gathered intelligence, identify patterns or trends within the organization, and assess the potential impact on national security. The analyst's role involves continuous monitoring, validation of information, and adapting analytical approaches to evolving circumstances.

They contribute valuable insights to the operational team and help shape the overall strategy based on the intelligence gathered.

In summary, the Undercover Infiltration scenario presents challenges from the perspectives of the undercover agent, the operational commander, and the intelligence analyst. Each perspective involves distinct challenges and considerations, ranging from the agent's on-the-ground experiences to strategic decision-making and analytical assessments. The successful outcome of the scenario relies on effective collaboration and communication among these key roles.

4 Conclusion

GenAI tools can provide supports for intelligence analysis that enable analysts to think more broadly.

When providing guidance or creating hypotheses in the early stages of analysis in the RAMPAGE framework, GenAI can do the easy work, providing reminders of considerations when evaluating evidence or generating the most obvious hypotheses for a given scenario. This frees the analyst to focus on the tasks that require greater creativity or more careful and specific analysis that GenAI cannot use big data to complete.

GenAI also has a particular strength in providing different ways to understand or see the same information. ChatGPT was able to literally provide multiple perspectives that may be relevant to a given scenario. It can also be used to explore counterfactuals and assist in multi-path reasoning activities, such as pre-mortem analysis. Even using different GenAI tools leads to very differently formatted answers. For example, ChatGPT provided answers in lists and bullet points based on our prompts, while Bing's Copilot would create summarizing paragraphs with sources cited. Copilot also has the option to control the type or tone of answers, selecting between more creative, more precise, or balanced between the two (the default setting). GenAI visualization tools, such as Taskade, provide novel visualizations that do not require analysts to have any visual art talents or familiarity with visualization software. These visualizations can highlight factors and relationships between pieces of evidence that might not otherwise be considered.

As a note of caution, currently available GenAI tools may not be entirely suitable for intelligence analysis, given their open availability and access to only generally available information. However, by highlighting examples of how GenAI could be incorporated into an intelligence analysis process, these lessons could be used to develop GenAI systems with similar capabilities that meet the needs of the intelligence and defense communities. They can be used as tools to develop intelligence analysis training and assist students in practicing analysis skills.

Disclosure of Interests. The authors have no competing interests to declare that are relevant to the content of this article.

References

1. McDermott, A.F., et al.: Developing an adaptive framework to support intelligence analysis. In: Sottilare, R.A., Schwarz, J. (eds.) Adaptive Instructional Systems. Design and Evaluation. HCII 2021. LNCS, vol. 12792, pp. 550–558. Springer, Cham (2021). https://doi.org/10.1007/978-3-030-77857-6_39
2. Kim, M., Adlof, L.: Adapting to the future: ChatGPT as a means for supporting constructivist learning environments. TechTrends **68**(1), 37–46 (2024)
3. Wang, N., Wang, X., Su, Y.S.: Critical analysis of the technological affordances, challenges and future directions of Generative AI in education: a systematic review. Asia Pac. J. Educ. (2024)
4. Asare, S.K., Wright, A.M.: A note on the interdependence between hypothesis generation and information search in conducting analytical procedures. Contemp. Account. Res. **20**, 235–251 (2003)
5. Bailey, C.D., Daily, C.M., Phillips, T.J., Jr.: Auditors' levels of dispositional need for closure and effects on hypothesis generation and confidence. Behav. Res. Account. **23**, 27–50 (2011)
6. Sprenger, A., Dougherty, M.R.: Generating and evaluating options for decision making: the impact of sequentially presented evidence. J. Exp. Psychol. Learn. Mem. Cogn. **38**, 550 (2012)
7. Thomas, R., Dougherty, M.R., Buttaccio, D.R.: Memory constraints on hypothesis generation and decision making. Curr. Dir. Psychol. Sci. **23**, 264–270 (2014)
8. Sprenger, A.M., et al.: Implications of cognitive load for hypothesis generation and probability judgment. Front. Psychol. **2**, 129 (2011)

Integrating ChatGPT into the ELEVATE-XR Adaptive Instructional Framework

Ashley F. McDermott[1] [ID] and Sarah J. Stager[2]([✉]) [ID]

[1] Hilltown Engineering, Chesterfield, MA 01012, USA
[2] The College of Information Science and Technology, Pennsylvania State University, University Park, PA 16801, USA
profstager@psu.edu

Abstract. With the fast and radical transformation of how both students and instructors can access information brought on by the development of generative AI, it is critical to develop best practices around how generative AI is used to support instruction and learning rather than being a detriment. In this paper, we examine how generative AI could be integrated into the Exercisable Learning-theory and EVidence-based Andragogy for Training Effectiveness (ELEVATE) framework developed by Stanney, Skinner, and Hughes [1] to inform adaptive approaches to instruction. Originally developed for designing eXtended Reality (XR) training experiences, the ELEVATE framework incorporates learning theories from behaviorism, cognitivism, and constructivism into a cohesive framework based on the Dreyfus and Dreyfus [2] skill acquisition model and Bloom's Revised Taxonomy [3]. The ELEVATE framework offers guidance for developing appropriate expectations and forms of instruction for students at 5 proficiency levels: novice, advanced beginner, competent, proficient, expert. The ELEVATE framework identifies language for appropriate learning objectives and types of learning activities that would be appropriate for students of different proficiency levels.

Keywords: Generative AI · adaptive learning · adaptive instruction

1 Introduction

The COVID-19 pandemic brought the quality of teaching and learning into focus and increased demand, both immediately to help with emergency remote teaching and longer-term improvement in pedagogical and andragogical practices [4]. This has placed a greater demand on instructional designers at learning and teaching centers at colleges and universities, as well as on individual instructors as they develop new courses or improve existing ones. Many instructors have never received training in instructional design and are unfamiliar with the process an instructional designer uses to help develop a course. Instructional designers are, of course, not experts in the material being taught in each course they help develop. Both instructional designers and course instructors need tools that can reduce the time it takes to develop a course, while maintaining or improving the quality of those courses.

R. A. Sottilare and J. Schwarz (Eds.): HCII 2024, LNCS 14727, pp. 250–260, 2024.
https://doi.org/10.1007/978-3-031-60609-0_18

Generative AI (GenAI), including ChatGPT, is having a significant impact on education, revolutionizing the way teaching and learning take place. ChatGPT specifically can generate coherent and contextually relevant responses, which makes it a valuable tool for instructional designers and course instructors seeking to engage with students in a more dynamic and interactive manner. ChatGPT can provide instant explanations, answer questions, and offer personalized feedback, enhancing the learning experience for students. Moreover, it serves as a versatile resource for instructors, helping them create interactive lessons, generate educational content, and offer additional support outside the traditional classroom setting. As ChatGPT continues to evolve, it has the potential to foster a more inclusive and adaptive learning environment, catering to individual needs and promoting a collaborative approach to education.

Tools like Chat GPT can be used to extend the relationship between the teaching and the learning by articulating the scaffolding of ready relevant instruction. For this paper, we are focusing on supporting and facilitating human thought and work in the context of designing instructional resources. While GenAI, such as ChatGPT is useful for automation, this paper will demonstrate a path for enhancing the connection between developing instructional resources by integrating specific language compiled by Stanney et al. in the ELEVATE Framework [1].

2   ELEVATE Framework

The ELEVATE framework incorporates learning theories from behaviorism, cognitivism, and constructivism into a cohesive framework based on the Dreyfus and Dreyfus [2] skill acquisition model and Bloom's Revised Taxonomy [3]. The ELEVATE framework offers guidance for developing appropriate expectations and forms of instruction for students at 5 proficiency levels: novice, advanced beginner, competent, proficient, expert. The ELEVATE framework identifies language for developing/applying appropriate learning objectives and types of learning activities that would be appropriate for students of different proficiency levels.

3   ChatGPT

The generative AI, ChatGPT, is a web-based tool with a usable interface including a task bar to input a prompt, the option to generate alternate responses, copy generated responses, and like/dislike the generated text, all with clearly identified icons. ChatGPT is a large language model (LLM) that uses vast amounts of available data to process language and respond to requests or prompts from users. ChatGPT also introduces itself:

Prompt: What is chatgpt and how can it be integrated in higher education to design instruction, specifically developing learning objectives?

Response:

ChatGPT is a language model developed by OpenAI, based on the GPT (Generative Pre-trained Transformer) architecture. It is designed to generate human-like text based on the input it receives. The model has been trained on a diverse range of internet text, allowing it to understand and generate coherent responses across various topics.

Integrating ChatGPT into higher education for instructional design, especially in developing learning objectives, involves leveraging its natural language understanding and generation capabilities. Here are some ways it could be used:

1. **Generating Learning Objectives:**
   - ChatGPT can assist educators in generating clear and concise learning objectives for their courses. Educators can provide information about the subject, target audience, and desired outcomes, and ChatGPT can suggest well-phrased learning objectives.

2. **Providing Examples and Context:**
   - When developing learning objectives, providing concrete examples and contextual information is crucial. ChatGPT can help in generating illustrative examples and contextual explanations to enhance the understanding of learning objectives.

3. **Reviewing and Refining Objectives:**
   - Educators can use ChatGPT as a tool for reviewing and refining learning objectives. The model can offer suggestions, identify potential issues, and help in ensuring that the objectives are measurable, specific, and aligned with educational goals.

4. **Creating Instructional Materials:**
   - ChatGPT can assist in generating content for instructional materials, such as course outlines, lesson plans, and supplementary materials. This can save educators time and provide additional resources to support the learning objectives.

5. **Answering Student Queries:**
- In an online learning environment, ChatGPT can be integrated to answer students' questions related to the learning objectives. It can provide additional explanations, clarify concepts, or direct students to relevant resources.

6. **Feedback and Assessment:**
- ChatGPT can assist in developing feedback for assessments and assignments. It can provide model answers, suggest improvement areas, and offer constructive feedback aligned with the learning objectives.

When integrating ChatGPT or similar language models into higher education, it's important to consider the following:

- **Ethical Use:** Ensure that the use of AI models aligns with ethical guidelines and respects privacy and data security.

- **Human Oversight:** While AI can be a valuable tool, it should not replace human oversight. Educators should review and adapt the suggestions provided by the model to ensure they align with educational goals and standards.

- **Iterative Process:** Instructional design is often an iterative process. Educators should view ChatGPT as a tool that aids in the process, allowing for refinement and improvement over time.

- **Explainability:** Ensure that learning objectives and content generated by the model are easily understandable by students.

Overall, ChatGPT can be a valuable assistant in the instructional design process, providing support in generating, refining, and implementing learning objectives in higher education.

While a number of articles discuss the extensive problems presented by GenAI, including ChatGPT, in education some positive integrations have been identified. Kasneci, E. et al. [5] identify the benefit large language models can have in "assisting teachers in the creation of [inclusive] lesson plans and activities... Generate questions and prompts that encourage the participation of people at different knowledge and ability levels, and elicit critical thinking and problem-solving, ... generate targeted and personalized practice problems and quizzes, which can help to ensure that students are mastering the material" (p.3).

In a systematic review, Wang, Wang and Su [6] reported a critical analysis of technological affordances, challenges, and future directions of GenAI. Extending the evaluation of the technological affordances of chatbots by Huang et al. [7], Wang, Wang and Su [6] conducted a content analysis of 27 articles, revealing four categories of technological affordances of GenAI. These include accessibility, personalization, automation, and interactivity. Of the four areas, personalization and interactivity are two areas that require specific language to generate the appropriate prompt. This paper addresses the gap of language to include for better fit for personalization and interactivity.

4   Incorporating ChatGPT in Instructional Design

The role of technology in teaching and learning continues to evolve. While learning objectives may exist in a course, the state of the knowledge, skills, and abilities (KSA) of individual students is largely unknown. This may be due to exposure, practice, time between exposure and practice, memory, cognitive load, injury, or even the language used to sequence mapped content. Given well informed language, Generative AI can be integrated as a tool to provide scaffolding for learners, enabling more individualized/adaptive instruction.

4.1   Role of ChatGPT

GenAI LLMs are particularly good at tasks that require access to large amounts of information and can be used to assist instructors in their course design. ChatGPT can be an excellent source for developing or improving learning objectives, instructional activities, and even assessment resources. For example, ChatGPT can write learning objectives clearly, since it is able to pull from many examples of learning objectives.

However, ChatGPT certainly cannot replace a human instructor in course design. The role of the instructor comes into play when reviewing the content of these learning objectives or other GenAI responses. LLMs are great at using language but do not have any inherent ability to apply logic. Instructors need to make sure any responses from a GenAI program make sense and are accurate. GenAI should be treated as a tool to support instructors.

4.2   Prompt Engineering

Appropriate prompt engineering is the key to unlocking the power of GenAI tools like ChatGPT. Prompts are the requests you make to the chatbot, and how well formulated they are will determine the usefulness of the answers. Bad prompts have a few characteristics [8]. The first one is assuming the GenAI is capable of completing any intelligent activity. GenAI tools are built on language models, which means they know how to create coherent linguistic responses. They do not have mathematical or logical reasoning capabilities, so creating mathematical prompts or logical puzzles are unlikely to produce accurate responses. The second is providing vague prompts that are either too big of a question to answer (such as "What is the meaning of life?") or too specific (such as "What are the exact details of OpenAI's business plan?"). The third is providing misleading prompts or prompts that are actually asking multiple questions at the same time (such as "Please explain the principles of generative AI, describe its history, and discuss its applications in business and education").

According to Microsoft, a prompt has five components:

These components include instructions, primary content, examples, cues, and supporting content. Instructions guide the model on what to do, primary content is the main text being processed, examples demonstrate desired behavior, cues act as jumpstarts for the output, and supporting content provides additional information to influence the output [9].

The first and most obvious component is the instructions for the GenAI program. These instructions must be clear and specific, minimizing room for different interpretations [9, 10]. It is important to remember that these are language models, so using analogies and descriptive language improve the models ability to understand the instructions, unlike more traditional search or chatbots. Instructions can be supplemented with cues and examples to indicate where the model should start and what form the response should take. It's also important to include with the discussion, a way for the GenAI to respond if it does not find a satisfactory response. Adding this to the instructions helps to reduce the chances of getting false or nonsensical answers, although it does not eliminate that possibility [9].

The primary content is the content that you want answers about, such as learning objectives for learning a specific topic. In this case, creating learning objectives is the instruction and the primary content is whatever the topic is. This content can be expounded upon using supporting content, such as the important concepts within the specified topic. Just like with humans, these language models can interpret the same requests differently, depending on the wording or order of the content and instructions. Given this attribute, it can be beneficial to make the same request multiple times with different wording and either use the response that best captures what you desired or pull from multiple responses. Here are two generic prompts for developing learning objectives for a course using the ELEVATE [1] framework:

Generic prompt for an existing course:

```
Develop learning objectives to differentiate teaching and
learning using the following learning objectives [insert
current LO] that address [LO terms by Proficiency Level]
for [specific population].
```

Generic prompt for a new course:

```
Develop learning objectives for [identify topic] that
differentiate teaching and learning that address [LO
terms by Proficiency Level] for [specific population].
```

These two prompts are examples of "zero-shot prompting" that does not include supporting information or cuing. A "few-shot prompting" would add examples of learning objectives. On their own, both zero-shot and few-shot prompting have very limited application [11]. The LLM GenAI programs are designed to be conversational, and this should guide your approach to interacting with them. One key aspect of this is that complex requests should be broken down into smaller requests, but those requests can be in context, so the GenAI can understand that the multiple requests are related [10]. The prompts provided above would be most useful within a larger context, using "chain-of-thought" reasoning, where you walk through chains of reasoning [11]. In this example, the prompts would be part of a conversation that already established what the population of interest knows and what the proficiency level means.

4.3 Developing ELEVATE Learning Objectives

Creating prompts that include the affordances curated by the ELEVATE framework will assist designers of learning to vary teaching and learning with the support of GenAI. Using this technique to identify language and craft appropriate prompts to create learning objectives and activities for every proficiency level is a game changer. It can be difficult and time consuming to design courses or learning activities for students at different proficiency levels. This can lead to "one size fits all" design or developing the course for only the lowest proficiency levels. While this can be a valid approach for introductory courses, all courses will include students with different KSAs and different interests or reasons for taking the course. Using the ELEVATE framework to understand the possible proficiency levels and how to design learning activities for them and combining this with GenAI's power to produce large numbers of examples, can bring differentiated, adaptive teaching approaches into reach.

The table below includes other criteria provided by ELEVATE that can be used to develop prompts that may improve the opportunity to efficiently and effectively improve connections between students and the desired content (Table 1).

Table 1. Terms for Prompt Engineering. (Extracted From Table 3 ELEVATE-XR Stanney et al., 2023)

	Learning Objectives	Learning Activities
Novice	Remember: Facts, Concepts, Procedures	Define, Describe, Identify, Arrange, Label, List, Match, Name, Recall
Advanced Beginner	Understand: Facts, Concepts, Procedures	Compare, Classify, Differentiate, Imitate, Explain, Translate, Interpret
Competent	Apply: Facts, Concepts, Procedures	Demonstrate, Employ, Examine, Execute, Illustrate, Manipulate, Use
Proficient	Analyze: Facts, Concepts, Procedures	Deconstruct, Differentiate, Deduce, Organize, Relate, Structure
Expert	Evaluate/Create: Facts, Concepts, Procedures	Assess, Contrast, Critique, Invent, Hypothesize, Validate

Below is an example of a prompt to generate differentiated learning objectives for different levels of proficiency and the response.

Prompt:

Write versions of the learning objective "Design and develop a mobile application to provide users with some usable and useful capability." using

•Novices are focused on absorbing declarative knowledge, but they lack schemas and thus are inclined to extraneous processing. As a result, they tend to find it difficult to determine the essential parts of learning material. Novices thus learn best with focused, guided (e.g., step-by-step instruction) scenarios that support following rules, discriminating relevant situational features, prioritizing important information, and organizing knowledge.

•Advanced beginners start to understand the context of a situation, can quickly access the particular rules that are relevant to a specific task and/or context, and formulate organizing principles into maxims. However, they still lack an understanding of how to filter incoming information based on relevance and thus can become easily overwhelmed. Those at the advanced beginner stage thus learn best with increasingly complex scenarios that require application of rules and choosing a course of action, connecting new knowledge with existing knowledge, integration of extraneous information, provision of specific and targeted feedback, and techniques to manage stress.

•At the third stage, the competent individual no longer struggles with basic rules, is able to develop intuition to guide their decision-making, and has the ability to devise personalized rules to formulate plans. Learning at the competent stage can be facilitated by the provision of authentic, complex scenarios and inverse problems (i.e., those that require calculating from a set of observations and causal factors that produced them) that foster self- produced them) that foster self-reflection, consideration of "why" decisions are made, and autonomy but with supportive feedback when needed.

•At the fourth level, the proficient individual can see the big picture, know which cues and information to focus on, and have an intuitive sense of the goal given the situation. Learning can be facilitated at this stage through provision of complex, unique scenarios that foster goal-setting, solving problems in novel and imagina-

tive ways, managing multiple distractions and emotional stimuli, self-reflection, and use of intuition.
•The expert no longer needs rules and can work intuitively, knowing what's important and what's not, what to do, and what the expected outcome is in any given situation. Learning at the expert stage can be facilitated by providing scenarios that deal with uncertainty during decision-making and challenge one's understanding, reflecting on a situation while engaging metacognitive knowledge and regulatory control processes that foster the discovery of new knowledge and sharing knowledge with others

ChatGPT:

Novice Level:
Design and develop a mobile application aimed at novices, emphasizing the absorption of declarative knowledge through focused, guided scenarios. Implement step-by-step instructions, support rule-following, and facilitate the discrimination of relevant situational features. Prioritize the organization of knowledge and the identification of essential learning material. Provide novices with a learning environment that helps them understand and follow rules effectively.

Advanced Beginner Level:
Create a mobile application tailored for advanced beginners, focusing on scenarios that enhance their understanding of contextual situations. Develop features that allow quick access to relevant rules and encourage the formulation of organizing principles into maxims. Help users at this level filter incoming information based on relevance, manage overwhelming situations, and apply rules to choose appropriate courses of action. Provide targeted feedback, connect new knowledge with existing knowledge, and introduce techniques for stress management.

Competent Level:
Design a mobile application for competent individuals, acknowledging their proficiency in basic rules. Foster the development of intuition for decision-making and the ability to devise personalized rules and plans. Facilitate learning through authentic, complex scenarios and inverse problems that promote self-reflection, considera-

tion of "why" decisions are made, and autonomy. Offer
supportive feedback when needed, creating an environment
that challenges learners while respecting their competence.

Proficient Level:
Develop a mobile application for proficient individuals
who can see the big picture and have an intuitive sense
of goals in various situations. Provide complex, unique
scenarios that foster goal-setting, creative problem-
solving, and the ability to manage multiple distractions
and emotional stimuli. Encourage self-reflection and the
use of intuition to make informed decisions. Create a
learning environment that challenges users at the proficient level and stimulates novel and imaginative thinking.

Expert Level:
Create a mobile application tailored for experts who no
longer rely on rules and can work intuitively. Design
scenarios that deal with uncertainty during decision-making, challenging their understanding and reflecting on
situations. Engage metacognitive knowledge and regulatory
control processes to foster the discovery of new
knowledge. Encourage the sharing of expertise with others
through collaborative features. Provide an environment
that respects the expertise of users at the expert level
and stimulates continuous learning and knowledge sharing.

ChatGPT and other GenAI programs can be used to generate large numbers of
examples and suggestions for learning objectives and learning activities, a task that can
be quite time consuming for an instructor. Now, the instructor can choose from the
responses provided by ChatGPT and bring their expertise to bear in selecting which
activities best align with the class and the learning goals, even altering the suggestions
to better suit the class without having to come up with activities.

5   Conclusion

This paper provides an example of how to use GenAI tools to support human efforts to
develop effective learning experiences. Rather than treating GenAI as a tool for automation or for replacing human effort, this paper aimed to demonstrate that GenAI can be
used to enhance and extend human capabilities, increasing the effectiveness of course
authors. Incorporating the terms identified by the ELEVATE framework into specific
prompts in ChatGPT enables course authors to extend the opportunity for learning to

more students with greater ease. The ELEVATE framework from Stanney and colleagues [1] neatly combines several learning theories into an actionable tool to develop ready, relevant learning. It provides guidance for understanding students with different levels of experience and how to develop learning objectives and activities that are appropriate for each experience level. ChatGPT brings the power of LLM GenAI to produce many examples of such learning objectives and activities. By combining their power, course authors can pick from many options the learning objectives and activities that best fit their course. Rather than creating a course or activity from scratch or based on a few results from an internet search or polling colleagues, the course author can essentially choose from a menu of options generated according to their specifications.

References

1. Stanney, K.M., Skinner, A., Hughes, C.: Exercisable learning-theory and evidence-based andragogy for training effectiveness using XR (ELEVATE-XR): elevating the ROI of immersive technologies. Int. J. Hum.-Comput. Interact. **39**(11), 2177–2198 (2023)
2. Dreyfus, S.E., Dreyfus, H.: A five-stage model of the mental activities involved in directed skill acquisition. Distribution **22** (1980)
3. Anderson, L.W., et al.: A Taxonomy for Learning, Teaching, and Assessing: A Revision of Bloom's Taxonomy of Educational Objectives. Addison Wesley Longman, Boston (2001)
4. Schlesselman, L.S.: Perspective from a teaching and learning center during emergency remote teaching. Am. J. Pharm. Educ. **84**(8), ajpe8142 (2020)
5. Kasneci, E., et al.: ChatGPT for good? On opportunities and challenges of large language models for education. Learn. Individ. Differences **103**, 102274 (2023)
6. Wang, N., Wang, X., Su, Y.S.: Critical analysis of the technological affordances, challenges and future directions of Generative AI in education: a systematic review. Asia Pac. J. Educ. (2024)
7. Huang, W., Hew, K.F., Fryer, L.K.: Chatbots for language learning—are they really useful? A systematic review of chatbot-supported language learning. J. Comput. Assist. Learn. **38**(1), 237–257 (2021)
8. Heston, T.F., Khun, C.: Prompt engineering in medical education. Int. Med. Educ. **2**(3), Article 3 (2023)
9. Tips to Become a Better Prompt Engineer for Generative AI. (n.d.). TECHCOMMUNITY.MICROSOFT.COM. https://techcommunity.microsoft.com/t5/ai-azure-ai-services-blog/15-tips-to-become-a-better-prompt-engineer-for-generative-ai/ba-p/3882935. Accessed 2 Feb 2024
10. Prompt Engineering for Generative AI | Machine Learning. (n.d.). Google for Developers. https://developers.google.com/machine-learning/resources/prompt-eng. Accessed 2 Feb 2024
11. What is prompt engineering? | IBM. (n.d.). https://www.ibm.com/topics/prompt-engineering. Accessed 2 Feb 2024

Bootstrapping Assessments for Team Simulations: Transfer Learning Between First-Person-Shooter Game Maps

Benjamin D. Nye[✉][iD], Mark G. Core[iD], Sai V. R. Chereddy, Vivian Young, and Daniel Auerbach

Institute for Creative Technologies, University of Southern California, 12015 Waterfront Drive, Playa Vista, CA 90094, USA
{nye,core}@ict.usc.edu

Abstract. Assessing teams and providing feedback on scenario-based training typically requires human observers or scenario-specific metrics crafted by experts, due to the complexity of general-purpose automated tools to assess team performance. Machine learning can help infer team performance patterns, but labeled data for a specific training scenario is often sparse. To address this issue, the Semi-Supervised Learning for Assessing Team Simulations (SLATS) project investigated the feasibility of semi-supervised learning and transfer learning which leverages training data from related scenarios to classify performance on a target scenario with the same metrics but a different terrain context. To this approach, we analyzed performance of teams in the first-person shooter Team Fortress 2 (TF2). TF2 teams for the "Capture Point" mode were classified into archetypes based on the performance of the team and the performance of individual members of the team across the corpus: novice, weak link, team of experts, and expert team. To investigate the feasibility of transfer learning, we isolated matches from two of the most frequent maps/terrains. Results found that leveraging data from the source map always improved classification F1-scores compared to relying solely upon target (test) map training data. The greatest benefits were observed when target data was limited (0 to 42 target examples). While further research is required to explore the effectiveness of transfer learning across training scenarios that are more dissimilar (e.g., different simulations, rather than just different maps), these results offer a promising direction to help bootstrap team assessments on new training scenarios by leveraging data from earlier, comparable scenarios. However, efficiently calculating reusable metrics for model features based on low-level scenario events and logs remains a challenge that requires further research.

Keywords: Semi-supervised learning · Transfer learning · Team training

1 Introduction

Assessing teams and providing feedback on scenario-based training is traditionally ad-hoc, due to a lack of general-purpose automated tools to assess team

R. A. Sottilare and J. Schwarz (Eds.): HCII 2024, LNCS 14727, pp. 261–271, 2024.
https://doi.org/10.1007/978-3-031-60609-0_19

performance. In many cases, the gold standard remains live observers with a virtual control panel or physical scorecard to record performance outcomes. While standards such as xAPI have facilitated the development of general-purpose data analytics [1], the patterns that represent expert versus novice performance can vary substantially based on scenario difficulty or objectives. Machine learning can help infer these patterns, but labeled data for a specific training scenario is often sparse.

To address this issue, the Semi-Supervised Learning for Assessing Team Simulations (SLATS) project investigated the feasibility of transfer learning which leverages training data from related scenarios to classify performance on a target scenario (i.e., same metrics but different conditions). This work aligns to efforts such as the Army's Synthetic Training Environment, which should enable scenario-based team training such as battle drills to be conducted in simulated environments with individual and team actions logged using the xAPI standard [4]. The SLATS project was driven by three goals:

1. **Classify Team Performance:** Automate or semi-automate activities that an observer-trainer might need to perform during or after a training scenario, to enable greater opportunities for team training (e.g., reducing cost and expertise bottlenecks). While not all abilities of a human observer or trainer can be replicated in an automated system, sufficient data should exist in a simulation to identify common errors that should be flagged as areas for improvement.
2. **Diagnose Performance Issues:** Develop a set of key team metrics and data views that aggregate lower-level scenario-specific assessments into actionable and interpretable insights. The use-cases of strongest interest are to produce metrics for: individual feedback, team feedback, scenario adaptation (for a future simulation), and instructor review/assessment.
3. **Generalized Framework:** Develop a re-usable set of metrics and tools that can be applied to assess team training in a variety of scenarios, leveraging industry and standards for recording performance and learning events.

In this paper, we investigate the potential of transfer learning to help achieve these goals. Analyses on an existing large corpus of team game scenarios (Team Fortress 2) are presented. We also present context on the overall machine learning pipeline used by SLATS, with an emphasis on how effectively these models could facilitate diagnostic feedback and generalization to new types of simulations. The results presented indicate that transfer learning offers an effective way to improve assessment for scenarios in a similar training system, but that the ability to generalize models broadly remains a challenge.

2   Background

2.1   Assessment Methodologies for Team Training

While scenario-based assessments have been explored in many educational contexts, assessments to support both team and individual learning remain challenging and are traditionally scenario-specific and labor intensive. Given the difficulty to develop such computer-based assessments, they are frequently not used even when the training itself delivered using computer-based training. For example, large organizations such as the Army still rely primarily on live observer-trainers to watch the exercise and manually determine feedback and after-action-review items (e.g., sustain vs. improve priorities), limiting training feedback to times where human experts are available. However, effective training requires many practice opportunities; it is not feasible for a large number of teams to practice toward expert performance when experts must facilitate each session.

Team training is substantially more complex than individual training, because learners may vary not just by skill level but also by the types of skills they are expected to know (e.g., specialization). Research on team assessment and performance has proposed role-based models for team behavior to address these issues [2,3]. There may also be differences in team versus individual performance. Metrics development has looked at distinguishing between i) individual tasks, ii) team tasks (outcomes), and iii) teamwork (process), in projects such as the Surveillance Scenario Team Tutor [3] and Squad Overmatch [5].

Smith-Jentsch, Johnston, and Payne [9] break teamwork into four categories: information exchange (domain-relevant content of communications), communication delivery (e.g., clarity, brevity, using proper terminology and language), supporting behavior (back-up behavior to correct errors or fill gaps), and leadership (adapting priorities and guidance to changes in the situation). Integrating across these frameworks, an ideal-world team assessment would account for: a) propagation of errors (e.g., inability to complete a task due to a teammate's failure), b) external influences (e.g., good process/bad outcome), and c) back-up behavior (e.g., assigning credit for successful performance to the proper individual). The models should also distinguish between task work (e.g., performance) vs. teamwork (e.g., coordination). These behaviors imply that assessing teams meaningfully requires capturing both team and individual metrics, with some structure or data-derived inferences to determine how individual metrics relate to higher-level team processes and outcomes.

2.2   Learning: Semi-supervised and Transfer Leaning

To address the cold start problem for scenario-based training data, we are using a semi-supervised approach to build a classifier to detect engagement archetypes. Given that labeled data is often hard to collect, semi-supervised methods leverage a small amount of labeled data to make better use of a larger set of unlabeled data [10]. In earlier work by our group, the SMART-E project (Service for Measurement and Adaptation to Real-Time Engagement) applied semi-supervised learning for generalized, automated assessment of engagement by individual learners. Research with SMART-E found that metrics were able to generalize across systems for engagement [7] and that semi-supervised learning offered advantages for classifying and interpreting engagement archetypes such as distracted learners versus those racing through the content [8]. As such, a goal for SLATS was to generalize this technique to assessing teams in scenarios.

However, while investigating semi-supervised techniques, we recognized that our semi-supervised approach was primarily helpful for an initial scenario where archetypes were not yet well understood. Later scenarios should be much faster to classify accurately if transfer learning can boost new scenario assessments based on patterns in earlier well-analyzed scenarios. However, the benefit of transfer learning depends on the similarity between the tasks [6]. Even for different maps or variations of scenarios with the same objectives, different team behaviors might be more successful overall.

3   Approach

The SLATS architecture is designed to process data in stages as shown in Fig. 1, such that each subsequent stage only relies on the prior stage as a data source. Raw events and logs are first produced by a training scenario, which are either directly recorded as xAPI statements or processed through a log-file converter to generate xAPI records. A log-cleaner function then fixes these raw logs to produce a second canonical xAPI log for processing (meaning that the raw xAPI statements always exist for alternate cleaning or record checks). In the second stage, the raw xAPI logs are analyzed to produce two types of metrics: direct metrics and intermediate metrics. Direct metrics require xAPI log data to perform their calculations (e.g., number of deaths for a player that session), while intermediate metrics can be calculated only based on other metrics (no xAPI data needed). Metrics may be individual or team, with certain team metrics being more likely to be intermediate (i.e., derived from the individual players). Metrics may either be custom functions or they may be determined by a lightweight markup file which specifies certain functions and aggregations (e.g., average, min/max, etc.).

As shown in Stage 3, a team session vector can be specified, which specifies the set of metrics that will be available as features for classifying team performance. In the example analysis below, teams are classified only on team-level metrics for easier interpretation, but this is not a requirement. A session vector is calculated for each scenario sessions, both for labeled data (known archetypes) and unlabeled data. In a multi-team match, each team will have its own session.

Classification occurs during the final stage. Following the approach described for the SMART-E semi-supervised model [8], unlabeled sessions are clustered based on their feature vectors. An alignment algorithm calculates the global best-match between each cluster and the labeled data for each archetype. Then, data for each cluster is assigned a candidate label based on the archetype which aligned to it. This pooled data set includes both truly labeled data and cluster-aligned data, which are used to train a machine learning model. Different clustering algorithms and classifier types may be selected using parameters, with Gaussian Mixture Models (GMM) clustering and Logistic Regression classification used by default. This approach to pooling labeled and unlabeled data for the classifiers increases accuracy versus using only unlabeled data for training and exploratory analysis indicates benefits up to about four times as much unlabeled data as labeled data (e.g., 20 labeled vs. 80 unlabeled).

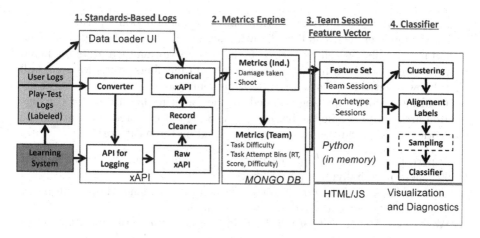

Fig. 1. SLATS Architecture Diagram

3.1 Team Archetypes

SLATS classifies teams into "archetypes" of performance that represent their performance level and use such classifications to areas to practice next. Unlike a traditional 0 to 100 score, we are instead interested in the development stage of a team from a poorly-coordinated set of novices to a highly-effective expert team. This is important for the ability to re-use metrics across different scenarios and simulations. For example, in one scenario it might be reasonable for an expert team to have only 10 communicative actions, while in another, an expert team might require 100. Moreover, metrics are not necessarily linear between archetypes: an expert gaming team might have fewer kills, because they win decisively without an extended conflict. Manually filling in and updating scenario-specific parameters and weights would be time consuming for scenario authors. To avoid this requirement, models were leveraged to estimate and update parameters with the goal of being able to distinguish between different classes of team behavior. Different archetype categories may be specified per-system that is registered in the SLATS framework. In the current work, we focused on classifying:

1. Expert Team: Team is effective and composed of successful individuals.
2. Team of Experts: Team members are good at individual tasks, but the team is not successful, such as due to poor communication or coordination.
3. Weak Link: Team members are good at individual tasks effectively, but the team is not successful, such as due to poor communication or coordination.
4. Novice: Team performances is poor, which is also reflected by lack of success or experience of its individual members.

Typically, a ground truth data set would be established based on expert labeling of a small set of sessions of each category. However, in this case due to the very large corpus of Team Fortress 2 (TF2) matches, we inferred labels based on knowledge about team performance and the individual performance of each team member across all their known matches. Each team (unique combination of individuals) was characterized by its team performance and its predicted performance based on a linear regression of team members' statistics across all their known matches (shooting, support, and survival).

Gold labels for teams were defined by the following heuristics, for the designated archetypes. While the broader SLATS project explored other archetype categories, research on transfer learning focused on these categories.

1. Expert Team: Over 75th percentile team performance and all members over 60th percentile individual performance
2. Team of Experts: Under 75th percentile team performance despite all members over 60th percentile individual performance.
3. Weak Link: Under 75th percentile team performance with at least one but not all more members under 40th percentile individual performance.
4. Novice: Under 25th percentile team performance and all members under 40th percentile individual performance

3.2 SLATS Diagnostics

While not the main focus of this paper, after a team was classified by SLATS this result could be visualized in a web interface as shown in Fig. 2. When providing diagnostic feedback, we consider the generalizable metrics collected and differentiate them by the individual vs. the team [9] and also the team expertise level. Based on the anchor points of Novice and Expert Team as the lowest and highest archetypes, respectively, a rank-order was inferred for the next-better archetype to advance toward.

The team's performance on each feature was shown as a bar chart. A green bar indicates the team's performance on a metric exceeds the typical team in their archetype (red-dotted line) or the next-better archetype (yellow-dotted line). A red bar indicates falling short of the typical standard for the current archetype on that performance feature (e.g., worse than other novice teams). As shown in the third bar "Survive", the next-better archetype might be worse than the current one on certain performance features. Suggested "Sustains" and "Improves" recommendations are displayed below the bar chart. For more expert team, areas to improve will typically be team metrics. However, for more novice teams, areas to improve will more commonly be individual skills to practice.

3.3 Transfer Learning Analysis

To evaluate this approach, TF2 data was used as a proxy for future synthetic battle training. Teams post log files of their matches to public online repositories and TF2 scenarios require balancing individual competencies (e.g., shooting accuracy, taking cover) with teamwork competencies (e.g., capturing positions, healing/support). We collected a corpus of TF2 matches using the same "mode" (e.g., goal and rules), and classified teams in our corpus of TF2 matches into

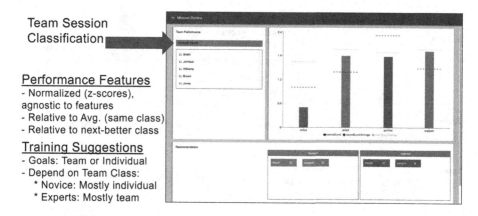

Fig. 2. SLATS Session Diagnostics User Interface (Color figure online)

four archetypes based on the performance of the team and the performance of individual members of the team across the corpus as explained above.

The relationships between team and individual metrics are outlined in Fig. 3, with individual metrics in orange and team metrics in blue. The feature vector for a session had four values: Move (capture and hold points), Shoot (kill or damage opponents), Support (heal or assist team member in a kill), and Survive (heal self, avoid damage, and avoid death). The Move metrics included each capture point as a distinct lower-level metric, so that performance was based on the percentage of the session each point was held and the maximum number of points they held at the same time. Team metrics were also normalized to average across the number of players and converting to z-scores for each team session metrics so they would be on comparable scales.

To investigate the feasibility of transfer learning, we isolated matches from two of the most frequent maps/terrains for a game mode called "Control Point" (Snakewater and Process). These matches require capturing and holding a set of control points on the map to win. In each analysis, test data was drawn solely from the target map and we explored the use of varying mixtures of training data from the source and target maps. The SLATS architecture was configured to use default classifier settings (GMM and Logistic Regression), with the expectation that if transfer learning assists simpler models it should also benefit more data-intensive models. A stratified random sample of sessions was selected from each map, which ensured that all archetypes were represented and that only one session per match was selected (i.e., avoiding two sessions from same match but different teams). A total of 319 sessions were processed to generate session feature vectors, with 246 Snakewater (Sn) and 73 Process (Pr) sessions prepared.

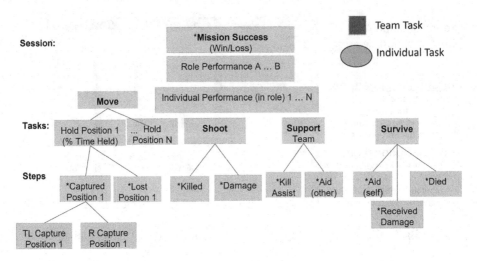

Fig. 3. TF2 Metrics Aggregation Diagram

4 Results

Cross validation on entire session corpus of 319 sessions showed strong classification results (5-fold CV; F1 = 0.97 ± 0.02). Table 1 shows the average F-1 scores for each additional 14 training sessions from either a source map or the target map (which also provides sessions used as test data).

Table 1. F-1 Scores (Avg. of 5-fold CV) for Team Classification based on Source and Target map sessions

Source Training N_{Sn}	Target Training N_{Pr}				
	0	14	28	42	56
0	N/A	0.867	0.865	0.895	0.962
14	0.912	0.917	0.945	0.954	0.969
28	0.900	0.959	0.969	0.949	0.987
42	0.936	0.989	0.976	0.967	0.980
56	0.942	0.969	0.980	0.960	0.960
70	0.933	0.939	0.949	0.969	0.987
84	0.962	0.980	0.960	0.987	0.980
94	0.953	0.969	0.966	0.939	0.967

Particularly when data is limited, including training samples from both maps improves classification performance on the Target test sessions. These benefits are most pronounced with fewer than 56 Target sessions (F-1 below 0.9 without

Source sessions, but 0.912–0.954 with even just 14 Source sessions). A follow-up analysis with greater randomization of samples confirmed these results, showing that the average best-performance tended to be about F1 = .966 and that it typically plateaued at approximately 72 samples (28 Source/42 Target).

5   Discussion

This research found that both semi-supervised learning and transfer learning can improve classification of team performance. As shown in Fig. 2 for diagnosis, archetype analysis enabled by the semi-supervised approach is helpful because team performance is not just on a monotonic scale but in cases where some metrics may decrease as teams improve overall. Transfer learning also showed benefits for overcoming the cold start problem of limited data. However, generalizable metrics pipelines were challenging to design when relying on xAPI standards-based approaches, which likely requires more specialized research in this area.

Transfer Effectiveness. Leveraging data from the source map always improved classification f1-scores compared to relying solely upon target map training data. The greatest benefits were observed when target data was limited (0 to 42 target examples). Although it was not always the case that more source data results in higher performance, validation data could be used to find the ideal mixture of source and target training data. While further research is required to explore the effectiveness of transfer learning across training scenarios that are more dissimilar (e.g., different simulations, rather than just different maps), these results offer a promising direction to help bootstrap team assessments on new training scenarios by leveraging data from earlier, comparable scenarios.

Improving Metrics Pipelines. Our work is complementary to research that improves underlying assessment metrics, such as research on multi-modal assessment of training scenarios [11]. Since SLATS archetypes are derived from aligning small amounts of labeled data with larger bottom-up clusters, the specific assessment metric components can be replaced with more advanced measures while following the same pipeline. In addition to more advanced metrics, more efficient calculations of standards-based metrics are also needed. The sheer volume of data for a highly-logged scenario (e.g., TF2) posed challenges in this research. Attempting to apply a standards-based approach for xAPI conversion of each low-level action (e.g., every shot fired) resulted in very large xAPI learning stores. Processing metrics on such records required optimized queries and database caching of results, which undermined the goal of easily generalizing team assessment across different training systems. Research groups have investigated data streams and other techniques to optimize metrics [1], which may offer a foundation for future work on reusable metrics.

Tradeoffs of Archetypes. The SLATS approach depends on interpretable team archetypes, which can be benchmarked against real teams rather than heuristic rules or cutoffs. However, training experts may not know or recognize distinct team archetypes for all training scenarios. In particular, the assumption of an "expert" category assumes that as experts gain skills, they tend to behave more and more similarly in comparable situations (i.e., converging on the best approach to situations such as landing a damaged aircraft). However, expert teams may also become more diverse in their behavior (e.g., developing their own patterns of communication, developing teamwork patterns unique to the strengths/weaknesses of individuals in the team). Thus, it may be difficult to recognize expert teams in some scenarios because how they coordinate and work together may differ, implying that additional archetypes might be required when these distinctions are relevant for training.

6   Conclusions and Future Directions

Research on the SLATS framework indicates that transfer learning offers advantages for team assessment, particularly when data is limited and relatively simple models are leveraged. Future research is needed however to replicate these findings with more advanced models, particularly models that could incorporate data streams more directly for a high volume of data. For example, new classes of neural network transformer models may be able to directly ingest event data streams and produce meaningful assessments.

These findings suggest that outcomes-based assessment for training scenarios and simulations may someday be automated usefully, not just for assessing team performance but also for tracing individual poor task performance that may need further practice. However, the current work did not model more complex processes or delayed consequences that may occur in other scenarios (e.g., the appropriate skill to practice if a small mistake early-on results in a large failure later). These types of assessments are important, as simulated assessments should distinguish between a good process versus a good (or bad) outcome when suggesting skills to study.

Finally, research on automatically generated formative assessments and diagnoses for training scenarios warrants further pilot studies and evaluation research to indicate how much these insights help guide and improve learning outcomes and study processes. Despite growing interest in automated or partially automated assessment, data on effectiveness remains limited. As such, future work should conduct studies to identify the benefits and limits of such feedback compared to control conditions such as a Wizard-of-Oz model (i.e., feedback controlled by a hidden human expert) or a system without formative assessments.

Acknowledgments. This research was sponsored by U.S. Army through the USC ICT University Affiliated Research Center (W911NF-14D0005). However, all statements in this work are the work of the authors alone and do not necessarily reflect the views of sponsors, and no official endorsement should be inferred.

References

1. Blake-Plock, S., Hoyt, W., Casey, C., Zapata-Rivera, D.: Data analytics and visualization for xAPI learning data: considerations for a GIFT strategy. In: Design Recommendations for Intelligent Tutoring Systems, vol. 8: Data Visualization, pp. 163–171 (2020)
2. Brawner, K., Sinatra, A.M., Gilbert, S.: Lessons learned creating a team tutoring architecture. In: Design Recommendations for Intelligent Tutoring Systems, vol. 6: Team Tutoring, pp. 201–220. US Army (2021)
3. Gilbert, S.B., et al.: Creating a team tutor using GIFT. Int. J. Artif. Intell. Educ. **28**(2), 286–313 (2018)
4. Goldberg, B., et al.: Forging competency and proficiency through the synthetic training environment with an experiential learning for readiness strategy. In: Interservice/Industry Training, Simulation, and Education Conference (I/ITSEC) (2021)
5. Johnston, J.H.: Team performance and assessment in GIFT - research recommendations based on lessons learned from the squad overmatch research program. In: Proceedings of the 6th Annual GIFT Users Symposium (GIFTSym6). US Army (2018)
6. Neyshabur, B., Sedghi, H., Zhang, C.: What is being transferred in transfer learning? In: Advances in Neural Information Processing Systems 33, pp. 512–523 (2020)
7. Nye, B.D., Core, M., Auerbach, D., Ghosal, A., Jaiswal, S., Rosenberg, M.: Integrating an engagement classification pipeline into a GIFT cybersecurity module. In: Proceedings of the 8th Annual Generalized Intelligent Framework for Tutoring (GIFT) Users Symposium (GIFTSym8), pp. 49–67. US Army (2020)
8. Nye, B.D., Core, M.G., Jaiswa, S., Ghosal, A., Auerbach, D.: Acting engaged: leveraging play persona archetypes for semi-supervised classification of engagement. Int. Educ. Data Min. Soc. (2021)
9. Smith-Jentsch, K.A., Johnston, J.H., Payne, S.C.: Measuring team-related expertise in complex environments. In: Cannon-Bowers, J.A., Salas, E. (eds.) Making Decisions Under Stress: Implications for Individual and Team Training. American Psychological Association (2018)
10. Triguero, I., García, S., Herrera, F.: Self-labeled techniques for semi-supervised learning: taxonomy, software and empirical study. Knowl. Inf. Syst. **42**(2), 245–284 (2015)
11. Vatral, C., Mohammed, N., Biswas, G., Goldberg, B.S.: GIFT external assessment engine for analyzing individual and team performance for dismounted battle drills. In: Proceedings of the Ninth Annual GIFT Users Symposium (GIFTsym9), pp. 107–127. US Army (2021)

Enhancing E-Learning Experience Through Embodied AI Tutors in Immersive Virtual Environments: A Multifaceted Approach for Personalized Educational Adaptation

Fatemeh Sarshartehrani[✉][ID], Elham Mohammadrezaei[ID], Majid Behravan[ID], and Denis Gracanin[ID]

Virginia Tech, Blacksburg, VA 24061, USA
{fatemehst,elliemh,behravan,gracanin}@vt.edu

Abstract. As digital education transcends traditional boundaries, e-learning experiences are increasingly shaped by cutting-edge technologies like artificial intelligence (AI), virtual reality (VR), and adaptive learning systems. This study examines the integration of AI-driven personalized instruction within immersive VR environments, targeting enhanced learner engagement-a core metric in online education effectiveness. Employing a user-centric design, the research utilizes embodied AI tutors, calibrated to individual learners' emotional intelligence and cognitive states, within a Python programming curriculum-a key area in computer science education. The methodology relies on intelligent tutoring systems and personalized learning pathways, catering to a diverse participant pool from Virginia Tech. Our data-driven approach, underpinned by the principles of educational psychology and computational pedagogy, indicates that AI-enhanced virtual learning environments significantly elevate user engagement and proficiency in programming education. Although the scope is limited to a single academic institution, the promising results advocate for the scalability of such AI-powered educational tools, with potential implications for distance learning, MOOCs, and lifelong learning platforms. This research contributes to the evolving narrative of smart education and the role of large language models (LLMs) in crafting bespoke educational experiences, suggesting a paradigm shift towards more interactive, personalized e-learning solutions that align with global educational technology trends.

Keywords: Adaptive Learning · Artificial Intelligence in Education · Immersive Virtual Environments · Learner Engagement · Personalized Instruction

1 Introduction

The landscape of education has undergone a seismic shift in the past decade, with e-learning emerging as a dominant force, reshaping how knowledge is accessed,

© The Author(s), under exclusive license to Springer Nature Switzerland AG 2024
R. A. Sottilare and J. Schwarz (Eds.): HCII 2024, LNCS 14727, pp. 272–287, 2024.
https://doi.org/10.1007/978-3-031-60609-0_20

imparted, and absorbed. This transformation has been significantly accelerated by technological advancements, particularly in artificial intelligence (AI) and virtual reality (VR). However, while these innovations have democratized access to education, they have also brought to the fore new challenges, especially in terms of learner engagement and the personalization of the educational experience.

Traditional e-learning systems often fall short in replicating the nuanced and interactive dynamics of a traditional classroom, leading to issues such as reduced student engagement, lack of personalized attention, and a one-size-fits-all approach to education. The advent of immersive virtual environments, coupled with the emergence of AI, offers a promising avenue to address these challenges [1]. By integrating embodied AI tutors within these environments, there is potential to revolutionize the e-learning experience, offering a more personalized, engaging, and interactive learning journey for students.

The concept of embodied AI tutors in virtual environments is grounded in the idea of creating an interactive, responsive educational experience that closely mirrors human tutoring. These AI tutors are designed to adapt to individual learners' emotional and cognitive states, offering a customized learning path that considers factors such as the learner's pace, preferences, and current understanding. This approach holds the promise of not only enhancing learner engagement but also significantly improving learning outcomes.

Furthermore, the integration of AI tutors in immersive virtual environments opens new horizons for e-learning. These environments can simulate real-world scenarios, provide immersive experiences, and enable interactive learning activities that were previously unattainable in traditional e-learning setups. The virtual setting, governed by an AI tutor, can dynamically adjust to the learner's needs, providing a rich, contextual, and adaptive learning environment.

This paper aims to delve into the feasibility, design, and impact of incorporating embodied AI tutors in immersive virtual environments for e-learning. We explore the transformative potential of this technology in creating personalized learning experiences and address the challenges and opportunities it presents. Our research is particularly focused on understanding how these AI tutors can adapt to various emotional, cognitive, and environmental factors to enhance the overall learning experience.

As the digital age continues to evolve, the role of technology in education becomes increasingly pivotal. This study contributes to the ongoing discourse in this field, proposing a multifaceted approach to e-learning that leverages the latest advancements in AI and VR to offer a more effective, engaging, and personalized educational experience.

2   Related Work

2.1   Virtual Reality in Education

In exploring the integration of advanced technological solutions in education, significant emphasis has been placed on the transformative potential of VR [2]. Studies by Lege and Bonner, Freina and Ott, and Christou have collectively underscored VR's capacity to enhance educational experiences through immersive, interactive learning environments [3–5]. These works highlight VR's role in facilitating deeper engagement, improving spatial memory, and enabling access to previously inaccessible learning scenarios. Despite the promising advancements, they also note critical challenges, including the development of VR-specific pedagogies, technological barriers, and the need for further research to optimize VR's educational impact. The insights from these studies provide a crucial backdrop to our investigation into the use of embodied AI tutors in immersive virtual environments, reinforcing the premise that while VR technologies offer substantial benefits in creating engaging and effective learning experiences, they necessitate careful integration with pedagogical strategies to realize their full potential in education.

2.2 Ethical Considerations in AI-Driven Education

Transitioning from the technological advances in VR, it is crucial to consider the ethical framework within which these innovations are applied, especially when creating AI tutoring systems in virtual reality environments. The importance of ethics in ensuring fair and responsible education outcomes cannot be overstated, as highlighted by studies from Borenstein and Howard, Remian, Garrett et al., and Huallpa et al. [6–9]. These studies illuminate the complex ethical challenges of AI education, including privacy, bias, accountability, and the need for ethical pedagogy, and emphasize the importance of designing AI tutoring systems with a strong ethical framework that prioritizes student data protection, fairness in adaptive learning, and a trustworthy and transparent environment. Incorporating these ethical considerations into the development of AI tutoring systems aligns with responsible AI development principles and enhances the quality and integrity of e-learning experiences.

2.3 Advancements in Intelligent Tutoring Systems (ITS)

Building on the ethical foundation, the quest for more effective and engaging e-learning environments has led to significant research dedicated to the integration of AI and ITS. For instance, a systematic review by Chien-Chang et al. highlights how AI and IT innovations within ITS are crucial for fostering sustainable education through personalized learning experiences and enhanced student engagement [10]. Similarly, meta-analyses by Ma et al. and Kulik & Fletcher provide compelling evidence that ITS, characterized by their adaptive learning capabilities, generally outperform traditional instructional methods, leading to improved learning outcomes [11,12]. Ennouamani et al.'s work on adaptive e-learning systems further underlines the transformative potential of personalization in web-based education, a principle that underpins our exploration of embodied AI tutors in immersive virtual environments [13].

2.4   Emotional and Cognitive Adaptation in Learning

Further enriching the discourse on adaptive learning technologies, significant research has explored the intricate dynamics between learners' cognitive, motivational, and emotional states and their learning experiences. For instance, studies on digital educational games have demonstrated the potential of inferring and adapting to learners' states to enhance engagement and learning effectiveness [14]. Similarly, the AutoTutor system's ability to detect and respond to affective and cognitive states underscores the viability and value of affect-sensitive ITS in personalizing learning experiences [15]. Further, empirical investigations into the impact of cognitive-affective states on learning with computer-based environments reveal the critical importance of addressing these states to improve educational outcomes [16]. These studies collectively highlight the foundational role of adaptive technologies in creating more personalized, responsive, and effective e-learning environments.

2.5   Personalized Learning Pathways

In the evolving landscape of e-learning, personalized education emerges as a critical driver for enhancing engagement and learning outcomes. Studies by Reddy et al., Shaw et al., Villatoro Moral and Crosseti, and Salinas and De-Benito illustrate the diverse methodologies and significant potential of personalized learning pathways, ranging from intervention tracking applications to asynchronous platforms guided by learning recommendation algorithms [17–20]. These pioneering approaches lay the groundwork for our exploration into AI-driven personalization within immersive virtual environments, where the dynamic adaptation to learners' emotional and cognitive states promises to further individualize learning experiences. By integrating real-time data on learner engagement and performance, our AI tutoring system aims to not only accommodate individual learning preferences and needs but also to foster a deeper, more effective learning process.

2.6   Large Language Models (LLMs) in Education

Moreover, the advent of LLMs has marked a significant shift in the educational landscape, offering novel methods for achieving personalized learning, intelligent tutoring, and effective educational assessments. These advanced AI systems extend beyond traditional educational tools, offering personalized learning experiences, enhancing student engagement, and providing innovative solutions to long-standing challenges in education. By leveraging the capabilities of LLMs, educators and learners alike can explore new horizons in knowledge acquisition, critical thinking, and creative problem-solving. The integration of these models into educational frameworks signifies a shift towards more dynamic, interactive, and learner-centered approaches, underscoring the profound impact of artificial intelligence on the future of education.

Gan et al. delve into the realm of LLMs within the sphere of digital and smart education, identifying them as pivotal in addressing traditional educational challenges such as individual learner differences and the allocation of teaching resources. The paper envisions EduLLMs as evolving tools that offer novel methods for achieving personalized learning, intelligent tutoring, and effective educational assessments. By providing a comprehensive review of current research status and applications, the work serves as a roadmap for educators, researchers, and policymakers, guiding them towards a deeper understanding of LLMs' potential to revolutionize the education sector while also acknowledging the technical, ethical, and practical challenges that lie ahead [21].

Mollick et al. scrutinize the dual-edged nature of LLMs in the educational landscape, acknowledging their transformative capabilities alongside their inherent risks and limitations. The discourse revolves around the delicate balance between exploiting AI's adaptive learning potentials and mitigating risks associated with AI's imperfections. The paper highlights the necessity for educators to actively guide students in the judicious use of AI tools, fostering a learning environment where AI assists rather than replaces the learning process. Through practical guidelines and theoretical insights, the authors propose a framework for integrating AI in education that encourages critical engagement with AI outputs, ensuring students remain critical thinkers and active participants in their learning journey [22].

2.7 Emotional Intelligence and Prompt Engineering in LLMs

The exploration of Emotional Intelligence (EQ) in LLMs represents a significant advancement in the field of artificial intelligence, particularly in educational applications. Recent studies, have begun to investigate the capacity of LLMs to understand and process human emotions, suggesting that these models can potentially engage in more empathetic and nuanced interactions. This development holds promise for creating more engaging and supportive learning environments, where AI tutors can respond to the emotional states of learners, thus facilitating a more personalized and effective educational experience. Wang et al.'s research marks a significant leap in evaluating the EQ of LLMs, specifically GPT-4. By conducting a psychometric evaluation, the study reveals GPT-4's ability to understand and interact with human emotions, outperforming the majority of human participants in EQ assessments. This breakthrough underscores the potential for LLMs to engage in emotionally intelligent dialogues, paving the way for more empathetic and effective communication in educational settings. The findings suggest a future where LLMs can provide not only academic support but also emotional guidance, enhancing the overall educational experience by fostering a sense of understanding and connection between AI tutors and learners [23].

Prompt engineering is a cutting-edge practice in the realm of artificial intelligence, especially within the context of LLMs. This technique involves crafting inputs (prompts) to an AI model to elicit specific outputs or behaviors, optimizing the interaction between humans and AI [24]. In educational settings, prompt engineering transcends mere technical manipulation; it becomes an art form that blends content knowledge, critical thinking, and iterative design. By effectively leveraging prompt engineering, educators and developers can tailor

LLMs to provide personalized learning experiences, encourage creative problem-solving, and facilitate a deeper engagement with the material. This approach not only enhances the efficacy of AI in education but also democratizes the learning process, allowing for a more inclusive and equitable educational landscape.

Cain extensively explores the innovative practice of prompt engineering within the context of Large Language Models (LLM AI) in education. The research delineates prompt engineering as a pivotal methodology, incorporating content knowledge, critical thinking, and iterative design. This approach is posited as essential for harnessing LLM AI's transformative potential, aiming to foster a learning environment that is not only personalized and engaging but also equitable. The paper advocates for the integration of LLM AI tools in educational settings, emphasizing their capacity to transform students and educators from passive recipients to active creators of learning content, thereby promoting a more dynamic and interactive educational experience [25].

In exploring the enhancement of learner engagement within e-learning environments, our research aligns with innovative methodologies such as gamification and data-driven strategies for engagement optimization. Studies by Rebelo and Isaías, Moubayed et al., and Atkins et al. have highlighted the effectiveness of engaging students through gamification elements and the significant correlation between engagement levels and academic performance [26–28]. These findings validate our pursuit of personalized AI tutoring as a means to deepen engagement and suggest that the adaptive capabilities of AI could represent the next frontier in cultivating engaging and effective e-learning experiences. By integrating these insights, our study contributes to the evolving landscape of e-learning technologies, where personalization and continuous engagement strategies are paramount for enhancing learning outcomes.

3   Method

This study explores the use of adaptive AI tutoring systems in immersive virtual environments, specifically in the context of Python programming education. The research was conducted with ten graduate students from Virginia Tech and used an experimental design to assess the effectiveness of our personalized AI tutoring framework. The study focuses on the deployment of AI tutors, which are designed to adapt their instructional strategies based on the unique emotional and cognitive profiles of learners. This is achieved through the use of survey data. We employed a data-driven approach to classify participants into distinct emotional and cognitive clusters. This classification was pivotal for tailoring the educational content to align with the unique needs and predispositions of learners, thereby maximizing the effectiveness of the AI tutoring system within an immersive virtual environment.

The goal of the research is to evaluate the effectiveness of AI tutors in fostering deeper learner engagement and enhancing knowledge acquisition. Additionally, the study aims to investigate the capacity of these tutors to provide a personalized learning experience by adapting to the psychological profile of each student. The research seeks to contribute important insights into the potential of AI and VR technologies to revolutionize e-learning, offering a more adaptive, engaging, and personally resonant educational experience.

This study sought participants who were either current students or staff members at Virginia Tech, ensuring a diverse range of experiences and perspectives within the educational context. A key requirement for participation was proficiency in English to guarantee clear understanding and communication throughout the study. Importantly, prior programming experience was not a prerequisite, allowing for a broad spectrum of knowledge levels among participants. This inclusivity aimed to mirror the varied backgrounds found in typical e-learning environments, providing valuable insights into the adaptive AI tutoring system's effectiveness across different learner profiles.

3.1 Pre-study Procedures

Initial Survey: Before commencing the experimental sessions, an initial survey was administered to gather essential data on the participants' emotional and mental states. This survey was designed to measure various psychological dimensions, including excitement, motivation, stress, anxiety, and overall engagement with the prospect of attending a VR class. The objective was to establish a baseline understanding of each participant's psychological readiness and predisposition toward learning in a virtual environment. This preliminary step was crucial for tailoring the AI tutoring system to address the needs and emotional states of learners, ultimately aiming to enhance the personalization and effectiveness of the e-learning experience.

Clustering and Group Assignment: After the initial survey, which collected self-reported data on emotional and mental states, we applied the k-means clustering algorithm ($k = 3$) to categorize participants into three primary groups: engaged and motivated, anxious and stressed, and distracted and disengaged. The clustering was informed by an analysis of the survey responses, aiming to capture a comprehensive snapshot of each participant's readiness and suitability for different learning approaches.

Pre-assessment and Content Adaptation: After clustering, participants were subjected to a Python programming knowledge quiz. The design of this quiz was intricately linked to their assigned cluster, ensuring that the questions were appropriately challenging and aligned with their emotional and cognitive state. For the engaged and motivated cluster, the quiz featured challenge-based and problem-solving questions, demanding a deeper engagement with the material. Meanwhile, participants in the anxious and stressed cluster encountered questions with straightforward instructions, minimizing additional stress and focusing on clarity. Those identified as disengaged or distracted were presented with interactive and relatable questions, designed to spark interest and foster engagement through relatable scenarios.

3.2   Teaching Sessions

Tailored Instructional Design: The AI tutor, central to our study, was programmed to adjust its teaching strategies based on the cluster assignments. Leveraging large language models, the tutor was capable of demonstrating personalized Python programming lessons in the virtual reality environment. This adaptation extended to the depth of content delivery, the complexity of tasks, and the nature of interactions, ensuring that each learner received an educational experience that was not only personalized but also conducive to their learning style and emotional disposition.

Python Lecture: The core of our experimental design involved conducting 30-min Python programming lectures tailored to the specific needs and emotional states of participants, segmented into three distinct groups. Each group received instruction through a virtual reality headset, ensuring an immersive learning experience. The AI tutor, a central component of our study, was intricately designed to deliver content in a manner that resonated with the emotional and cognitive profiles of the learners. This approach was premised on the hypothesis that personalized teaching strategies, informed by the understanding of learners' states, could significantly enhance engagement and learning effectiveness [29]. Consequently, the AI tutor dynamically adjusted its instructional methods, varying the complexity of the material and the interaction style to suit the designated cluster of each participant group. This customization aimed to optimize the educational encounter, enhancing both the absorption of Python programming concepts and the overall user experience within the virtual environment.

3.3   Post-session Assessment

Feedback Collection: Following the conclusion of the teaching sessions, participants were invited to share their feedback through a survey. This feedback mechanism was designed to capture participant perceptions, including overall satisfaction with the teaching session, the relevance and utility of the course content, and the degree to which the session met the learning objectives. Additionally, the survey sought to gauge the appropriateness of the course's pace, ensuring that the delivery was neither too rapid nor too sluggish for effective comprehension. This feedback was instrumental in assessing the qualitative impact of the AI tutoring system, offering insights into its strengths and areas for refinement.

Knowledge Assessment: To quantitatively measure the educational outcomes of our intervention, participants were subjected to pre- and post-session assessments. These assessments were crafted to evaluate the participants' Python programming knowledge before and after exposure to the AI tutor's personalized instruction. By comparing the initial competencies with the post-session understanding, we aimed to ascertain the learning gains attributed to the adaptive tutoring approach. This comparison not only served as a direct indicator of the AI tutor's effectiveness in conveying programming concepts but also provided a metric for assessing improvements in knowledge acquisition among participants.

4   System Design

Our system, developed using the Unity Game Engine, delivers an adaptive AI tutoring experience for Python programming tailored to learners' emotional states. By recognizing states such as engagement, stress, or distraction, the system provides personalized educational content within immersive virtual environments. Utilizing Unity's extensive capabilities, we've crafted a versatile platform that supports interactive learning through VR. The core of our system is an Adaptive AI Tutor, which dynamically adjusts teaching strategies based on the learner's performance and emotional state, leveraging AI for responsive content delivery. The Python programming curriculum is structured to cater to a wide range of learners, featuring summary descriptions, illustrative coding examples, and interactive exercises. Our adaptive mechanisms ensure content presentation is modified in real-time, optimizing learning efficiency and engagement based on the learner's current state. However, creating a system that accurately identifies and adapts to the learner's emotional state presented unique challenges, including the integration of AI for emotion recognition and maintaining a balance between educational depth and engaging interactivity.

Anxious and Stressed Scenario: To alleviate learner anxiety, this setting employs a straightforward design, emphasizing clarity and support to encourage confidence in coding (Fig. 1a).

Engaged and Motivated Scenario: This environment is vibrant and dynamic, designed to foster curiosity and sustained interest in learning through interactive challenges and rewards (Fig. 1b).

Disengaged and Distracted Scenario: Targeting re-engagement, this environment utilizes captivating narratives and relatable contexts to draw the learner's focus back to the educational content (Fig. 1c).

5   Result

The post-survey results depicted in Table 1 provide a comprehensive overview of participants' sentiments across various dimensions of the course experience. The table categorizes respondents' feelings into three main groups: Stressed, Disengaged, and Engaged, allowing for a nuanced understanding of their perceptions. By utilizing abbreviated labels such as OS (Overall Satisfaction), CC (Course Content), LO (Learning Outcomes), and Pace (Pace of the Course), the table streamlines data presentation without sacrificing clarity.

(a) Disengaged Group Learning Environment

(b) Engaged Group Learning Environment

(c) Stressed Group Learning Environment

Fig. 1. The three environments for the three clusters.

Examining the data reveals intriguing patterns. Notably, respondents who reported feeling "Very satisfied" displayed a remarkable level of contentment across multiple aspects within the Engaged category. For instance, approximately 66.7% expressed high satisfaction levels with Overall Satisfaction, Course Content, and Learning Outcomes. Conversely, participants indicating dissatisfaction ("Dissatisfied") exhibited noteworthy discontentment, particularly within the Stressed and Disengaged categories. Here, substantial proportions of respondents reported dissatisfaction with Overall Satisfaction and Pace of the Course, suggesting areas that may require improvement.

Furthermore, the table illustrates variations in respondent perceptions based on their reported levels of satisfaction. For instance, while those expressing "Satisfied" sentiments demonstrated relatively balanced perceptions across all categories, individuals who reported feeling "Neutral" showcased mixed feelings. Despite indicating neutrality in some aspects, such as Overall Satisfaction and Learning Outcomes, a considerable number of respondents within this group expressed dissatisfaction with Course Content, suggesting potential areas for enhancement.

Table 1. Post survey results. The numbers represent percentages.

Response Options	Stressed				Disengaged				Engaged			
	OS	CC	LO	Pace	OS	CC	LO	Pace	OS	CC	LO	Pace
Very satisfied	66.7	66.7	33.3	0	0	0	0	50	20	60	20	60
Satisfied	0	33.3	33.3	33.3	0	100	0	0	60	20	60	20
Neutral	0	0	33.3	0	50	0	100	0	20	20	20	0
Dissatisfied	33.3	0	0	66.7	50	0	0	50	0	0	0	0
Very dissatisfied	0	0	0	0	0	0	0	0	0	0	0	20

The assessment conducted before and after the teaching course reveals significant improvements in student engagement and understanding (Table 2). Prior to the intervention, while all students were found to be correct in their responses when stressed, only 50% of the responses were correct when students were disengaged. Notably, engagement was high with 86.7% of correct responses, suggesting that a majority were motivated despite some missing responses (13.3%).

Following the teaching intervention, there was a marked improvement in the students' responses. Under stress, the rate of correct responses remained constant at 100%. However, the most significant improvement was observed in the previously disengaged cohort, where correct responses increased from 50% to 100%. This suggests that the intervention was highly effective in capturing and maintaining the students' attention. Furthermore, engagement levels increased slightly but notably, with correct responses rising to 93.3%. Correspondingly, the rate of missing responses decreased to 6.7%, indicating fewer instances of inattention or lack of motivation post-intervention.

Table 2. Summary of responses before and after intervention.

State	Stressed	Disengaged	Engaged
Before			
Correct	100	50	86.7
Missing	0	50	13.3
After			
Correct	100	100	93.3
Missing	0	0	6.7

These results suggest that the teaching course was successful in enhancing student engagement and reducing the rates of disengagement and missing responses (Table 2). Such improvements may be attributed to the pedagogical strategies employed during the course, which appeared to resonate well with the students, thereby increasing their attentiveness and participation. According to the table, the stressed group scored 100% on both pre and post-assessments. However, it is important to note that the assessments for the stressed group were designed to be less challenging to avoid inducing additional stress.

5.1 Positive Aspects

Participants appreciated various elements of the course, which contributed to a positive learning experience:

- The use of straightforward and easy-to-understand coding examples was highly valued.
- The relaxing, green environment was likened to learning Python in a park, enhancing the enjoyment of the sessions.
- The structured progression from simple to complex concepts, along with encouragement at the end of each topic, was commended.
- The novelty and interesting approach of the course/session were highlighted as key strengths.

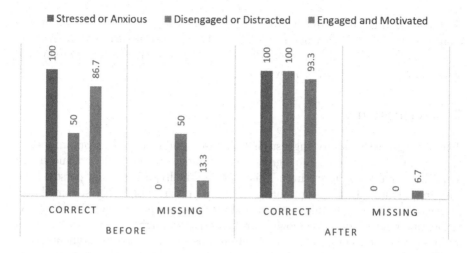

Fig. 2. Responses Improvement

- Participants appreciated the ability to learn at their own pace and the availability of additional code samples.
- The combination of visual slides, audible explanations, and helpful code snippets effectively facilitated learning.

5.2 Areas for Improvement

Feedback also identified several areas where the course could be improved to enhance the learning experience:

- The slide design was seen as monotonous; adding visuals, cartoons, or colors could increase engagement.
- A sense of isolation was reported due to the lack of a classroom atmosphere, indicating a need for more interactivity.
- The user interface's limitations, specifically the absence of a "back" button, hindered the ability to review content.
- The resolution and uniformity of code snippets need attention to maintain consistency across the course.
- Suggestions for increased interactivity included features allowing participants to actively engage with the content.
- Obstructions by the robot arm and overly wordy slides were mentioned, along with a recommendation for a preliminary screener test.
- The desire for more examples, particularly in unused screen areas, was expressed.
- Incorporating gamification and adjusting the speech pace were suggested to cater to diverse learning preferences.

Incorporating the feedback from participants will be crucial in refining the course to better meet learner needs and enhance overall satisfaction. Adjustments to the course design, content delivery, and interactive elements are among the key areas targeted for improvement.

6   Discussion

The findings of this study suggest that the use of adaptive AI tutors in immersive virtual environments can significantly enhance engagement and knowledge acquisition in e-learning settings. This supports the hypothesis that personalized tutoring, attuned to the emotional and cognitive states of learners, contributes positively to the educational experience. Participants who interacted with the AI tutors tailored to their psychological profiles reported higher levels of satisfaction and demonstrated notable improvements in post-intervention assessments.

The observed increase in correct responses and engagement metrics post-intervention highlights the potential of adaptive AI tutors to address issues related to disengagement and cognitive overload in e-learning. The results align with prior research [30] suggesting that personalized, emotionally intelligent AI can create more effective learning environments by providing customized support and feedback.

The study's implications extend beyond the immediate educational outcomes. By demonstrating the feasibility and benefits of personalized AI tutors, this research paves the way for developing more sophisticated, responsive e-learning systems that cater to the diverse needs of students. The positive response to the immersive, tailored environments reinforces the importance of situating e-learning within contexts that resonate with learners' psychological states. However, the study is not without limitations. The sample size was relatively small and restricted to a specific demographic, which may limit the generalizability of the findings. Moreover, the study was conducted over a short period (one session), and therefore, the long-term effects of AI tutoring on learning efficiency and satisfaction remain unknown.

Feedback regarding the system's design revealed a need for greater interactivity and visual appeal. Future development will focus on incorporating more dynamic elements and personalization options to enhance user engagement and reduce feelings of isolation. Moreover, the system's user interface will be improved to provide learners with more control over their learning experience, such as the ability to review previous content easily.

Ethical considerations in AI-driven education remain paramount. This study adhered to ethical standards by ensuring data privacy and bias mitigation. Practically, the deployment of AI tutors on a larger scale will necessitate addressing technological barriers and ensuring educators are equipped to integrate these systems effectively into their curricula.

Further research should explore the impact of adaptive AI tutoring across diverse populations and educational contexts. Longitudinal studies are needed to assess the long-term effectiveness of AI tutors and their impact on learners' motivation and self-efficacy. Additionally, future work should investigate the scalability of AI tutoring systems and their integration with various pedagogical approaches.

The integration of adaptive AI tutors in e-learning represents a promising advancement in educational technology. While the preliminary findings are encouraging, ongoing research and development are essential to fully realize the potential of AI tutors in enhancing educational outcomes and learner satisfaction.

7 Conclusion

This study has highlighted the potential of adaptive AI tutors within immersive virtual environments to enhance learner engagement and knowledge acquisition. The tailored approach, responsive to the emotional and cognitive states of individuals, offers a significant improvement over traditional e-learning methods. However, the results, while promising, are preliminary and based on a limited sample from a single institution, suggesting the need for further research to establish broader applicability and long-term effectiveness. As educational technologies advance, the insights provided by this study can inform the development of more effective, personalized, and engaging e-learning experiences. The ongoing exploration and refinement of AI tutors in education continue to offer an exciting frontier for enhancing teaching and learning practices.

References

1. Mohammadrezaei, E., Gračanin, D.: Extended reality for smart built environments design: smart lighting design testbed. In: Streitz, N.A., Konomi, S. (eds.) HCII 2022. LNCS, vol. 13325, pp. 181–192. Springer, Cham (2022). https://doi.org/10.1007/978-3-031-05463-1_13
2. Mohammadrezaei, E., Ghasemi, S., Dongre, P., Gračanin, D., Zhang, H.: Systematic review of extended reality for smart built environments lighting design simulations. IEEE Access **12**, 17058–17089 (2024)
3. Lege, R., Bonner, E.: Virtual reality in education: the promise, progress, and challenge. Jalt Call J. **16**(3), 167–180 (2020)
4. Freina, L., Ott, M.: A literature review on immersive virtual reality in education: state of the art and perspectives. In: The International Scientific Conference Elearning and Software for Education, vol. 1, pp. 133–141 (2015)
5. Christou, C.: Virtual reality in education. In: Affective, Interactive and Cognitive Methods for E-learning Design: Creating an Optimal Education Experience, pp. 228–243. IGI Global (2010)
6. Borenstein, J., Howard, A.: Emerging challenges in AI and the need for AI ethics education. AI Ethics **1**, 61–65 (2021)
7. Remian, D.: Augmenting education: ethical considerations for incorporating artificial intelligence in education. Instr. Des. Capstones Collect. **52**, 1–57 (2019)
8. Garrett, N., Beard, N., Fiesler, C.: More than "if time allows" the role of ethics in AI education. In: Proceedings of the AAAI/ACM Conference on AI, Ethics, and Society, pp. 272–278 (2020)

9. Huallpa, J.J., et al.: Exploring the ethical considerations of using chat GPT in university education. Period. Eng. Nat. Sci. **11**(4), 105–115 (2023)

10. Lin, C.-C., Huang, A.Y.Q., Lu, O.H.T.: Artificial intelligence in intelligent tutoring systems toward sustainable education: a systematic review. Smart Learn. Environ. **10**(1), 41 (2023)

11. Ma, W., Adesope, O.O., Nesbit, J.C., Liu, Q.: Intelligent tutoring systems and learning outcomes: a meta-analysis. J. Educ. Psychol. **106**(4), 901 (2014)

12. Kulik, J.A., Fletcher, J.D.: Effectiveness of intelligent tutoring systems: a meta-analytic review. Rev. Educ. Res. **86**(1), 42–78 (2016)

13. Ennouamani, S., Mahani, Z.: An overview of adaptive e-learning systems. In: 2017 Eighth International Conference on Intelligent Computing and Information Systems (ICICIS), pp. 342–347. IEEE (2017)

14. D'Mello, S., et al.: AutoTutor detects and responds to learners affective and cognitive states. In: Workshop on Emotional and Cognitive Issues at the International Conference on Intelligent Tutoring Systems, pp. 306–308 (2008)

15. Bedek, M., Seitlinger, P., Kopeinik, S., Albert, D.: Inferring a learner's cognitive, motivational and emotional state in a digital educational game. Electron. J. E-Learn. **10**(2), 172–184 (2012)

16. Baker, R.S.J., D'Mello, S.K., Rodrigo, M.M.T., Graesser, A.C.: Better to be frustrated than bored: the incidence, persistence, and impact of learners' cognitive-affective states during interactions with three different computer-based learning environments. Int. J. Hum.-Comput. Stud. **68**(4), 223–241 (2010)

17. Reddy, V.K., et al.: Personalized learning pathways: enabling intervention creation and tracking. IBM J. Res. Dev. **59**(6), 4–1 (2015)

18. Shaw, C., Larson, R., Sibdari, S.: An asynchronous, personalized learning platform–guided learning pathways (GLP). Creat. Educ. **5**, 1189–1204 (2014)

19. Moral, S.V., de Benito Crosseti, B.: Self-regulation of learning and the co-design of personalized learning pathways in higher education: a theoretical model approach. J. Interact. Media Educ. **1**, 2022 (2022)

20. Salinas, J., De-Benito, B.: Construction of personalized learning pathways through mixed methods. Comun. Media Educ. Res. J. **28**(65), 31–41 (2020)

21. Gan, W., Qi, Z., Wu, J., Lin, J.C.-W.: Large language models in education: vision and opportunities. In: 2023 IEEE International Conference on Big Data (BigData), pp. 4776–4785. IEEE (2023)

22. Mollick, E.R., Mollick, L.: Using AI to implement effective teaching strategies in classrooms: five strategies, including prompts. Including Prompts, 17 March 2023 (2023)

23. Wang, X., Li, X., Yin, Z., Yue, W., Liu, J.: Emotional intelligence of large language models. J. Pac. Rim Psychol. **17**, 1–12 (2023)

24. Meskó, B.: Prompt engineering as an important emerging skill for medical professionals: tutorial. J. Med. Internet Res. **25**, e50638 (2023)

25. Cain, W.: Prompting change: exploring prompt engineering in large language model AI and its potential to transform education. TechTrends **68**(1), 47–57 (2024)

26. Moubayed, A., Injadat, M., Shami, A., Lutfiyya, H.: Relationship between student engagement and performance in e-learning environment using association rules. In: 2018 IEEE World Engineering Education Conference (EDUNINE), pp. 1–6. IEEE (2018)

27. Atkins, A., Wanick, V., Wills, G.: Metrics feedback cycle: measuring and improving user engagement in gamified elearning systems. Int. J. Serious Games **4**(4), 3–19 (2017)

28. Rebelo, S., Isaías, P.: Gamification as an engagement tool in e-learning websites. J. Inf. Technol. Educ. Res. **19**, p833 (2020)
29. Grant, P., Basye, D.: Personalized Learning: A Guide for Engaging Students with Technology. International Society for Technology in Education (2014)
30. Pratama, M.P., Sampelolo, R., Lura, H.: Revolutionizing education: harnessing the power of artificial intelligence for personalized learning. Klasikal: J. Educ. Lang. Teach. Sci. **5**(2), 350–357 (2023)

Can Large Language Models Recognize and Respond to Student Misconceptions?

Francis Smart[1], Nathan D. Bos[2(✉)], and Jaelyn T. Bos[3]

[1] Michigan State University, East Lansing, USA
[2] Johns Hopkins University, Baltimore, USA
NBOS@mitre.org
[3] University of California, Santa Cruz, Santa Cruz, USA

Abstract. Expert human tutors can observe learner mistakes to understand their misconceptions and procedural errors. Highly capable, but opaque large language models have shown remarkable abilities across numerous domains, and may be useful for adaptive instruction in a variety of ways. Working with publicly available data from the National Assessment of Educational Progress, (388 questions selected from 4th, 8th and 12th grade math and science) we examined these three questions:

1) Do language models find the same problems difficult as students do? We found statistically significant, but small similarities in performance that differ somewhat by model.
2) Do language models have the same pattern of errors as students? Our findings reveal that, under the "minimal " prompts, the models often mirror students in choosing the same incorrect answers. However, this alignment decreases when prompt models used "chain of thoughts".
3) Can language models interpret and explain students' wrong answers? We presented frequently-chosen wrong answers to NAEP items to GPT-4 and an experienced science teacher, and compared their explanations. There was a good correspondence between these explanations, with 81% being fully or partially in agreement.

Discussion focuses on how these capabilities can be used for test design and adaptive instruction.

Keywords: Large Language Model · Difficulty Prediction · Distractor Prediction · Distractor Analysis · Item Analysis · Question Difficulty · Large Language Models · Machine Learning · Natural Language Processing

1 Introduction and Previous Research

In this paper, we assess the potential for large language models (LLMs) to fill several possible roles in adaptive instruction. LLMs produce human-like responses to many (although far from all) types of questions, and some have scored very well on standardized tests. There is potential for these models to fill various roles in test construction, such as serving as item-testing proxies and distractor writing and analysis tools. There is also great potential for these systems to aid in instruction and provide individualized feedback to students.

Prior research in the field of educational assessment and tutoring has focused on various approaches to understanding student errors and providing targeted interventions. Expert human tutors have long been recognized as valuable resources for identifying and explaining student misconceptions and procedural errors. These tutors possess the knowledge and experience to observe student mistakes and provide personalized feedback to enhance learning outcomes. [1] Computer-based tutors have also been in development for decades, and have demonstrated the value of immediate feedback and scaffolding from such systems [2–4].

Alternatively, there is growing interest in exploring the potential of LLMs as tools for understanding and explaining student error patterns. LLMs like GPT-4 and GPT 3.5 have demonstrated impressive capabilities in natural language processing and language generation, opening up new possibilities for LLM application in the educational domain.

A significant body of research has explored the use of LLMs in educational settings. These models have been widely explored in education for various purposes, including generating automated questions [5–9], creating educational materials [10–12], scoring student responses [13, 14], and providing feedback [15, 16].

Settles et al. [9] associated with the platform (Duolingo) successfully utilized LLMs to generate a large number of linguistic items and then employed a second model to predict item difficulty through the analysis of features such as word and sentence length, word corpus difficulty, as well as the predicted Common European Framework of Reference. They do not directly report on "true" compared with "predicted" difficulties but do report measures of internal consistency of 96% and test-retest reliability of 80%.

The potential of LLMs extends beyond these applications in the field of education. Rae et al. [17] found that LLMs excel in various knowledge-based and problem-solving tasks, even when they have not received specific training. Their study covered 152 tasks, including 57 tasks relevant to education, such as High School Chemistry and Astronomy. Similarly, Hocky and White [10] explored the proficiency of LLMs in solving specialized chemistry coding challenges, highlighting their potential in interactive tutoring settings.

While LLMs offer significant advantages, other researchers have raised concerns regarding their misuse. Rudolph et al. [18] emphasize the need for cautious implementation, as LLMs can be potentially misused to generate inauthentic student work or lead to insincere responses from teachers. The authors suggests that the overuse of LLMs in both generating and evaluating student work will increase the risk that students will learn little and teachers will not be able to evaluate work, understand individual failings, and adapt to student needs.

2 Methods

NAEP Item Selection. In this study, we analyzed 388 questions selected from the publicly accessible National Assessment of Educational Progress (NAEP) Question Tool (NAEP Question Tool, 2023). These questions are distributed into subsets consisting of

300 Mathematics questions and 88 Science questions for grades 4, 8, and 12. The Mathematics items originated from assessments conducted in 1990, 1992, 1996, 2003, 2005, 2007, 2009, 2011, and 2013, whereas the Science items were from 2005, 2009, 2011, and 2019. Our selection process started with an initial pool of 1171 Mathematics items and 393 Science items, from which only multiple-choice questions were considered.

Language models have very limited visual reasoning abilities, despite some recent progress in this area, [19] During the selection process, two raters reviewed and eliminated questions that were judged to be too dependent on visual information, such as references to figures, graphs, drawings, charts, and maps. All questions had Sect. 508-compliant descriptions intended to provide enough information to answer the questions without visual aids. When we thought these text descriptors were adequate the items were kept, but when using the visual description was judged to make the problem significantly more difficult the items were excluded. Following these criteria, the item pool was narrowed down to 369 Mathematics items and 93 Science items.

We also found an alarming number of mistakes in the 508 compliant descriptions, where the text provided was insufficient to solve the problem or even misleading. Our review uncovered 69 items (representing an error rate of 15%) that contained some inaccuracies. Two examples:

Question ID:2009-12M2 #11 M181601. "Which of the following expressions is equal to one over X plus 2 minus 2 over X plus 1?

- **Commentary:** There is dual ambiguity about what is in the numerator and what is the denominator; multiple answers are possible.

Question ID:1992-8M15 #11 M049201. Description: "Shaded triangle inside a square". Question: "If the area of the shaded triangle shown above is 4 square inches, what is the area of the entire square?".

- **Commentary:** The geometric figure description is simply incomplete, not enough information to answer the question.

Problematic Items Were Eliminated from Our Test Question Set. Responses to questions were obtained from models through an API interface. Each question was sent as a separate query. Each model was prompted to answer the same question three times under two different prompt models: the "Minimal" prompt and the "Chain of Thought" prompt.

Prompt Engineering. Using the right verbal prompts to frame problems and give instructions to LLMs can make a large difference in performance [20]. The process of finding high quality prompts is called prompt engineering. For all studies reported we pre- experimented with different prompting strategies, and selected a method that gave consistent and high-quality results. When prompt variants are present the results will be presented in terms of largest effect (under the best prompt).

A very common and effective type of prompt, often used with problems that require reasoning are "Chain of Thought" prompts. These instruct the models to solve problems step by step rather than immediately producing an answer, and were shown to greatly improve performance on problems such as math story problems. [21].

Here is an example of a 4th Grade Math question from the NAEP, with the basic prompts used in this study:

Mr. Garcia bought 5 dozen eggs at $0.89 per dozen. What was the total cost of the eggs?

A. $ 4.05   B. $ 4.45   C. $ 5.89   D. $ 10.68.

- **Minimal prompt:** "What is the correct answer? Answer with only a single letter in parenthesizes (letter)".
- **Chain of Thought prompt:** "Solve this item listing the minimum steps required to get to the solution. Finally list the solution as a letter in parentheses for example (Z). Be concise".

3   Results

Model Accuracy and Effects of Prompts. Table 1 Shows the language models we used. Score is the average percentage correct on the test questions. Note that the expected score in the case of random choice is 0.25. The Spearman correlation with student percentage correct is also shown, along with the p statistic of the correlation. Separate statistics are shown for Minimal prompting (Min) and Chain-of-Thought prompting (COT).

Overall models had comparable accuracy to students. See Appendix A for more detailed correlation results between models and students by grade and subject. Most, but not all of the models displayed a relationship with the difficulty ratings (Easy, Medium, Hard) given by NAEP. The main exception was GPT-4, which showed only a weak correlation with difficulty, because of its high accuracy, correctly answering 95% of the items.

Table 1. Model accuracy by prompt type, and correlation with item difficulty

Model	Min	COT	Min		COT	
	Score	Score	rho	p-stat	rho	p-stat
Llama 7b	0.430	0.484	0.130	0.011	0.152	0.053
Llama 13b	0.532	0.521	0.137	0.007	0.142	0.071
Llama 70b	0.568	0.708	0.094	0.064	0.093	0.239
GPT 3.5	0.649	0.874	0.125	0.014	0.171	0.030
GPT 4	0.764	0.961	0.067	0.191	−0.018	0.823
Gemini-Pro		0.832			0.134	0.091

There was also a positive correlation between model size and accuracy. Models are listed in the table in increasing size. We correlated the item scores with the known model parameters – 125 billion for GPT-3.5, an estimated 1 trillion for GPT-4, and 175 billion for Gemini-Pro, along with Llama's various sizes of 7, 13, and 70 billion parameters. There was a significant positive correlation of 0.15, indicating that larger models did tend to score better.

The prompting regime appeared to make a large difference in model accuracy, as can be seen in the increase in scores between columns 1 and 2. It also appears that larger models benefitted more from the Chain of Thought prompting than smaller models.

3.1 Study 1: Do LLMs Find the Same Problems Difficult as Do Students?

We investigated whether Large Language Models (LLMs) could be used to anticipate the difficulty level of test items by having the LLMs directly attempt to answer the questions. These items were presented to the LLMs with both the item body and answer options included.

There were several reasons to be interested in the results of this comparison. Language models have shown remarkably good performance on a number of tests designed for humans, but to what extent do they solve problems in the same way as humans? The study of error patterns is frequently used to try to understand underlying thought processes of humans; if language models showed similar error patterns this might indicate similar processes. We also wanted to know whether more sophisticated LLMs—those with a higher number of parameters—would behave like more adept test-takers. More pragmatically, if language models find the same problems easy or difficult, their performance could be used as indicators of item difficulty.

As was shown in Table 1, there was a small positive correlation between which problems models got correct and how difficult NAEP rated the problems. Tables 2 and 3 show correlations between model results and student results by problem, by age and subject area.

Table 2. Correlations between students and models with minimal prompting.

Model	Science	Math	Grade 4	Grade 8	Grade 12	All
llama7b	0.107	0.128*	0.14	0.126	0.086	0.110*
llama13b	0.370***	0.097	0.160*	0.119	0.118	0.137**
llama70b	0.249*	0.084	0.074	0.128	0.102	0.094
gpt35	0.331**	0.106	0.156*	0.059	0.167	0.127*
gpt4	0.172	0.076	0.178*	−0.039	−0.022	0.067

As we saw in the previous analysis, there is some relationship between model accuracy and student accuracy. In general, model results were closer to student results in science than math. Overall, however, this relationship was weaker than what would be needed to use language models as student proxies using this method. We also thought we might see a stronger interaction between model size and grade level—it would have been very convenient if, for example, Llama13b could serve as a proxy for an 8th grader, whereby GPT3.5 could serve as a proxy for a 12th grader. The relationships observed here are statistically meaningful, but practically limited.

Table 3. Correlations between students and models with chain of thought prompting.

Model	Science	Math	Grade 4	Grade 8	Grade 12	All
llama7b	0.260*	0.174**	0.152	0.185*	0.199*	0.181***
llama13b	0.259*	0.194***	0.142	0.212*	0.278**	0.186***
llama70b	0.277**	0.105	0.093	0.188*	0.072	0.123*
gpt35	0.153	0.175**	0.171*	0.171	0.144	0.166***
gpt4	0.017	0.076	−0.018	0.202*	−0.006	0.064
Gemini Pro	0.424***	0.059	0.134	0.133	0.004	0.105*

3.2 Study 2. Do Language Models Have the Same Pattern of Errors as Students?

Behavioral scientists have long used incorrect answers as windows into the unobservable processes of problem solving. This study focused on wrong answer choices made by students and language models on the NAEP. This analysis focuses on the choice of wrong answers, which often reflect common misconceptions or procedural errors. We initially conducted a correlation analysis of wrong answer choices between models and students. The results were similar to the accuracy results—mostly significant but quite small correlations. These pilot results are not shown.

We decided to focus on most often chosen distractors. We narrowed the question set to questions the models got wrong and selected questions where students got 20% or higher incorrect, to ensure there was some variance to study. We also eliminated a small number of questions where two items tied for most-chosen.

If these answers were randomly distributed, there would be about 33% in each distractor choice. Student answers were clearly not randomly distributed: the most-chosen distractors for each problem represented about 48% of the data. We then tested how often models chose the most-chosen student distractor choices. Results are in Table 4.

Table 4. Model choice of students' most-chosen distractor

Model	Min Mean	COT Mean	Min p-stat	COT p-stat	Min = COT
Llama 7b	0.372	0.230	0.035	0.000	0.000
Llama 13b	0.419	0.215	0.000	0.000	0.000
Llama 70b	0.443	0.241	0.000	0.000	0.000
GPT 3.5	0.415	0.245	0.000	0.014	0.000
GPT 4	0.406	0.234	0.014	0.131	0.031

The results of this analysis were more interesting than expected. With minimal prompting, language models resembled students. Each model, when it got an answer wrong, tended to pick the same wrong answer as students significantly more often than chance. The statistical test was a two-way test against chance level prediction of 0.33.

Unexpectedly, after Chain of Thought prompting, this pattern reversed: model wrong answers were significantly less likely to be the same as students' wrong answers. To confirm this difference, we did a paired test comparison, and confirmed that prompt type led to a significant difference on most-chosen distractors. This has a number of implications for adaptive instruction, which will be discussed.

3.3 Study 3. Can Language Models Interpret and Explain Students' Wrong Answers?

Can language models offer explanations for student errors akin to those provided by teachers? Large Language Models can do more than answer multiple-choice questions, they can also produce prose, often very lucid and articulate prose. An important job of a teacher or tutor is explaining concepts, and an expert tutor or teacher can explain concepts at the right level and from the right perspective to reach a learner who is having difficulty. How well could a large language model perform that function?

For this analysis we narrowed the set of wrong answer choices to what we will call "dominant distractors". The criteria for a dominant distractor are that the problem accuracy has to be less than 80% (has enough wrong answers to support analysis) and one distractor has to have received at least 40% of those incorrect answers. We hypothesized that dominant distractors should represent common misconceptions or a specific procedural error. Addressing these specific gaps can be the key to helping students avoid similar errors.

For this analysis, we recruited an experienced middle school science teacher, and a current Ph.D. student in Ecology and Evolutionary Biology (the 3rd author) to review 72 dominant distractors and provide a best-guess explanation. We obtained GPT-4's explanations for the same set of dominant distractors.

Here are two examples of typical student mistakes.

Example Problem #1. What two gases make up most of the Earth's atmosphere?

A. Hydrogen and oxygen
B. Hydrogen and nitrogen
C. Oxygen and carbon dioxide
D. Oxygen and nitrogen

The correct answer is D, but C was a dominant distractor. Clearly this represents a knowledge gap about the composition of the atmosphere, but why was that particular option chosen so often?

Teacher's explanation	GPT-4 explanation
Nitrogen makes up the majority of earth's atmosphere. Students may recognize that CO_2 is biologically important, but it comprises a tiny fraction of the atmosphere	Students commonly know that both oxygen and carbon dioxide are present in the Earth's atmosphere. Since they are aware that carbon dioxide is considerably talked about due to its role in global warming and environmental issues, they might have presumed it to be one of the major components, thereby choosing it alongside oxygen

Example Problem #2. In a bag of marbles, 1/2 are red, 1/4 are blue, 1/6 are green, and 1/12 are yellow. If a marble is taken from the bag without looking, is it most likely to be:

A. Red
B. Blue
C. Yellow
D. Green

The correct answer is A, the dominant distractor was C, which is the opposite of correct. Why?

Teacher's explanation	GPT-4 explanation
Reversed understanding of numerator and denominator of fraction and how they relate to probabilities	Students might have chosen 'yellow' as the answer due to a misconception about how fractions work. In this case, they may not fully understand that a smaller denominator in a fraction does not imply a larger value. Thus, they may have seen '1/12' and assumed that because '12' is the largest number, this must mean there are more yellow marbles. This is a fundamental misunderstanding about what the fraction represents specifically, that the denominator represents the total number of parts, and a larger denominator means each part is smaller

After obtaining explanations from the teacher and GPT-4, the second author reviewed these explanations and coded them as 2-full agreement, 1-partial agreement, or 3-mismatch. In the mismatch category, both the teacher and the model might both be correct or incorrect, there is no way to know without a source of ground truth, but their explanations were different. Here is the observed agreement:

Full Agreement	56% (N = 40)
Partial	25% (N = 18)
Mismatch	19% (N = 14)

We interpret the 81% of full or partial agreement as providing some evidence that language models can 'understand' student mistakes and provide helpful explanations of them. A few more observations follow.

The GPT-4 explanations did, infrequently, contain procedural errors themselves, or scientific mis-statements, where the teachers' did not appear to contain these. For example, at one point in the second explanation listed, GPT-4 said 'smaller denominator' when it should have said 'larger'. Using an LLM for feedback might itself require a quality-enhancing prompt procedure.

One difference between the explanations that was noteworthy: the GPT-4 explanations were considerably longer, as in the given examples, with an average word count of 11 for the teacher and 65 for the LLM. Sometimes these longer explanations contained nuances that the teacher explanations did not. In problem 1, GPT elaborates on why carbon dioxide might be more familiar to the student than nitrogen. In problem 2, GPT notes that in addition to the procedural error, the student may simply be keying on the larger number (12 vs 3), even though it is in the denominator.

The teacher was not asked to produce long responses for this coding task, often opting for a minimal explanation that conveyed the correct idea. Partially this was because they knew there was no actual student involved. But it is also a near-universal fact of education that teacher attention is a scarce commodity, and teachers do not have time or attention to provide the fullest, most complete feedback to every student on every problem. Attention is less of a concern for a language model, which can produce longer and sometimes more complete explanations very quickly and on-demand. This points to some potential use cases for LLMs, which will be discussed.

As a second part of the explanation analysis, we asked both the teacher and the LLM to categorize the mistakes as misconceptions, procedural errors or knowledge gaps. We found that the categorization scheme needs work: there was a lot of ambiguity in the ratings (as reported by the 3rd author) and little agreement. In particular, there was considerable disagreement as to what answers might represent a true misconception, which would benefit from being directly addressed, versus a knowledge gap, which would best be addressed by focusing on the correct answer.

4   Discussion

Large Language models, used in a standard way, appear to be mediocre proxies for students. There are statistically significant similarities in both which problems they find difficult, and what types of errors they make. These similarities might indicate that, at least some of the time, large language models are not only reproducing human performance, but using similar methods.

Improving these abilities would provide useful tools for test development. Estimating problem difficulty is an important part of test design. Distractor analysis is an important part of test evaluation. Error patterns often offer clues to internal problem-solving processes, and doing this in an automated way would have a number of applications. However, we also note that the most advanced LLM used in this study, GPT-4, showed the lowest correlation with student performance by reason of significantly outperforming the students at all levels. This means that more advanced LLMs are not necessarily better proxies for estimating problem difficulty. Another possible application of LLMs in test design would be to use language models to generate effective distractors.

We also, as a tangential finding, found a large number of errors in the text alternatives to visual figures in NAEP problems. These were mistakes that would prevent a student from answering a problem correctly, or at all. Vision-impaired students depend on alternative texts to be both accurate and helpful. Large Language models might be employed to both help evaluate, and even help produce these text descriptions.

The most intriguing and unexpected finding was that language model distractor choice are similar to students with minimal prompting, but statistically dissimilar with chain of thought prompting. This finding suggests several follow-ups.

First, we wonder whether, if the step-by-step processes produced by models improve the model answers, would these same steps be useful for students? Could adaptive instruction systems mine model self-explanations for useful student hints?

Similarly, are there other methods of prompting that would lead models to be even more similar to students? Are there methods that could usefully mimic students of different ages, or different strengths and weaknesses?

Our study of model-generated feedback also suggests useful methods for automated tutoring, similar to some that are already being employed. GPT-4 gave feedback similar to an experienced science teacher. It also gave more feedback and was able to provide it immediately. Rapid feedback given 'in the moment' may have great value, as opposed to waiting for grading results. The task of returning comments with grading is also a significant burden for teachers, and language models could make this process easier and perhaps better.

Disclosure of Interests. The authors have no competing interests to declare that are relevant to the content of this article.

Appendix A. Detailed Language Model Performance by Subject and Grade

Accuracy of Models Answering NAEP Items, Miminal Prompting

Grade	Subject	students	llama7b	llama13b	llama70b	gpt35	gpt4
4	Math	55.2%	40.6%	54.3%	52.7%	62.0%	75.9%
4	Science	60.7%	72.0%	80.0%	86.6%	92.0%	100.0%
8	Math	53.5%	36.6%	42.1%	45.9%	55.7%	70.9%
8	Science	54.4%	60.0%	65.3%	78.6%	85.3%	100.0%
12	Math	53.2%	45.9%	45.3%	48.0%	57.3%	57.3%
12	Science	45.1%	70.2%	71.6%	80.7%	82.4%	92.1%

Accuracy of Models Answering NAEP Items, Chain of Thought Prompting

Grade	Subject	students	llama7b	llama13b	llama70b	gpt35	gpt4	Gemini pro
4	Math	55.2%	45.1%	48.4%	68.1%	85.9%	96.6%	81.0%
4	Science	60.7%	65.3%	73.3%	85.3%	96.0%	93.3%	95.8%
8	Math	53.5%	36.4%	38.8%	59.2%	82.7%	96.4%	76.8%
8	Science	54.4%	52.0%	65.3%	76.0%	84.0%	92.0%	97.3%
12	Math	53.2%	33.7%	49.7%	56.3%	82.0%	93.4%	76.4%
12	Science	45.1%	54.0%	74.6%	78.8%	86.8%	94.7%	82.9%

References

1. D'Mello, S.: Expert tutors feedback is immediate, direct, and discriminating (2010)
2. McKendree, J.: Effective feedback content for tutoring complex skills. Hum.-Comp. Interact. **5**, 381–413 (1990). https://doi.org/10.1207/s15327051hci0504_2
3. Koedinger, K.R., Aleven, V.: Exploring the assistance dilemma in experiments with cognitive tutors. Educ. Psychol. Rev. **19**, 239–264 (2007). https://doi.org/10.1007/s10648-007-9049-0
4. Kantack, N., Cohen, N., Bos, N., Lowman, C., Everett, J., Endres, T.: Instructive artificial intelligence (AI) for human training, assistance, and explainability. In: Artificial Intelligence and Machine Learning for Multi-domain Operations Applications IV, pp. 45–54. SPIE (2022). https://doi.org/10.1117/12.2618616
5. Bezirhan, U., Davier, M.: Automated reading passage generation with OpenAI's large language model (2023)
6. Raina, V., Gales, M.: Multiple-choice question generation: towards an automated assessment framework (2022)
7. Wang, Z., Valdez, J., Basu Mallick, D., Baraniuk, R.G.: Towards human-like educational question generation with large language models. In: Rodrigo, M.M., Matsuda, N., Cristea, A.I., Dimitrova, V. (eds.) AIED 2022. LNCS, vol. 13355, pp. 153–166. Springer, Cham (2022). https://doi.org/10.1007/978-3-031-11644-5_13

8. von Davier, M.: Training Optimus Prime, M.D.: Generating medical certification items by fine-tuning OpenAI's gpt2 transformer model (2019). http://arxiv.org/abs/1908.08594, https://doi.org/10.48550/arXiv.1908.08594

9. Settles, B., LaFlair, G.T., Hagiwara, M.: Machine learning-driven language assessment. Trans. Assoc. Comput. Linguist. **8**, 247–263 (2020). https://doi.org/10.1162/tacl_a_00310

10. Hocky, G.M., White, A.D.: Natural language processing models that automate programming will transform chemistry research and teaching. Digit. Discov. **1**, 79–83 (2022)

11. Moore, S., Nguyen, H.A., Bier, N., Domadia, T., Stamper, J.: Assessing the quality of student-generated short answer questions using GPT-3. In: Hilliger, I., Muñoz-Merino, P.J., De Laet, T., Ortega-Arranz, A., Farrell, T. (eds.) EC-TEL 2022. LNCS, vol. 13450, pp. 243–257. Springer, Cham (2022). https://doi.org/10.1007/978-3-031-16290-9_18

12. Walsh, J.: Lesson plan generation using natural language processing: prompting best practices with openai's GPT-3 model (2022)

13. Mizumoto, A., Eguchi, M.: Exploring the potential of using an AI language model for automated essay scoring. Res. Methods Appl. Linguist. **2**, 100050 (2023)

14. Wu, X., He, X., Liu, T., Liu, N., Zhai, X.: Matching exemplar as next sentence prediction (MeNSP): zero-shot prompt learning for automatic scoring in science education. In: Wang, N., Rebolledo-Mendez, G., Matsuda, N., Santos, O.C., Dimitrova, V. (eds.) AIED 2023. LNCS, vol. 13916, pp. 401–413. Springer, Cham (2023). https://doi.org/10.1007/978-3-031-36272-9_33

15. Matelsky, J.K., Parodi, F., Liu, T., Lange, R.D., Kording, K.P.: A large language model-assisted education tool to provide feedback on open-ended responses (2023)

16. Peng, B., Galley, M., He, P., Cheng, H., Xie, Y., Hu, Y., Gao, J.: Check your facts and try again: Improving large language models with external knowledge and automated feedback (2023)

17. Rae, J.W., et al.: Scaling language models (2021)

18. Rudolph, J., Tan, S., Tan, S.: War of the chatbots: Bard, Bing Chat, ChatGPT, Ernie and beyond. The new AI gold rush and its impact on higher education. J. Appl. Learn. Teach. **6** (2023)

19. Lu, P., et al.: Mathvista: evaluating mathematical reasoning of foundation models in visual contexts. arXiv preprint arXiv:2310.02255 (2023)

20. White, J., et al.: A prompt pattern catalog to enhance prompt engineering with chatgpt. arXiv preprint arXiv:2302.11382 (2023)

21. Wei, J., et al.: Chain-of-thought prompting elicits reasoning in large language models (2023). http://arxiv.org/abs/2201.11903, https://doi.org/10.48550/arXiv.2201.11903

Examining the Role of Knowledge Management in Adaptive Military Training Systems

Robert A. Sottilare$^{(\boxtimes)}$ (iD)

Soar Technology LLC, Orlando, FL 32817, USA
bob.sottilare@soartech.com

Abstract. This paper explores the pivotal role of knowledge management (KM) in revolutionizing adaptive military training systems (AMTSs). As military organizations confront increasingly complex and dynamic operational environments, traditional training methodologies must evolve to cultivate the adaptivity, cognitive flexibility, and real-time decision-making skills required to operate successfully. KM is the process of capturing, organizing, storing, and sharing information and expertise, and involves the creation of systems and practices to identify, create, represent, distribute, and enable the adoption of insights and experiences. In the case of adaptive training, the goal of knowledge management is to improve efficiency, promote innovation, and enhance decision-making through knowledge transfer from training experiences to application in operational (work) experiences. The integration of advanced technologies, such as artificial intelligence, virtual reality, and machine learning, necessitate a cohesive KM framework. Case studies demonstrate methods used in successful implementations, including cyber warfare preparedness, urban warfare simulations, and personalized leadership development. Challenges such as resistance to cultural change, security concerns, and resource limitations are also discussed, alongside strategies for overcoming these barriers. This paper emphasizes the intersection of adaptive training technologies and KM, showcasing how AI-driven adaptivity tailors training experiences for efficient skill acquisition. The paper also outlines future directions for research, including the integration of emerging technologies, enhanced personalization, blockchain for data security, and ethical considerations in responsible AI implementation. The broader implications for human-computer interaction research and military preparedness underscore the significance of KM in fostering a culture of continuous learning, collaboration, and informed decision-making within military organizations.

Keywords: Knowledge Management · Adaptive Military Training · Cognitive Flexibility · Artificial Intelligence

1 Introduction

We begin by providing an overview of the evolving landscape of military training systems, emphasizing the increasing complexity of contemporary warfare and the critical need for adaptive training approaches. We highlight the challenges faced by traditional training methodologies in keeping pace with rapidly changing technology and dynamic operational environments encountered by military personnel.

R. A. Sottilare and J. Schwarz (Eds.): HCII 2024, LNCS 14727, pp. 300–313, 2024.
https://doi.org/10.1007/978-3-031-60609-0_22

1.1 The Evolving Landscape of Military Training Systems

The landscape of military training systems has undergone a profound transformation in response to the multifaceted demands of modern warfare. Traditional training methods, characterized by static curricula and standardized exercises, are proving insufficient in preparing armed forces for the intricacies of contemporary conflicts. The integration of advanced technologies, such as artificial intelligence, virtual reality, and data analytics, is reshaping the training paradigm, enabling a more dynamic and adaptive approach that supports the tailoring of training scenario difficulty to match the capabilities of trainees and maintain trainee engagement [1]. This evolution necessitates a reevaluation of traditional structures to harness the full potential of emerging tools and methodologies, emphasizing the need for a cohesive and knowledge-driven framework.

1.2 The Complexity of Contemporary Military Operations

The nature of contemporary military operations is marked by unprecedented complexity, featuring asymmetrical threats, rapid technological advancements, and dynamic operational landscapes. In this environment, military forces must possess a diverse skill set that extends beyond traditional military tactics. Adaptive training approaches become imperative as they enable military personnel to cultivate skills such as cognitive flexibility [2], rapid decision-making [3], and interdisciplinary collaboration [4]. The intricacies of modern military operations demand training systems that can accurately simulate the diverse scenarios representing real-world conditions, adapt to evolving threats, and prepare soldiers for the unpredictable nature of 21st-century military operations. It is critically important for adaptive training systems to efficiently transfer required knowledge gained during training to the operational or work environment.

1.3 Challenges Faced by Trainers Using Traditional Training Methodologies

Traditional military training methodologies face significant challenges in meeting the demands of contemporary warfare. Rigidity in curricula and the inability to simulate the complexity of evolving real-world scenarios limit the effectiveness of these approaches in supporting operational mission objectives [5]. Trainers and simulation developers often struggle to provide realistic and dynamic training environments, hindering the development of critical skills such as adaptivity and rapid decision-making. Additionally, the static nature of traditional training fails to address the individualized learning needs of military personnel, making it difficult to prepare them for the diverse challenges they may encounter in modern operational settings.

1.4 Rapidly Changing Technology and Dynamic Operational Environments

The accelerating pace of technological advancements and the ever-changing nature of operational environments pose formidable challenges to military training systems. Traditional methodologies, which may have once been effective, struggle to keep pace with the introduction of cutting-edge technologies and the evolving tactics employed by adversaries [6]. The dynamic nature of contemporary conflicts requires training systems to adapt swiftly, incorporating the latest advancements to ensure that military personnel are equipped with the most relevant skills. This rapid evolution underscores the necessity for adaptive training approaches that integrate KM and emerging technologies to maintain a state of readiness in the face of unpredictable and rapidly changing circumstances.

2 Military Training Paradigms

This section explores the historical progression of military training paradigms, examining traditional models and their limitations in addressing the diverse skill sets required by modern armed forces. We emphasize the growing importance of adaptivity, cognitive flexibility, and real-time decision-making skills, underscoring the necessity for innovative approaches to training that integrate cutting-edge technologies.

2.1 Historical Progression of Military Training Paradigms

The historical evolution of military training paradigms reflects the shifting nature of warfare and the evolving needs of armed forces. Traditionally, training methods were often rooted in rigid discipline, emphasizing standardized drills and hierarchical structures. Over time, as warfare became more complex, training paradigms adapted to incorporate a broader range of skills. The transition from classical formations and linear tactics to the development of combined arms tactics marked a significant shift in training approaches. The historical progression reflects a continual effort to align training methodologies with the strategic and technological advancements of the times. In other words, as threats evolve so should training methods to ensure high efficiency in transferring knowledge (successful experiences) to application in operational environments.

2.2 Traditional Training Models and Their Limitations

Traditional military training models, often characterized by fixed curricula and repetitive exercises, face limitations in preparing modern armed forces for the diverse and dynamic challenges of contemporary warfare. These models tend to focus on specific skill sets and scenarios, lacking the adaptivity required for the unpredictable nature of 21st-century conflicts. The rigid nature of traditional training can result in a disconnect between the skills emphasized in training and the multifaceted demands of real-world missions. As armed forces increasingly engage in asymmetrical warfare and complex peacekeeping operations, the inadequacies of traditional models become more pronounced, highlighting the need for a more versatile and adaptive training paradigm.

2.3 The Growing Importance of Adaptiveness, Cognitive Flexibility, and Real-Time Decision-Making Skills

In the face of evolving threats and operational landscapes, there is a growing recognition of the importance of certain skills that extend beyond traditional battlefield tactics. Adaptiveness, cognitive flexibility, and real-time decision-making skills have become critical components of military preparedness.

Adaptiveness is the ability of an entity or system to adjust and thrive in changing conditions or environments by modifying its behavior, structure, or functions in response to varying conditions with the goal of achieving better performance or outcomes [7]. Cognitive flexibility is the ability to switch between concepts, and to think about multiple concepts simultaneously [8] and involves adapting to changing conditions, adjusting to new information, and shifting one's thinking between different tasks or perspectives [9]. Real-time decision-making skills involve the ability to make effective and timely decisions in dynamic and rapidly changing conditions. Individuals with strong real-time decision-making skills can quickly analyze information, assess the situation, and choose the most appropriate course of action based on their ability to rapidly process information, think critically, prioritize options, and assess risk [10].

Modern armed forces must navigate unpredictable scenarios that demand quick and informed decisions. The ability to adapt to changing circumstances, think critically, and make rapid decisions is paramount. Recognizing the significance of these skills has led to a paradigm shift in training philosophies, with an increased emphasis on cultivating mental agility and flexibility in addition to physical prowess.

2.4 Innovative Approaches to Training that Integrate Cutting-Edge Technologies

The contemporary military landscape necessitates innovative approaches to training that seamlessly integrate cutting-edge technologies. Traditional models struggle to incorporate the complexities of modern warfare, where technological advancements play a pivotal role. The integration of artificial intelligence, virtual reality, and simulation technologies offers the potential to create realistic and dynamic training environments. These technologies enable armed forces to simulate diverse scenarios, from cyber threats to urban warfare, providing a more holistic approach to skill development. The necessity for innovation lies in the ability to bridge the gap between theoretical training and the practical application of skills in complex and evolving operational settings.

3 KM in Military Training

Focusing on the pivotal role of KM, this section delves into how effective information capture, organization, and dissemination can enhance military training outcomes. We discuss the potential of KM systems in preserving institutional knowledge, fostering collaboration, and providing timely access to relevant information, thereby contributing to the adaptivity of military training programs.

3.1 What is KM?

KM is a multidisciplinary approach to identifying, capturing, organizing, and utilizing an organization's collective knowledge to achieve its objectives more effectively [11]. KM at the organizational level seeks to transfer knowledge to its members so that knowledge can be applied in training and the resulting firsthand knowledge and skills can be transferred to effective use in operational environments encountered in the future. KM involves the systematic process of creating, acquiring, organizing, storing, and disseminating information and expertise within an organization. The goal of KM is to enhance the organization's ability to learn, adapt, and innovate by making the right information available to the right people at the right time. Key components of KM include [12]:

- *Knowledge Creation:* The generation of new knowledge through innovation, research, and experience. This involves converting individual and collective insights into explicit knowledge that can be shared and utilized.
- *Knowledge Capture:* The process of identifying and collecting relevant information and expertise within an organization. This can include documentation, lessons learned from past experiences, and insights gained through various activities.
- *Knowledge Organization:* Structuring and categorizing knowledge in a meaningful way to facilitate easy retrieval and understanding. This can involve creating taxonomies, databases, and other classification systems.
- *Knowledge Storage:* The secure and accessible storage of explicit knowledge within databases, repositories, or other information systems. This ensures that information is readily available to those who need it.
- *Knowledge Retrieval:* The ability to search for and retrieve relevant knowledge when needed. This involves having efficient systems in place to access and apply the knowledge stored within an organization.
- *Knowledge Sharing:* The process of disseminating knowledge among individuals and teams. This can include formal training programs, mentorship, collaborative platforms, and communication channels that encourage the exchange of information.
- *Knowledge Application:* The practical use of knowledge to solve problems, make decisions, and drive innovation. This involves translating knowledge into action for the benefit of the organization.

Effective KM contributes to improved decision-making, increased organizational agility, enhanced innovation, and the preservation of institutional memory. It is particularly crucial in dynamic and knowledge-intensive environments where staying competitive requires constant learning and adaptation. KM is applied in various sectors, including military operations, business, healthcare, education, and government, to leverage intellectual capital and create a culture of continuous improvement.

3.2 The Pivotal Role of KM in Training

KM plays a pivotal role in revolutionizing military training by facilitating the efficient capture, organization, and dissemination of critical information. In contemporary warfare, where information is a valuable asset, effective KM systems are essential for ensuring that military personnel are equipped with the most relevant and up-to-date knowledge. These systems provide a structured approach to handling information, promoting seamless collaboration, and enabling the integration of new insights into training methodologies and curricula. The role of KM in training extends beyond simple data storage. KM serves as the backbone for adaptive training approaches, enabling armed forces to stay agile in response to evolving threats and operational requirements. KM as a tool in operational environments should also be duplicated in training environments to provide necessary scenario realism.

3.3 Information Capture, Organization, and Dissemination

The effectiveness of military training outcomes is intrinsically linked to the efficiency of information capture, organization, and dissemination. KM systems enable the systematic collection of data from various sources, ranging from historical mission reports to real-time intelligence feeds. The organized structuring of this information ensures that it is readily accessible to trainers and trainees alike. By facilitating the dissemination of relevant knowledge, training programs can be tailored to address specific challenges, incorporating lessons learned from past experiences. This targeted and informed approach enhances the overall effectiveness of military training, fostering a more responsive and adaptive force capable of meeting the demands of modern warfare.

3.4 Preserving Institutional Knowledge, Fostering Collaboration, and Enhancing Access to Relevant Information

KM systems offer immense potential in preserving institutional knowledge within military organizations. These systems act as repositories for accumulated wisdom, lessons learned, and best practices from past operations. By capturing and codifying this institutional knowledge, military training benefits from the collective experiences of seasoned personnel, ensuring that valuable insights are not lost over time. Additionally, KM fosters collaboration by providing a platform for sharing expertise and facilitating communication among dispersed units. Timely access to relevant information becomes a cornerstone of decision-making, allowing military personnel to make informed choices based on a comprehensive understanding of the operational landscape. In essence, KM systems not only preserve the legacy of institutional knowledge but also empower armed forces to continuously adapt and excel in dynamic and challenging environments.

4 Integration of KM with Adaptive Training Technologies

Here, we examine the intersection of adaptive technologies and KM within military training systems. We discuss the incorporation of artificial intelligence (AI) and machine learning (ML) to create personalized and dynamic training environments. This section explores how these technologies can leverage KM to tailor training experiences, allowing for more efficient skill acquisition and improved readiness.

4.1 The Nexus of Adaptive Training Technologies and KM

The intersection of adaptive training technologies and KM represents a paradigm shift in military training methodologies. Adaptive training technologies, encompassing artificial intelligence (AI), are increasingly integrated with KM systems to create a symbiotic relationship. KM serves as the backbone for these technologies, providing a structured repository for relevant military intelligence, historical data, and lessons learned. This integration enables adaptive training systems to dynamically adjust scenarios based on real-world insights, ensuring that training experiences are not only immersive but also informed by the collective knowledge of the armed forces. This is especially true in adaptive training systems that seek to use generative AI such as large language models (LLMs) like ChatGPT to create streams of military communications or drive the real-time perception-action cycle of intelligent computer-generated forces (CGFs) [13].

4.2 The Roles of AI and ML in Training Environments

The incorporation of AI and machine learning marks a transformative era in military training. AI algorithms analyze vast amounts of data to identify patterns and generate realistic simulations. ML adapts training scenarios based on individual and collective performance, creating personalized experiences. This integration enhances the realism of training exercises, allowing personnel to engage with diverse and dynamic scenarios, from complex battlefield situations to cyber warfare challenges. The result is a training environment that not only replicates the complexities of modern warfare but also leverages cutting-edge technologies to optimize skill development and readiness.

4.3 Leveraging KM to Enable Tailored Training Experiences

Adaptive training technologies leverage KM to tailor training experiences, optimizing skill acquisition and enhancing overall readiness. KM systems act as repositories for organizational intelligence, historical mission data, and expert insights. By integrating this wealth of information with adaptive technologies, training programs can dynamically adjust content based on the specific needs and performance of individual trainees or units. This tailoring allows for a more efficient allocation of resources, focusing on areas where improvement is needed and accelerating the development of crucial skills. The result is a training ecosystem that is not only adaptive to emerging threats and scenarios but also finely tuned to the unique requirements of each military unit, fostering a more agile and prepared armed forces.

Figure 1 illustrates a cycle of training transfer that is facilitated by KM. In the figure, required operational missions (green box) are driven by evolving threats which are documented through organizational intelligence and in turn drive benchmarking of training scenarios (yellow box) that represent a large set of possible missions and threats. Knowledge from training scenarios is captured from trainee communications and simulator data include after action reviews (AARs). Benchmarking of operational scenarios (blue box) represent knowledge gained during actual missions, and this knowledge is captured from operational communications, telemetry, and mission debriefs. Both AARs and mission debriefs capture lessons-learned (trainee or operator errors).

Knowledge and skills gained (proficiency; orange box) during training or operational experiences are the result of training or mission assessments respectively. The accuracy of these assessments is critical to accurately representing the current proficiency of trainees/operators, and in determining next steps in their progression of training and operational missions.

The transfer of training to operations is the process of applying knowledge and skills that have been learned in one domain to another domain. For example, if you learn how to play chess, you might transfer some of the strategic thinking and problem-solving skills learned in chess to other games or tasks. In Fig. 1, transfer between training and operations (gold boxes) can be largely dependent on experience in a particular environment but may also be highly dependent on the semantic similarity of the environments or the missions. Semantic similarity means the degree of likeness or similarity between two items. In this case, we are examining the similarity in the training and operational environments or missions. For example, similarity in the physical conditions of the training and operational environments can promote transfer of knowledge, and similarity in tasks (even in very different physical environments) can also promote transfer.

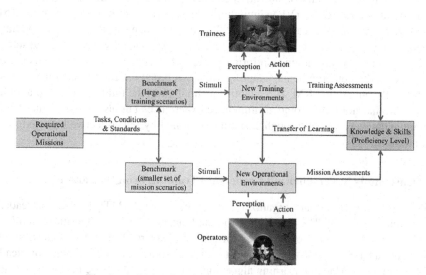

Fig. 1. A Cycle of Training Transfer facilitated by KM

5   Case Studies: Implementing AMTSs

Drawing on three real-world examples, this section presents case studies showcasing successful implementations of AMTSs with a strong emphasis on KM. We analyze the impact of these systems on trainee performance, mission success, and long-term strategic goals, providing empirical evidence of the benefits derived from integrating KM into adaptive training methodologies.

5.1   Case Study 1: Cyber Warfare Training Enhancement with AI and KM

In this case study, an armed forces branch implemented an Adaptive Military Training System focused on cyber warfare preparedness [14]. The system seamlessly integrated artificial intelligence (AI) algorithms and KM solutions to enhance training effectiveness. KM systems captured and organized the latest cyber threat intelligence, historical attack patterns, and best practices. The AI-driven training platform utilized this knowledge to dynamically generate realistic cyber scenarios, adapting difficulty levels based on individual trainee performance. This adaptive approach not only optimized skill development but also allowed continuous updates to the training curriculum based on evolving cyber threats, ensuring that military personnel were well-prepared for rapidly changing cyber warfare landscapes.

5.2   Case Study 2: Virtual Reality Simulation for Urban Warfare Training

In this case study, a military organization employed an Adaptive Military Training System utilizing virtual reality (VR) simulations to support urban warfare scenarios [15]. KM played a critical role in aggregating intelligence on urban terrain, past urban warfare missions, and lessons learned. This information was systematically organized and stored in a KM system. The VR training platform leveraged this repository to dynamically recreate realistic urban environments, adapting scenarios based on trainee performance and feedback. By integrating KM with adaptive technologies, the military achieved a training system that not only provided immersive experiences but also ensured that urban warfare skills were honed through scenarios informed by the collective knowledge of past missions.

5.3   Case Study 3: Personalized Leadership Training Using ML and KM

In this case study, a military organization implemented an AMTS focused on leadership development [16]. The system incorporated machine learning algorithms and KM solutions to tailor training experiences for aspiring leaders. The KM system captured historical leadership success stories, organizational strategies, and lessons from leadership challenges. Machine learning algorithms then analyzed trainee performance and identified specific leadership skills requiring improvement. The training program dynamically adjusted scenarios, incorporating real-world leadership challenges based on the collective knowledge stored in the system. This adaptive approach not only personalized leadership training but also ensured that future leaders were equipped with the insights and skills needed to navigate complex operational environments effectively.

6   Challenges

In this section, we address the challenges and potential barriers associated with implementing KM in AMTSs. We propose strategies to overcome these obstacles and discuss future directions for research and development in the pursuit of more effective and responsive training approaches. Five barriers to effective KM solutions in AMTSs follow:

6.1   Resistance to Cultural Change Challenge

One major challenge in implementing KM in AMTSs is resistance to cultural change within the armed forces. Military organizations often have established traditions and hierarchical structures that may resist the adoption of new KM practices. Overcoming this challenge requires a comprehensive change management strategy. It involves fostering a culture that values continuous learning and knowledge sharing. Leadership endorsement, clear communication about the benefits of KM, and incentives for participation can encourage personnel to embrace the cultural shift toward a more collaborative and knowledge-driven environment.

6.2 Security Concerns and Data Protection Challenges

The sensitive nature of military information poses security concerns and challenges in implementing KM systems. Ensuring data protection and preventing unauthorized access to classified information is crucial. Implementing robust cybersecurity measures and encryption protocols is essential. Developing access controls based on security clearances, implementing secure cloud solutions, and conducting regular security audits can help mitigate these concerns. Additionally, providing comprehensive training on data security and reinforcing the importance of confidentiality can contribute to a secure KM environment.

6.3 Integration with Legacy Systems Challenges

Military organizations often rely on legacy systems that may not easily integrate with modern KM technologies. The challenge lies in seamlessly incorporating new systems without disrupting existing operational workflows. Adopting an incremental approach to integration is key. This involves identifying interoperability standards, developing APIs (Application Programming Interfaces), and gradually migrating data from legacy systems to more adaptive platforms. Collaboration with technology experts and phased implementation can help ensure a smooth transition without compromising existing operational capabilities.

6.4 Lack of User Engagement Challenge

If military personnel are not adequately trained on KM systems, or if the systems are not user-friendly, there may be a lack of engagement, hindering the successful implementation of these systems. Prioritizing user-friendly interfaces and conducting thorough training programs are essential. Providing ongoing support and emphasizing the practical benefits of KM in daily operations can increase user engagement. Feedback mechanisms should also be implemented to continuously refine and improve the usability of the systems based on user experiences.

6.5 Limited Resources and Budget Constraints

Military organizations often operate within budget constraints and resource limitations. Implementing KM systems may require significant financial investment and allocation of resources. Strategic resource allocation and phased implementation can help address budget constraints. Leveraging open-source solutions, seeking partnerships with technology providers, and prioritizing key functionalities can optimize resource usage. Making a compelling case for the long-term cost-effectiveness and operational benefits of KM can also garner support for the necessary financial investments.

7 Future Directions

Future directions for KM research and development in the context of adaptive military training approaches should focus on advancing technologies, refining methodologies, and addressing emerging challenges. Here are key areas for consideration:

7.1 Integration of Advanced Technologies

As technology continues to evolve, research should explore the integration of emerging technologies such as augmented reality (AR), natural language processing (NLP), and advanced analytics into KM systems for adaptive training. AR can enhance realistic simulation experiences, while NLP can improve the understanding and extraction of information from unstructured data sources. Advanced analytics can provide valuable insights into training effectiveness and identify areas for improvement.

7.2 Enhanced Personalization and AI-Driven Adaptivity

Future research should aim to enhance the personalization of adaptive training experiences through artificial intelligence (AI). By leveraging machine learning algorithms, KM systems can analyze individual and collective performance data to tailor training scenarios based on specific needs and learning styles. This approach ensures that training adapts in real-time, optimizing skill development and readiness for each trainee.

7.3 Blockchain for Secure KM

Exploring the potential applications of blockchain technology in KM within military training systems is crucial. Blockchain can enhance data security, integrity, and transparency, addressing concerns related to unauthorized access and data tampering. Implementing decentralized and immutable ledgers for knowledge repositories can ensure the trustworthiness of information critical for training scenarios and decision-making processes.

7.4 Human-Centric Design and User Experience (UX)

Research efforts should focus on improving the user experience of KM systems to encourage greater adoption and engagement among military personnel. Human-centric design principles, usability studies, and feedback mechanisms can contribute to creating intuitive interfaces and workflows. Understanding the needs and preferences of end-users is paramount for successful implementation and sustained use.

7.5 Interoperability and Collaboration Platforms

Future research should emphasize the development of interoperable KM systems that facilitate seamless collaboration both within and across military organizations. Integration with existing military databases, communication platforms, and information-sharing networks is essential for creating a cohesive and interconnected knowledge ecosystem. Collaboration platforms should enable real-time communication and information exchange among military personnel, fostering a culture of continuous learning and shared expertise.

7.6 Ethical Considerations and Responsible AI

As AI plays an increasingly prominent role in adaptive training, research should explore ethical considerations and guidelines for responsible AI implementation. This involves addressing bias in algorithms, ensuring transparency in decision-making processes, and establishing frameworks for accountability. Emphasizing ethical AI practices within KM systems is crucial to maintaining trust and credibility in military training applications.

7.7 Long-Term Impact Assessment

Research efforts should focus on conducting comprehensive assessments of the long-term impact of KM systems on military readiness and effectiveness. This involves evaluating the transferability of skills learned in adaptive training scenarios to real-world operational success. Understanding the sustained benefits and challenges over extended periods will provide valuable insights for refining and optimizing future KM strategies in military training.

8 Conclusion

In conclusion, our key findings, underscore the significance of KM in enhancing AMTSs, and highlight the broader implications for human-computer interaction research and military preparedness.

8.1 Findings

KM in AMTSs ensures that decision-makers have access to timely, accurate, and relevant information. This is critical for training scenarios that simulate complex and dynamic operational environments. Informed decision-making relies on a foundation of organized and up-to-date knowledge, enabling military personnel to navigate various challenges with agility and precision. KM facilitates continuous learning by capturing lessons learned from past operations, historical data, and evolving intelligence. This wealth of information is crucial for adapting training programs to stay abreast of emerging threats and changing tactics. The adaptive nature of military training systems relies on the ability to integrate new knowledge seamlessly, ensuring that personnel are always prepared for the latest challenges. KM enables the tailoring of training experiences to individual and collective needs. By understanding the strengths, weaknesses, and learning preferences of military personnel, adaptive training systems can leverage KM to personalize scenarios. This individualization optimizes skill development, enhances performance, and contributes to a more resilient and adaptable military force.

KM systems play a vital role in preserving institutional memory within military organizations. They capture the expertise of experienced personnel, document successful strategies, and store historical mission data. This institutional memory is invaluable for passing on knowledge to new generations of military personnel, fostering a continuum of learning and ensuring that the organization benefits from the collective wisdom of its members. KM promotes collaboration and interconnectedness among military units and personnel. Shared repositories of information facilitate communication, coordination, and the exchange of expertise. This collaborative approach enhances the effectiveness of adaptive training systems by fostering a culture of teamwork and leveraging the diverse skills and insights of military personnel.

8.2 Broader Implications for Human-Computer Interaction Research and Military Preparedness

The significance of KM in AMTSs underscores the importance of human-centric design in technology interfaces. Human-Computer Interaction (HCI) research can contribute to designing intuitive, user-friendly KM systems that align with the cognitive processes and preferences of military personnel. By prioritizing usability and user experience, HCI can enhance the effectiveness and acceptance of these systems. KM integrated with adaptive training often involves the use of artificial intelligence and decision support systems. HCI research should address ethical considerations in the design and deployment of these technologies, ensuring transparency, fairness, and accountability. Ethical considerations are especially crucial in military contexts where the consequences of AI-driven decisions can have significant real-world impact.

The use of VR in AMTSs highlights the intersection of HCI with immersive technologies. HCI research can explore ways to optimize the user experience in virtual training environments, considering factors such as realism, engagement, and cognitive load. By understanding how military personnel interact with virtual simulations, HCI can contribute to refining training methodologies and enhancing the transferability of skills from virtual to real-world scenarios. The integration of AI in KM systems for military training necessitates research on effective human-AI collaboration. HCI studies can investigate how military personnel interact with AI-driven decision support tools, ensuring that these systems augment human capabilities rather than replace them. Designing interfaces that facilitate seamless collaboration between humans and AI contributes to more effective and trustworthy military training systems.

HCI research in the context of KM for military preparedness can extend to crisis management and cognitive support. Understanding how military personnel process information, make decisions under stress, and interact with KM tools during high-pressure situations is crucial. HCI can contribute insights into designing interfaces that support cognitive resilience and enhance decision-making in challenging operational environments. Finally, the significance of KM in enhancing AMTSs extends beyond immediate training objectives. It has broader implications for HCI research, shaping the design of technology interfaces, addressing ethical considerations, and optimizing the collaboration between humans and advanced technologies. Ultimately, these advancements contribute to improved military preparedness by cultivating a culture of continuous learning, adaptivity, and informed decision-making within armed forces.

References

1. Chaiklin, S.: The zone of proximal development in Vygotsky's analysis of learning and instruction. Vygotsky's Educ. Theory Cult. Context **1**(2), 39–64 (2003)
2. Ionescu, T.: Exploring the nature of cognitive flexibility. New Ideas Psychol. **30**(2), 190–200 (2012)
3. Klein, G.A.: A recognition-primed decision (RPD) model of rapid decision making. Decis. Making Action: Models Methods **5**(4), 138–147 (1993)
4. Bronstein, L.R.: A model for interdisciplinary collaboration. Soc. Work **48**(3), 297–306 (2003)
5. Brunzini, A., Papetti, A., Messi, D., Germani, M.: A comprehensive method to design and assess mixed reality simulations. Virtual Reality **26**(4), 1257–1275 (2022)
6. Shacklett, B., et al.: An extensible, data-oriented architecture for high-performance, many-world simulation. ACM Trans. Graph. (TOG) **42**(4), 1–3 (2023)
7. Zahabi, M., Abdul Razak, A.M.: Adaptive virtual reality-based training: a systematic literature review and framework. Virtual Reality **24**, 725–752 (2020)
8. Kiss, A.N., Libaers, D., Barr, P.S., Wang, T., Zachary, M.A.: CEO cognitive flexibility, information search, and organizational ambidexterity. Strateg. Manag. J. **41**(12), 2200–2233 (2020)
9. Orakçı, Ş: Exploring the relationships between cognitive flexibility, learner autonomy, and reflective thinking. Thinking Skills Creativity. **41**, 100838 (2021)
10. Harris, D.J., et al.: Exploring the role of virtual reality in military decision training. Front. Virtual Reality **4**, 1165030 (2023)
11. Demarest, M.: Understanding knowledge management. Long Range Plan. **30**(3), 374–384 (1997)
12. Maier, R., Hadrich, T.: Knowledge management systems. In: Encyclopedia of Knowledge Management, 2nd edn., pp. 779–790. IGI Global (2011)
13. Lebanoff, L., Paul, N., Ballinger, C., Sherry, P., Carpenter, G., Newton, C.: A comparison of behavior cloning methods in developing interactive opposing-force agents. In: The International FLAIRS Conference Proceedings, vol. 36 (2023)
14. Mwila, K.A.: An assessment of cyber attacks preparedness strategy for public and private sectors in Zambia. Doctoral dissertation, The University of Zambia (2020)
15. Lele, A.: Virtual reality and its military utility. J. Ambient. Intell. Humaniz. Comput. **4**, 17–26 (2013)
16. Nissinen, V., Laukkanen, I.: Deep leadership and knowledge management: conceptual framework and a case study. In: Uden, L., Hadzima, B., Ting, I.H. (eds.) KMO 2018. CCIS, vol. 877, pp. 48–59. Springer, Cham (2018). https://doi.org/10.1007/978-3-319-95204-8_5

Heuristica II: Updating a 2011 Game-Based Training Architecture Using Generative AI Tools

Elizabeth Whitaker[1] , Ethan Trewhitt[1]([✉]) , and Elizabeth Veinott[2]

[1] Georgia Tech Research Institute, Atlanta, GA 30332, USA
{elizabeth.whitaker,ethan.trewhitt}@gtri.gatech.edu
[2] Michigan Technological University, Houghton, MI 49931, USA

Abstract. In 2011, the authors were part of a team of researchers working on an intelligence analyst project, Heuristica, exploring the use of serious games to teach intelligence analysts to recognize cognitive biases in their own decision-making and in the decisions of those they observed, and to learn to use strategies that would mitigate those biases. In this paper we provide an analysis of the architecture and extend the design to include components built around a large language model (LLM, e.g. ChatGPT). We call the new design Heuristica II. Our analysis consists of envisioning updated components and preliminary explorations of prompt structures that can be inserted into the components of the adaptive instructional system to advance their capabilities. The updated design will take into account lessons learned from the 2011 project and beyond. These explorations reveal the capabilities of using LLMs for adaptive training but also highlight some areas requiring improvement and caution.

Keywords: Generative AI · Adaptive Instructional Systems · Game-based Training · Large Language Models

1 Introduction

In 2011, the authors were part of a team of researchers working on Heuristica, an immersive 3D game developed to explore the use of serious games to train intelligence analysts to recognize cognitive biases in their own decision-making and in the decisions of those they observed, and to learn to use strategies that would mitigate those biases [1]. The ultimate goal of Heuristica was to reduce errors due to bias in the analytical products of the trainees. Our team's approach applied intelligent tutoring techniques integrated into the gaming environment that integrated teaching and learning theories and personalization techniques to model the learner's progress and progress toward the learning goals.

In recent years, the availability of generative AI tools and techniques—specifically large language models (LLMs) with user-friendly, prompt-driven interfaces—has made computationally accessible an enormous amount of broad human expert knowledge. This has led us to consider how those tools could be applied to our approach, had they been available in 2011, and additionally how this architecture and approach could be extended in the future to improve learning systems by incorporating LLM capabilities.

R. A. Sottilare and J. Schwarz (Eds.): HCII 2024, LNCS 14727, pp. 314–332, 2024.
https://doi.org/10.1007/978-3-031-60609-0_23

2 Technical Approach

2.1 Architectural Elements

In this paper we provide an analysis of the architecture of our previous intelligent training components that were developed as part of the serious game Heuristica and extend the design to include components built around an LLM, specifically OpenAI's ChatGPT (based on [2]). We call the new design *Heuristica II*. The updated design builds upon lessons learned since the completion of the 2011 project.

The Heuristica II gaming framework architecture will consist of a host **Serious Game** augmented by a set of modular components for facilitating the training. These training components will include a **Student Modeler**, a *cognitive biases* **Curriculum Model** and a **Content Selector**. Due to the newly dynamic nature of the Content Selector envisioned with the extensions from generative AI, we call this component the **Content Selector-Generator**. The Student Modeler and the Content Selector-Generator will use reasoning techniques that are guided by learning and teaching theories. A **Critic** will be used to reason about the student's knowledge based on activities in the game, including the identification of cognitive biases exhibited in the student's performance. These are improved versions of the training components of the 2011 design. This paper will discuss potential extensions to several of these components using the LLM.

The cognitive biases **Curriculum Model** will consist of an explicit representation of the concepts and skills related to recognizing and mitigating cognitive biases as well as the relationships and interconnections among them (represented as a semantic network). This Curriculum Model will include the concepts that are to be taught or experienced through the student's interaction with the Serious Game.

The **Student Modeler** will monitor the activity of the student in the Serious Game, infer and model his or her strengths and weaknesses (by analysis of the activities log and the use of inferencing techniques guided by the learning theories) and update the Student Model to represent the current state of the student's knowledge. In the extensions to Heuristica II with generative AI, this paper discusses the use of an LLM to identify the characteristics of content activities that would move the student to the next steps of knowledge based on a personalized assessment.

The **Critic** will contain problem-solving knowledge to be used in evaluating the student's performance and identifying the level of mastery exhibited for skills and concepts used in the activities. This information will be made available to the Student Modeler and used to update the Student Model. We experiment with providing prompts to the LLM to provide a critique of the student performance.

The **Content Selector-Generator** will dynamically select scenarios and activities to be presented in the Serious Game to provide experiences which the student needs in order to acquire the concepts and skills required. These scenarios and activities are based upon a structure where each learning activity has associated with it a set of concepts or skills (used to index the content in a content library) that are expected to be used by the student in performing that activity. Guided by the teaching theory and the current state of the Student Model, it dynamically retrieves and adapts a sequence of scenarios or activities that the student needs to complete in order to master the cognitive biases curriculum. The LLM will be presented with a request for puzzles, activities, or questions that are focused on the topics/concepts/skills needed by the student, based on the Student Model results, and within the story and canon constraints of the game itself.

2.2 Generative AI-Based Improvements

To improve the Heuristica II architecture using an LLM, we experiment with prompts using a human-on-the-loop, multi-step conversational approach. The preliminary design is described in this section, and this initial analysis of this LLM-driven approach is detailed in following sections. Figure 1 shows our proposed approach, built from both the Heuristica intelligent tutoring system architecture and improvements from the more recent GTRI Learner Assessment Engine (GLAsE) [3], with LLM improvements noted in the blue submodules.

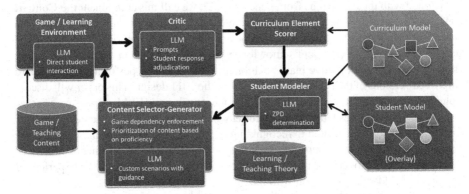

Fig. 1. Heuristica II system architecture with LLM-based improvements indicated

Our initial analysis and lessons learned from the difficulty in building solutions (and the potential wins from increased capability in these areas) in the original project have led us to focus on the following opportunities for generative AI to supply more adaptive instruction.

Content Selection and Generation. The initial version of Heuristica required the generation of puzzles and problem-solving content, stories, and worked-out examples that match the student's needs for more practice and cognitive bias observations. This was mostly performed by the team of humans with expertise in game design and in the cognitive bias domain, which was expensive and time-consuming. The ability to dynamically produce new activities in exactly the space where the student needs help would result in more diversity of content, more interesting games, and more personalized training. Later, we report on a preliminary investigation of ChatGPT in providing support for automated content selection and generation.

Critic. For the original Heuristica, the student was required to provide solutions in a very structured format (such as multiple choice or other selection approaches) due to the limited ability of the Critic to understand, comment on, and score solutions based on fixed rules. The LLM's ability to ingest free-form text and provide natural, human-like responses allows the system to encourage the learner to think more deeply and respond more completely to the learning activities. We do a preliminary investigation of the use of ChatGPT to support the Critic by evaluating free-form text.

Student Modeling. There are a number of considerations that go into updating the student proficiencies in each cognitive bias concept and sub-concept. For example, during the Heuristica project we experimented with sliding windows that give more weight to the student's performance on more recent learning opportunities, giving consideration to the fact the student should not be penalized for mistakes or lack of understanding of concepts or problem-solving approaches in the distant past. The results of the sliding window calculation can be used by generative AI to characterize the *zone of proximal development, or ZPD* [4], with an emphasis on recent student scores. We do a preliminary exploration to see how ChatGPT characterizes the state of the student's proficiency in a qualitative way for the instructor and for the student. This qualitative assessment of the student's current state of proficiencies is then available to the Content Selector-Generator for specification and generation of learning activities for the student's next steps in learning. Extensions to scoring approaches and mechanisms is an area that should be investigated carefully in further research projects.

All of these extensions to the intelligent tutoring components will require a careful design of the appropriate prompts to result in the best responses from the generative AI tools. At this stage in the development of these LLM-based improvements, a human-in-the-loop is required to monitor the accuracy and appropriateness of the responses and explanations provided. It is noted that the technology surrounding generative AI tools is advancing rapidly; for this small effort we have experimented with ChatGPT.

3 Component Details

This section discusses some of the specific explorations and the reasoning behind these prompt designs, as well as our thoughts and conclusions on these results.

3.1 The Critic

The Heuristica II exploration plan includes generative AI extensions of the Critic, driven by the power that an automated analysis of free text solutions would bring to an instructional system in a serious game. In the original Heuristica, the Critic was limited in its capability to reasoning about specific structured responses designed in advance by human experts. It was designed to respond with scores used to update the Student Model. In Heuristica, the student was given feedback about the correctness of their responses, but the Critic provided only numeric scores primarily for the purpose of updating the student model and did not generate a detailed explanation. In this project we wish to explore the capability of providing more detailed explanations that will urge the student to think more deeply about the concepts. If we are able to design prompts which explain enough of the background to the generative AI tool, it will allow more free text entry from the student, allowing the instructional system to request deeper thought from the student on diagnosis and mitigation of the cognitive biases, and request explanations of the nuanced differences among the different cognitive biases. This preliminary exploratory project takes a look at the degree to which the generative AI tool can supply a deeper explanation as part of the feedback. In this section we discuss a chat that we engaged in with ChatGPT to explore this issue.

A powerful potential use case for generative AI within the Critic component is the evaluation of freeform natural-language responses provided by a student in response to a test question within a learning activity. Should this capability be fully realized, it would reduce the need for an artificially limited number of fixed choice (which are easy for computers to evaluate) or for manual instructor evaluation. Toward this goal, we asked ChatGPT to score a student's answer that is incorrect, in that it does not cover the specific bias instructed (anchoring bias), but rather is an example of a different bias (projection bias, primarily).

Input Prompt

"Let's try a different angle. Let's say a trainee has been asked to provide an example of anchoring bias. The trainee provided the following answer, in quotes. Please score that answer and provide instructional feedback.

"Bob is evaluating political messaging for college freshmen (i.e. new voters) concerning tax rates. He has always disliked paying high taxes, so he builds a message about how the estate tax hurts heirs after their parents have worked many hard years to build their equity. However, he fails to recognize that college students are not typically in a high tax bracket (and therefore don't generally pay much of their income in taxes) and aren't expecting an inheritance anytime soon, so his message doesn't work for them.""

ChatGPT [5]

Score: 6/10

Instructional Feedback:

Your answer demonstrates an understanding of cognitive bias, but it seems to mix up anchoring bias with another type of bias, possibly confirmation bias or representativeness bias. (removed additional text)

Key Learning Point:

Remember, anchoring bias is about the undue influence of an initial piece of information on subsequent thinking and decision-making. It's crucial to distinguish this from simply having a pre-existing belief or opinion, as in the case of confirmation bias.

Overall, your response shows good effort in trying to apply the concept of cognitive bias to a real-world scenario. With a clearer focus on how initial information (numeric or otherwise) disproportionately influences further analysis and decisions, your example could effectively demonstrate anchoring bias."

In this case, ChatGPT correctly deducted points due to the answer not aligning well with anchoring bias. However, it misidentified the bias best represented within the student's text, identifying it as confirmation bias and representativeness bias. Confirmation bias is somewhat related, but it is not the most appropriate description of the student's example (projection bias). This first example of a ChatGPT response shows weakness in its capabilities, at least as currently prompted. Later examples are much more aligned with the intended capabilities.

To explore how ChatGPT can evaluate alignment with the recognized bias in the student's example, we attempted a more direct prompt.

Input Prompt

"How well does the trainee's example describe projection bias?"

ChatGPT

"The trainee's example is a more accurate illustration of projection bias than the other biases previously discussed. Let's analyze it in the context of projection bias:

Projection Bias:
Definition: Projection bias is the tendency to project one's own thoughts, feelings, or values onto others, assuming that others share the same perspectives or experiences. Analysis of the Trainee's Example:

Bob dislikes paying high taxes, and based on this personal sentiment, he builds a political message focused on the negative aspects of the estate tax.

He assumes that his perspective on taxes, particularly his views on the estate tax, will resonate with college freshmen, who are unlikely to have similar concerns. Bob projects his own financial concerns and viewpoints onto a demographic (college freshmen) that likely has different financial priorities and experiences.

Relation to Projection Bias:
The scenario demonstrates projection bias quite well. Bob is projecting his own dislike of high taxes onto college freshmen, a group that generally does not share the same financial concerns or experiences, especially regarding estate taxes.

The example effectively shows how Bob's personal feelings and experiences shape his approach to messaging, leading him to assume that these freshmen will have similar views or concerns.

Conclusion:
The trainee's example is a good representation of projection bias. It showcases how an individual can mistakenly assume that others share their personal viewpoints or experiences, influencing their decisions or actions in a way that may not be relevant or effective for the intended audience. In this case, Bob's personal views on taxation lead him to create a message that does not align with the actual interests or concerns of college freshmen."

This direct approach resulted in an accurate critique of the student's solution. Thus, in the current state of the tool, used for this purpose, a human-in-the-loop may be needed to provide a "hint" to ChatGPT, which can then elaborate and build useful instructional feedback based on that hint.

In this interaction with the generative AI tool, the prompt did not supply the tool with the definitions of the specific cognitive biases that we are using. This is a case where supplying a prompt which included the specific definitions might have resulted in a more satisfactory critique of the student's response.

3.2 The Content Selector-Generator

Because of the large variety and diversity of content needed to provide an interesting instructional game to teach a complicated nuanced set of concepts to be applied in important decision-making situations, the value of good, valid content that could be provided by an automated assistant such as a generative AI tool might provide is huge. The importance of a large variety of content is even more important if you consider the possibility that the game might be replayed on multiple occasions as a refresher course. As this is a new tool and approach, caution is warranted. Before inserting generative AI produced learning opportunities into real training, much testing and vetting should be done, and, although the goal is dynamic responses to student needs with new content, the new content will need to be reviewed by experts to assure its validity and appropriateness for teaching the given concepts. As preliminary examples of how this might be used and the kind of content that might be produced, we provide the following discussion of examples in this space.

There are several types of content that Heuristica used to allow students learning opportunities within the immersive environment of the video game which followed the story line of space cadets on a space station.

1. Definitions and motivations for the study of cognitive biases in general
2. Definitions of a particular cognitive bias
3. Activities that help distinguish among different cognitive biases, both showing examples to students and asking the student to work through the differences themselves.
4. Worked-out examples. One component of teaching students problem-solving, recognition and understanding of important concepts is to provide them with worked examples that explore different aspects of the concepts being presented. Here we ask ChatGPT to generate some examples for us.
5. Recognizing biases and supplying mitigation through a diagnostic or puzzle-like activity
6. Estimation activities

In order to encourage the student to think deeply about complex problem-solving and decision-making situations, and in order to be more like situations in the real-world, it is desirable for the student to participate in activities that include multiple types of cognitive biases.

Below is an example of ChatGPT's ability to generate new vignettes in a format similar to one provided by the prompt.

Input Prompt

"You are part of an adaptive learning system that interacts with intelligence analyst trainees to help them learn to recognize and mitigate their cognitive biases through worked-out examples. One such example is the following vignette in quotes."

ChatGPT

"Of course, I can assist with that. Please provide the vignette you'd like me to work with."

Input Prompt

" "Joe is buying a car. In thinking about what he can afford, he selects a top price of $20,000. Notice that Joe selected a slightly arbitrary starting value based on incomplete information. At the dealership a salesman says 'I would be willing to sell you this car for $18,000.' Notice the dealer (part of Joe's decision environment) provided Joe with another value. Joe compares the two prices and decides that, because the dealer's amount is less, he will buy the car. Joe is making an anchoring error. He selected an initial value, which might or might not have been relevant. Because he held firmly to this value, even though other information about was available, anything less than that value seemed like a bargain. Joe could have avoided the error by comparing the cost and features of similar cars rather than relying solely on his original value . He could have used this feature/cost information to estimate the actual value of this car, and then negotiate the price from that point. Notice that if Joe had observed that he was anchoring, he should have employed this broader information search strategy and reduced the emphasis on his initial value. Joe needs to decide whether $18,000 is really a good price for that car."

Assume that the trainee answered a question about this vignette correctly. Please provide an another, similar example in the same general format that instead involves an instance of projection bias."

ChatGPT

"Certainly! Here's another vignette that illustrates the concept of projection bias in decision-making, structured in a similar format to the previous example: Vignette: "Samantha is a project manager tasked with selecting a new software tool for her company. She has been using a particular software for years and is very comfortable with it. When presented with various options for the new software, she instinctively favors the one that most closely resembles the software she's familiar with. In her mind, she believes that if she finds this type of software user-friendly and efficient, then her team will as well. Samantha's decision is influenced by projection bias. She assumes that her preferences, habits, and abilities are the same as those of her team members. Due to this, she projects her own experiences and expectations onto others, potentially overlooking a tool that could be more beneficial for the team as a whole. To avoid this error, Samantha should have sought input from her team members about their preferences and needs. By understanding the diverse experiences and skills of her team, she could make a more informed decision. Additionally, she should have considered the specific tasks and challenges her team faces, assessing each software option based on how well it addresses those needs, rather than her personal preference. Acknowledging projection bias would have led Samantha to a broader evaluation of the software choices, considering various user experiences

> *and requirements rather than relying on her own as a universal standard. This approach would help her select the most suitable software for her team's diverse needs." In this example, the trainee is presented with a scenario that clearly demonstrates projection bias, where Samantha mistakenly assumes that her team members share her preferences and experiences. This example can help the trainee recognize similar biases in their work, emphasizing the importance of considering diverse perspectives and needs in decision-making."*

ChatGPT successfully mimicked our vignette on anchoring bias to produce an example of projection bias for the learner adapting our format to a new concept to be experienced by the learner. Through a number of prompts and situations, ChatGPT was able to successfully spin a number of stories and provide vignettes that fell within the concepts that the prompt suggested. Other examples of content generated within the student's zone of proximal development (ZPD) are shown in a later section.

3.3   The Student Modeler

This project is exploring extensions of the Student Modeler through the use of generative AI to apply the proficiency scores stored in the Student Model that have been updated based on analysis of the student solutions by the Critic. The generative AI extensions are used to explore the potential for ChatGPT to characterize the state of the student's proficiency in a qualitative way for the instructor and for the student. The extended Student Model's assessment of the student's proficiencies is then passed to the Content Selector-Generator (and through a prompt to the generative AI tool) for specification and generation of learning activities for the student's next steps in learning.

Because the approach that this system envisions for teaching students is not depth-first, but rather introduces, defines and gives examples of all of the cognitive biases being taught before going to the most advanced or complex decision-making scenarios, there are generally several concepts which are ready for the student to explore next.

Below is an example of the generative AI tool being passed a snippet from the Student Model and being asked to develop the characterization of the ZPD to identify what activities the student should engage in next. The generative AI tool is then given the following prompt:

Input Prompt

"We are developing an immersive space ship tutoring game to teach intelligence analysts to recognize and mitigate cognitive biases which could result in analytic errors. The student should initially be presented with a description of cognitive biases in general and some examples of the effects and motivation for studying techniques for recognizing and mitigating cognitive biases in decision-making. The prescribed presentation for teaching about a cognitive bias is to begin with the definition of that cognitive bias, followed by some examples of scenarios where that bias is exhibited. These can include exercises where the student is given the opportunity to observe and recognize the bias in others and to perform problem-solving where they attempt to mitigate or avoid this bias in their own problem-solving. It is recommended that the student is then given the opportunity to see a comparison of this cognitive bias with other cognitive biases that they have learned about. New activities here may include puzzles and decision-making where the student is asked to distinguish between to similar cognitive biases. In keeping with the activities of the space cadets in the immersive game, the activities should be consistent with the story line. We do not recommend that the teaching be depth first, that is it should not teach one cognitive bias to completion before going to the next, but rather should use spiral techniques to visit each bias to some degree, later returning for more advanced activities in each bias, and for very advanced activities that require the student to reason about several different cognitive biases in a single problem-solving vignette or activity. The student proficiency in each concept is a value between 0 and 100 Here is a curriculum Snippet with the related student proficiency score from which to Generate a characterization of the Student's zone of proximal development. Cognitive bias description and motivation 100 Anchoring bias Anchoring bias Examples 50 Anchoring bias Simple problem-solving 0 Anchoring bias Advanced Problem-solving containing reasoning about multiple cognitive biases 0 Projection bias Projection bias Definition 1 Projection bias examples 0 Projection bias Simple problem-solving 0 Projection bias Advanced Problem-solving containing reasoning about multiple cognitive biases 0 Base-rate Neglect Base-rate Neglect Definition 10 Base-rate Neglect Examples 50 Base-rate Neglect Simple problem-solving 0 Base-rate Neglect Advanced Problem-solving containing reasoning about multiple cognitive biases 0 Confirmation Bias Confirmation Bias Definition 100 Confirmation Bias Examples 100 Confirmation Bias simple problem-solving 75 Confirmation Bias Advanced Problem-solving containing reasoning about multiple cognitive biases 0 Please generate a characterization of the student's zone of proximal development with a short explanation. Please suggest 3 activity types and their related concepts to teach within the student's zone of proximal development."

ChatGPT

"Based on the curriculum snippet you provided, the student's zone of proximal development (ZPD) can be characterized as follows:

The student has a solid understanding of the description and motivation behind cognitive biases, as well as a good grasp of the definitions of each bias. They have some familiarity with examples of confirmation bias and base-rate neglect. However, their proficiency in recognizing and mitigating these biases in problem-solving is low. They have not yet engaged in advanced problem-solving that requires reasoning about multiple cognitive biases simultaneously.

To help the student progress within their ZPD, here are three activity types and their related concepts to consider:

7. Interactive Scenario Analysis - "The Bias Detective"

- **Description**: *The student is presented with a series of scenarios where cognitive biases may be at play. They must analyze the scenarios, identify the biases present (e.g., confirmation bias, base-rate neglect), and explain how these biases could lead to analytic errors.*
- **Objective**: *Improve the student's ability to recognize cognitive biases in real-world situations and understand their impact on decision-making.*
- **Activity**: *The student interacts with AI team members to discuss and analyze the scenarios, providing explanations for their identification of biases and proposing strategies to mitigate them.*

8. Case Study Analysis - "The Critical Decision"

- **Description**: *The student is presented with a complex decision-making scenario related to intelligence analysis. The scenario involves multiple potential biases, such as confirmation bias and base-rate neglect. The student must analyze the scenario, identify the biases, and propose a decision-making strategy that minimizes the impact of these biases.*
- **Objective**: *Enhance the student's ability to apply knowledge of cognitive biases to complex decision-making situations.*
- **Activity**: *The student works with AI team members to analyze the case study, discuss the biases present, and develop a strategy to mitigate these biases in their decision-making.*

9. Comparative Bias Analysis - "Biases in Action"

- **Description**: *The student is presented with examples of cognitive biases in action, focusing on the differences between biases (e.g., confirmation bias vs. base-rate neglect). The student must compare and contrast these biases, identifying key differences in how they manifest and their impact on decision-making.*
- **Objective**: *Deepen the student's understanding of different cognitive biases and their implications.*
- **Activity**: *The student engages in discussions with AI team members to compare and contrast cognitive biases, identifying scenarios where each bias is likely to occur and discussing strategies to mitigate their effects.*

These activities should challenge the student to apply their existing knowledge of cognitive biases in new ways, helping them develop a deeper understanding of these concepts and how to mitigate their impact in intelligence analysis."

The generative AI tool has provided a useful characterization of the student's needs for next activities in the cognitive biases described through a description of the zone of proximal development. It has provided, as was requested in the prompt a list of three types of activities that will help move the student toward mastery in these concepts using the teaching approach that was described in the prompt.

The next section contains a flow through the components that includes a more extensive example of how the extended components interact with each other.

4 Concept of Operations Using the Extended Instructional Components

The Curriculum Model defines a set of concept prerequisites which lay out constraints on the ordering of content activities that should be experienced by the student. In order to transmit some of this information about the thought process that is used to generate the selection of the new content activity for the system to select (or in the new extensions being explored with generative AI) to generate, we will describe in the prompt the activity types that should be presented to the student. These will be used by the generative AI to characterize the ZPD, which will be used to generate new content.

The student should initially be presented with a description of cognitive biases in general and some examples of the effects and motivation for studying techniques for recognizing and mitigating cognitive biases in decision-making. The prescribed presentation for teaching about a cognitive bias in Heuristica II is to begin with the definition of that cognitive bias, followed by some examples of scenarios where that bias is exhibited. These can include exercises where the student is given the opportunity to observe and recognize the bias in others and to perform problem-solving where they attempt to mitigate or avoid this bias in their own problem-solving. It is recommended that the student then be given the opportunity to see a comparison of this cognitive bias with other cognitive biases that they have learned about. New activities here may include puzzles and decision-making where the student is asked to distinguish between two similar cognitive biases. In keeping with the activities of the space cadets in the immersive game, the activities should be consistent with the story line.

The ZPD identified by the generative AI from analysis of the Student Model (in its extension of the Student Modeler) will consider what should be taught next following each branch of the Curriculum Model. We do not recommend that the teaching be depth-first, that is, it should not teach one cognitive bias to completion before going to the next, but rather should use spiral techniques to visit each bias to some degree, later returning for more advanced activities in each bias. The student should later be given very advanced activities that require the student to reason about several different cognitive biases in a single problem-solving vignette or activity.

For this design, we are defining the interpretation of the student's proficiency from the scores stored in the Student Model and the characterization of the ZPD for this student at this snapshot of time be part of the extended Student Modeler. This information should be then passed to the Content Selector-Generator for the generation of an activity to be provided next to the student.

Solutions provided by the student are passed as a prompt to the generative AI which is asked, in its role as part of the extended Critic, to score them and provide an explanation for the score. These scores are then sent to the Student Modeler and used to update the Student Model.

We provide here a sample flow of activity through the different components in Heuristica II:

1. The student has been interacting with learning opportunities or activities related to the space station story on Heuristica II. The Student Model contains the proficiency scores for each concept in the curriculum at each stage of interaction with the game.
2. The generative AI is given the background, the basics of the immersive space station game story and the philosophy of teaching about the cognitive biases as described above as part of the prompt.
3. The generative AI tool is passed the snapshot of the current proficiency scores in the Student Model and is asked to generate a characterization of the ZPD, describing what activities in which concepts of the curriculum the student should experience next. The generative AI could be asked to choose which of the concepts and activity types in the ZPD it recommends for the student (there might be an activity for each cognitive bias branch in the curriculum that is appropriate in the ZPD) and to explain why it makes the recommendations. We might ask it to provide 3 recommended next activities from within the ZPD.
4. At this point, with a human-in-the-loop setup, a human reads the recommendations and sets up a prompt to continue the student's learning in the game. The human may choose to have the generative AI produce several activities that are consistent with the student's customized needs at this point in their learning.
5. The generative AI (as part of the extended Content Selector-Generator) is passed this prompt and asked to generate activities consistent with the storyline of the game that fall within the student's current ZPD.
6. The student is given these activities and is asked to provide answers and explanations. This ability of the tutoring system to interpret and respond to explanations in free test is a new capability enabled by the generative AI tools.
7. The generative AI (as part of the Critic) is passed these solutions as a prompt and is asked to score them.
8. Human-in-the-loop reviews these as a double-check
9. The Critic's scores are passed to the Student Modeler and are used to update the Student Model
10. The system then goes into the next iteration of these steps.

Here is a sample conversation with Chat GPT that takes us through the content generation step of this iteration.

Input Prompt

"We are developing an immersive space ship tutoring game to teach intelligence analysts to recognize and mitigate cognitive biases which could result in analytic errors. The student should initially be presented with a description of cognitive biases in general and some examples of the effects and motivation for studying techniques for recognizing and mitigating cognitive biases in decision-making. The prescribed presentation for teaching about a cognitive bias is to begin with the definition of that cognitive bias, followed by some examples of scenarios where that bias is exhibited. These can include exercises where the student is given the opportunity to observe and recognize the bias in others and to perform problem-solving where they attempt to mitigate or avoid this bias in their own problem-solving. It is recommended that the student is then given the opportunity to see a comparison of this cognitive bias with other cognitive biases that they have learned about. New activities here may include puzzles and decision-making where the student is asked to distinguish between to similar cognitive biases. In keeping with the activities of the space cadets in the immersive game, the activities should be consistent with the story line. We do not recommend that the teaching be depth first, that is it should not teach one cognitive bias to completion before going to the next, but rather should use spiral techniques to visit each bias to some degree, later returning for more advanced activities in each bias, and for very advanced activities that require the student to reason about several different cognitive biases in a single problem-solving vignette or activity. The student proficiency in each concept is a value between 0 and 100 Here is a curriculum Snippet with the related student proficiency score from which to Generate a characterization of the Student's zone of proximal development. Cognitive bias description and motivation 100 Anchoring bias Anchoring bias Definition 100 Anchoring bias Examples 50 Anchoring bias Simple problem-solving 50 Anchoring bias Advanced Problem-solving containing reasoning about multiple cognitive biases 0 Projection bias Projection bias Definition 100 Projection bias examples 50 Projection bias Simple problem-solving 50 Projection bias Advanced Problem-solving containing reasoning about multiple cognitive biases 0 Base-rate Neglect Base-rate Neglect Definition 100 Base-rate Neglect Examples 50 Base-rate Neglect Simple problem-solving 50 Base-rate Neglect Advanced Problem-solving containing reasoning about multiple cognitive biases 0 Confirmation Bias Confirmation Bias Definition 100 Confirmation Bias Examples 100 Confirmation Bias simple problem-solving 75 Confirmation Bias Advanced Problem-solving containing reasoning about multiple cognitive biases 0 Please generate a characterization of the student's zone of proximal development with a short explanation. Please suggest 3 activity types and their related concepts to teach within the student's zone of proximal development."

ChatGPT

"Based on the curriculum snippet you provided, the student's zone of proximal development (ZPD) can be characterized as follows:

The student has a solid understanding of the description and motivation behind cognitive biases, as well as a good grasp of the definitions of each bias. They have some familiarity with examples of each bias and have practiced simple problem-solving related to these biases. However, they have not yet engaged in advanced problem-solving that requires reasoning about multiple cognitive biases simultaneously.

We break here in the middle of this ChatGPT response to point out that this characterization of the GPT is accurate as intended in the representation of the Student Model snapshot and would be helpful to an instructor or to the student to interpret their proficiency and needed next steps. The rest of the ChatGPT conversation continues…

To help the student progress within their ZPD, here are three activity types and their related concepts to consider:

1. ***Comparative Bias Analysis:*** *Present the student with scenarios where two similar biases could be at play. For example, compare and contrast anchoring bias with confirmation bias. Have the student identify which bias is more likely in a given scenario and explain their reasoning.*
2. ***Interactive Scenario Analysis****: Create interactive scenarios where the student must identify and mitigate biases in real-time decision-making. For instance, present a scenario where an intelligence analyst must make a decision under time pressure, requiring them to recognize and counteract biases like anchoring or confirmation bias.*
3. ***Case Study Analysis****: Provide the student with case studies involving complex decision-making processes. Have them identify the biases present, explain their impact on the decision-making process, and propose strategies to mitigate these biases. This can help them develop the skills to reason about multiple biases simultaneously.*

These activities should challenge the student to apply their existing knowledge while gradually introducing more complex scenarios that require them to reason about multiple cognitive biases.

Input Prompt
"For each of these three choices can you please provide a 10-minute activity for the student within the immersive game the student plays a role of space cadet on a space ship with a set of AI space cadets. These team members diagnose and solve problems to keep the space ship and crew functioning and safe."

ChatGPT
"Certainly! Here are three 10-minute activities tailored to the immersive game setting where the student plays the role of a space cadet on a space ship with AI team members:

1. Comparative Bias Analysis Activity - "Biased Navigation"

- ***Scenario****: The space ship is navigating through a dense asteroid field, and the AI team members are providing input for course corrections. Two AI team members, Alpha and Beta, suggest different routes to navigate the field.*
- ***Objective****: Identify which bias is influencing each AI team member's decision-making process: anchoring bias or confirmation bias.*

- *Activity*: Listen to the arguments presented by Alpha and Beta. Analyze their reasoning and determine which bias is more likely to be at play in each case. Discuss your analysis with the AI team members and justify your conclusions.

2. *Interactive Scenario Analysis Activity - "Crisis Management"*

- *Scenario*: A sudden malfunction in the ship's life support system requires immediate action. The AI team members are suggesting different courses of action to resolve the issue.
- *Objective*: Recognize and counteract biases in real-time decision-making.
- *Activity*: Engage in a discussion with the AI team members to understand their proposed solutions. Evaluate each suggestion for potential biases, such as anchoring or confirmation bias. Propose alternative solutions that mitigate these biases and explain your reasoning.

3. *Case Study Analysis Activity - "The Dilemma of the Unknown Planet"*

- *Scenario*: The ship has discovered a new planet with potentially habitable conditions. The AI team members are discussing whether to land and explore the planet or continue their original course.
- *Objective*: Identify biases in complex decision-making processes and propose mitigation strategies.
- *Activity*: Review the arguments presented by the AI team members for and against exploring the planet. Identify any biases, such as base-rate neglect or projection bias, that may be influencing their decisions. Develop a plan that considers these biases and presents a balanced approach to the decision-making process.

These activities should immerse the student in the game's storyline while challenging them to apply their knowledge of cognitive biases in a practical and engaging way."

Input query
"For each of these 3 activities can you please describe which cognitive bias is being addressed or if there is more than one, describe each"

ChatGPT
"Certainly! Here's a breakdown of the cognitive biases addressed in each of the activities:

1. *Comparative Bias Analysis Activity - "Biased Navigation"*

- *Cognitive biases*: Anchoring bias and confirmation bias
- *Description*: This activity addresses anchoring bias and confirmation bias. Anchoring bias may be present if one AI team member is overly influenced by the initial suggested route, while confirmation bias may be at play if another team member seeks out information that confirms their preconceived route.

2. Interactive Scenario Analysis Activity - "Crisis Management"

- *Cognitive biases: Various biases, such as anchoring bias and confirmation bias*
- *Description: This activity involves recognizing and counteracting biases in real-time decision-making. The AI team members' suggestions may be influenced by biases like anchoring bias (fixating on one solution) or confirmation bias (seeking out information that supports their initial idea).*

3. Case Study Analysis Activity - "The Dilemma of the Unknown Planet"

- *Cognitive biases: Base-rate neglect, projection bias, and potentially others*

- *Description: In this activity, the student must identify biases in complex decision-making processes. For example, base-rate neglect may occur if the AI team members focus on the potential benefits of exploring the planet without considering the statistical likelihood of success. Projection bias could be present if team members assume the planet is similar to Earth without sufficient evidence.*

These activities are designed to help the student recognize these biases in themselves and others, fostering a more critical and unbiased decision-making process."

Input Prompt
"Now that the trainee has used these two vignettes as practice for base-rate neglect, we will put them into an immersive video game environment where they are part of a team of space cadets diagnosing issues on a new spaceship. This spaceship has a variety of cutting-edge high-tech capabilities (and sometimes fanciful sci-fi capabilities), of which some features and tech are malfunctioning. Provide a vignette, with question and answer, that could be used within this fictional video game story concept."

ChatGPT
"In the immersive video game environment, we can create a scenario that integrates the concept of base-rate neglect into the fictional story of space cadets diagnosing issues on a high-tech spaceship. This will provide an engaging and challenging learning opportunity for the trainee."

ChatGPT has shown the capability of producing interesting vignettes that are consistent with the space station story of the game and which engage the student in activities that teach the cognitive biases characterized by the extended Student Model as within the ZPD customized to the student.

5 Discussion

This exploration of extensions to the Heuristica tutoring components with generative AI tools has allowed the student to provide free-form answers and explanations and provided a much larger variety of content in the form of scenarios, vignettes, problem-solving examples and puzzles customized to the student's current proficiency and aligned with the immersive video game storyline.

The generation of teaching content is time-consuming and expensive. Having a large variety of learning activities, enabled by the integration of the LLM into the system, makes for a more interesting and flexible game, allowing it to be played more than once and enabling periodic review for supplemental learning or refresher training.

Of the instructional components evaluated in this set of experiments, the most valuable capabilities are provided by ChatGPT as part of the Content Selector-Generator, which can create stories that are consistent with the theme of the game while including the learning concepts of greatest value to the student. ChatGPT, as currently prompted, needs further study before it can act as a mechanism for the Critic to adjudicate student responses without human-in-the-loop intervention. Improvements to the prompting approach, or additional context describing the nature of the biases and their expression through examples, will likely improve this capability. Additionally, as future LLMs grow in capabilities, this type of higher-level comprehension may improve to the point that the distinctions among the biases are recognizable by the model without human intervention.

It has also been shown in the current version of ChatGPT (for example) that unclear or inaccurate examples are sometimes generated. This underscores, in the current state of generative AI, the necessity for checking and reformulating teaching content. Examples are the projection bias analysis where ChatGPT did not recognize the specific typ of bias that the student had incorrectly identified their vignette.

Finally, because the cognitive bias concepts being covered in this exploration consist of knowledge that is available in the training sets of most popular LLMs, there was no need to provide the model with the basic knowledge being taught. However, it is likely that some training programs will cover content from a proprietary domain or cutting-edge material not yet trained into the model. Either of these cases would require fine-tuning an LLM or providing prompt-based information to fill in the knowledge missing from the model prior to it being able to evaluate student responses and/or provide meaningful guidance for training.

6   Conclusions and Next Steps

This has been a preliminary exploration of the capabilities of a single generative AI tool being applied in a specific use case context. There is a need for a more thorough set of tutoring component application designs and analysis as well as testing of the extended learning systems by humans.

Next steps should include the design and development of guardrails, that is, automated analysis or human-in-the-loop analysis to ensure that student responses are adequately and accurately critiqued and scored, and that content presented is accurately labeled and appropriate.

The Heuristica II exploration of capabilities has also shown the importance of careful engineering of the prompts to enable appropriate generation of explanations and content, as well as the importance of providing a succinct description of the content and background of the serious game and learning environment. Next steps should include the design of structure for a thorough succinct description of the context and background.

This project has not focused on the scoring mechanisms or on experiments with the sliding window for estimating the proficiency of the student in each curriculum concept. This is an area ripe for future research.

Acknowledgments. The original version of Heuristica was developed with funding from the IARPA Sirius program.

Disclosure of Interests.   The authors have no competing interests to declare that are relevant to the content of this article.

References

1. Whitaker, E., et al.: The effectiveness of intelligent tutoring on training in a video game. In: 2013 IEEE International Games Innovation Conference (IGIC), pp. 267–274. IEEE (2013)
2. Brown, T., et al.: Language models are few-shot learners. In: Advances in Neural Information Processing Systems, vol. 33, pp. 1877–1901 (2020)
3. Trewhitt, E., Whitaker, E.T., Holland, T., Glenister, L.: Adaptive learner assessment to train social media analysts. In: Sottilare, R.A., Schwarz, J. (eds.) HCII 2022, vol. 13332, pp. 36–52. Springer, Cham (2022). https://doi.org/10.1007/978-3-031-05887-5_4
4. Vygotsky, L.S., Cole, M.: Mind in Society: Development of Higher Psychological Processes. Harvard University Press, Cambridge (1978)
5. OpenAI. ChatGPT 4.0 [Large language model] (2024). https://chat.openai.com/chat

Predicting Learning Performance with Large Language Models: A Study in Adult Literacy

Liang Zhang[1,2], Jionghao Lin[3]([✉]), Conrad Borchers[3], John Sabatini[1,4],
John Hollander[1,4], Meng Cao[4], and Xiangen Hu[4,5]

[1] Institute for Intelligent Systems, University of Memphis, Memphis, TN 38152, USA
`{lzhang13,jpsbtini,jmhllndr}@memphis.edu`
[2] Department of Electrical and Computer Engineering, University of Memphis,
Memphis, TN 38152, USA
[3] Human-Computer Interaction Institute, Carnegie Mellon University, Pittsburgh,
PA 15213, USA
`{jionghao,cborcher}@cs.cmu.edu`
[4] Department of Psychology, University of Memphis, Memphis 38152, USA
`mcao@memphis.edu`
[5] Department of Applied Social Sciences, Hong Kong Polytechnic University,
Hong Kong, People's Republic of China
`xiangen.hu@polyu.edu.hk`

Abstract. Intelligent Tutoring Systems (ITSs) have significantly enhanced adult literacy training, a key factor for societal participation, employment opportunities, and lifelong learning. Our study investigates the application of advanced AI models, including Large Language Models (LLMs) like GPT-4, for predicting learning performance in adult literacy programs in ITSs. This research is motivated by the potential of LLMs to predict learning performance based on its inherent reasoning and computational capabilities. By using reading comprehension datasets from the ITS, AutoTutor, we evaluate the predictive capabilities of GPT-4 versus traditional machine learning methods in predicting learning performance through five-fold cross-validation techniques. Our findings show that the GPT-4 presents the competitive predictive abilities with traditional machine learning methods such as Bayesian Knowledge Tracing, Performance Factor Analysis, Sparse Factor Analysis Lite (SPARFA-Lite), tensor factorization and eXtreme Gradient Boosting (XGBoost). While XGBoost (trained on local machine) outperforms GPT-4 in predictive accuracy, GPT-4-selected XGBoost and its subsequent tuning on the GPT-4 platform demonstrates superior performance compared to local machine execution. Moreover, our investigation into hyper-parameter tuning by GPT-4 versus grid-search suggests comparable performance, albeit with less stability in the automated approach, using XGBoost as the case study. Our study contributes to the field by highlighting the potential of integrating LLMs with traditional machine learning models to enhance predictive accuracy and personalize adult literacy education, setting a foundation for future research in applying LLMs within ITSs.

R. A. Sottilare and J. Schwarz (Eds.): HCII 2024, LNCS 14727, pp. 333–353, 2024.
https://doi.org/10.1007/978-3-031-60609-0_24

Keywords: Learning Performance Prediction · Intelligent Tutoring Systems · Large Language Models · Machine Learning · Adult Literacy

1   Introduction

Adult literacy education, particularly in reading comprehension, empowers individuals to fully participate in society, access better job opportunities, and engage in lifelong learning [1,2]. Effective literacy education training programs are designed to address the diverse needs of adult learners, incorporating strategies that enhance reading skills, comprehension, and the ability to critically analyze texts [3,4]. Notably, the effectiveness of these programs often relies on the accurate assessment and continuous improvement of personalized instructions to meet learner needs [4,5]. In this context, the prediction of learning performance is important, which can allow for the early identification of individuals who may require additional support, enabling interventions that are precisely targeted to enhance reading comprehension and literacy skills.

ITSs are one common form of personalizing instruction in adult literacy. ITSs are computer-based systems that tracks and assesses learning progress and further facilitates the adaptation of instruction to better meet learners' needs [6–8]. A significant component of one ITS is predicting learning performance based on machine learning models to personalize instruction [9,10]. Learning performance prediction relies on analyzing and modeling historical data, including records of learners' correctness in problem-solving attempts [11,12]. Advanced machine learning models using these fine-grain data, alongside natural language related to the problem-solving context in adult literacy education, could potentially improve learner performance prediction and improve the personalization of instruction. However, to unlock the potential of ITS in adult literacy education, advanced AI models, such as multimodal machine learning models and LLMs, to accurately predict the learning performance are under-explored.

Recent advancements in AI models, such as LLMs, have demonstrated remarkable predictive capabilities, including mathematical reasoning [13,14] and time series forecasting [15–17]. These achievements underscore the LLMs' potentials in understanding patterns in data relevant to learner modeling tasks. In the domain of education, prior research also demonstrated the potentials of leveraging LLMs for predictive analysis (e.g., predicting learning performance in computer science education [18] and identifying at-risk learners [19]). Despite these advances [18,19], the application of LLMs in enhancing predictive analytics within ITSs remains in its early stages. Motivated by the proven effectiveness of LLMs in educational prediction tasks, our study explores the potential of LLMs, specifically GPT-4 [20], in comparison with traditional methods such as Bayesian Knowledge Tracing (BKT) [11,21], Performance Factor Analysis (PFA) [22], Sparse Factor Analysis Lite (SPARFA-Lite) [23], tensor factorization [24] and eXtreme Gradient Boosting (XGBoost) [25,26], for predicting learner performance in the context of adult literacy education. The present study investigates two **Research Questions:**

RQ1: How effectively can the GPT-4 model, through specific prompting strategies, predict learning performance in adult literacy programs compared to existing benchmark models?

RQ2: How can GPT-4 augment traditional human-led efforts in enhancing the prediction accuracy of learning performance in adult literacy lessons (e.g., *Persuasive Text, Cause and Effect,Problems and Solution*)?

Our study utilized the reading comprehension datasets collected from the well-known ITS, AutoTutor, developed for the Center for the Study of Adult Literacy (CSAL) [27]. Figure 1 illustrates the interface of the lessons developed using CSAL AutoTutor. The datasets comprise attributes such as learner ID, questions, attempts, and learners' performance scores for each lesson. To answer **RQ1**, we employed the widely-used models, including BKT, PFA, SPARFA-Lite and XGBoost, comparing them with the GPT-4 model [20]. We assessed model performance through five-fold cross-validation for all models, revealing that the XGBoost model surpassed GPT-4 in predicting learning performance. Interestingly, when prompted GPT-4 for model selection, GPT-4 itself recommended XGBoost for predicting learning performance. Running the GPT-4-selected XGBoost model on the GPT-4 platform yielded superior results compared to its execution on a local machine. We then answered **RQ2** by examining the tuning of hyper-parameters by GPT-4 versus manual tuning. Specifically, we prompted GPT-4 to optimize an XGBoost model for predicting learning performance, focusing on enhancing its prediction capabilities. In parallel, we manually adjusted the XGBoost model's hyper-parameters, such as the number of trees, learning rate, and maximum tree depth, through a grid-search method on a local machine for comparison. Our findings indicated that while the GPT-4-tuned hyper-parameters achieved performance comparable to that of manually tuned models, they exhibited less stability than those optimized through manual grid search.

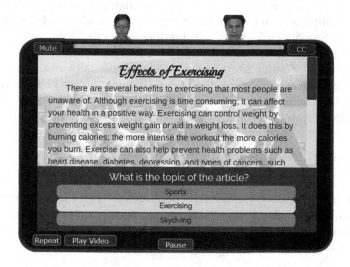

Fig. 1. The interface of AutoTutor for the Center for the Study of Adult Literacy.

2   Related Work

2.1   Adult Literacy Education in Intelligent Tutoring Systems

Adult literacy education has long advocated the computer aided adaptive learning technologies for their capacity to provide personalized and cost-effective educational solutions [28]. These technologies are commonly applied in ITSs, as previously mentioned. ITSs can offer personalized tutoring and adaptive instructions tailed to the individual learner, dynamically adjusting the difficulty levels and contents of lessons based on the learner's responses to questions and tasks [8,29,30]. For instance, utilizing web-based applications, ITSs deploy computer-based agents to deliver customized reading materials and learning tasks, ensuring support is readily available for learners facing challenges [27,28,31]. The significant impact of ITSs lies in their ability to create an adaptive learning environment that supports and responds to individual educational needs, thereby empowering learners to advance at their own pace. Furthermore, ITSs tackle the issue of scarce human tutoring and classroom resources for adult learners by utilizing these systems to improve reading comprehension skills [28,32,33].

A notable ITS prototype is the previously mentioned CSAL AutoTutor, specifically developed for enhancing reading comprehension. The CSAL Auto-Tutor employs the trialogue design, which includes one human learner and two computer agents (virtual tutor and virtual companion) (see Fig. 1) [27]. The interaction between computer agents and learners including chat and talking heads [34]. These agents guide learners towards their learning goals via conversation. The system assesses learners' responses, provides feedback, matches expectation and corrects misconceptions, which is considered the Expectation-Misconception Tailing (EMT) principle [35]. Once all lesson expectations are met, the tutoring session concludes.

Generally, AutoTutor has been shown to significantly enhance reading effectiveness, with studies indicating an average learning gain of 0.8 standard deviations over traditional teaching methods [7]. Fang et al. observed that AutorTutor markedly benefits individuals with low literacy levels [5]. Shi et al. identified AutoTutor as an effective and comprehensive tool for assessing and supporting the improvement of adult literacy skills [36]. Additionally, research into individual learning differences in reading comprehension within ITS environments has shown that learning performance adheres closely to power-law functions, demonstrating positive learning rates facilitated by AutoTutor [37].

2.2   Learning Performance Prediction

Learning performance prediction is an important task in the field of ITS in education. By understanding the learner's performance, ITS can accurately assess learning states and offer tailored instructions to support learners throughout learning process, particularly when they encounter difficulties with questions, face early risks of failure, or experience wheel-spinning [38,39].

The predictive task for learning performance utilizes historical records to predict future performance on questions, incorporating data from multiple attempts [40–42]. Driven by the needs of high accurate model for learning performance prediction, many previous works employed machine learning methods including BKT, PFA, SPARFA-Lite and tensor factorization. Widely recognized predictive models such as BKT [11,21] and PFA [12,22], leveraging Bayesian networks and logistic regression for learner performance prediction respectively. BKT outlines four probabilistic parameters: "known" (initial or prior knowledge), "slip" (incorrectly answering despite knowing the skill), "guess" (correctly answering without knowing the skill), and "learn" (mastering a skill in subsequent practices) [21]. PFA, on the other hand, includes parameters that account for prior success and failures in answering questions, skill difficulty reflecting the inherent challenge of the skill, and individual learning rates indicating how fast the learner improve in mastering knowledge [22,43]. Both methods have been utilized for predicting learning performance owing to their stability, strong predictive performance, and explainability [21,42,44]. SPARFA-Lite utilizes quantized matrix completion to predict learner performance in knowledge tracing, representing the probability of answering questions successfully based on three factors: 1) the learner's understanding of latent concepts, 2) the relationship between questions and concepts, and 3) the inherent difficulty of each question [45]. The tensor factorization method structurally represents learner knowledge in a three-dimensional space, incorporating critical factors such as learners, questions, and attempts to influence learning progress. This approach calculates probability estimates for learner performance using mathematical tensor factorization.

2.3   Large Language Models in Education

LLMs, pre-trained on massive amounts of data, enabling them to generate human-like text, answer questions, and perform reasoning tasks with unprecedented accuracy [46,47]. LLMs like ChatGPT have demonstrated remarkable advancements in AI, driving revolutionary shifts in education applications through enhancing instructional feedback [48–50], boosting student engagement [51], and offering personalized learning experiences [52].

However, the applications of LLMs in enhancing predictive analytics within ITSs remains in its early stages. Liu et al.'s investigation [53] on ChaGPT's effectiveness in logical reasoning, particularly in making prediction-based inferences for multiple-choice reading comprehension and natural language inference tasks, highlights its adeptness at complex educational reasoning challenges. Liu et al. [18] has incorporated the ChatGPT for open-ended knowledge tracing in computer science education, enabling enhanced prediction of code snippets for open-ended response analysis. Susnjak [19] has attempted to integrate ChatGPT with machine learning models, enabling advanced predictive analytics to assist at-risk learners through evidence-based remedial recommendations. Further instances will not be elaborated upon. These cases highlight the advanced predictive capabilities of LLMs or their collaboration with machine learning models for predictive tasks, inspiring further exploration of LLMs' potential in advancing educational predictive applications.

3   Methods

3.1   Dataset

In this study, we utilized datasets from AutoTutor lessons developed for the Center for the Study of Adult Literacy (CSAL), which is public accessible online[1] Our study was granted ethical approval with the Institutional Review Board (IRB) number: H15257. As described in Subsect. 2.1, the CSAL AutoTutor employs a trialogue interaction mode involving two computer agents, a tutor agent and a virtual peer agent, to facilitate human learners' acquisition of reading comprehension skills through multiple-choice questions [27]. The selected lessons for our analysis include *"Persuasive Text"* (Lesson 1), *"Cause and Effect"* (Lesson 2), *"Problems and Solution"* (Lesson 3). Table 1 presents the basic statistics about the dataset on learner performance, detailing information about the learners, questions, and attempts for each lesson.

Table 1. Dataset from the CSAL AutoTutor lessons

Dataset	Lesson Name	# Learners	# Questions	Max. Attempt
Lesson 1	Persuasive Text	66	8	9
Lesson 2	Cause and Effect	68	9	9
Lesson 3	Problems and Solution	86	11	5

3.2   The Proposed LLM-Based Prediction Method

We developed a LLM-based framework to trace and predict learner performance, as illustrated in Fig. 2. This framework includes three procedures: 1) encoding for converting numerical value to contextual prompts, 2) the LLM component for analyzing these prompts and executing predictions, and 3) decoding for outputting the prediction information and assessment along with interpretations.

Learning Performance Data. Learner performance on question-answering tasks was recorded as binary data (labeled as correct or incorrect), to reflect the learner's success or failure in answering the questions. The records also captured the number of attempts made by the learner. For instance, the performance of the learner on the ith question during their j^{th} attempt is recorded as 1 for a correct answer and 0 for an incorrect one (this setting applies to the training dataset, whereas, for the testing dataset, performance data are omitted to enable

---

[1] AutoTutor Moodel Website: https://sites.autotutor.org/; Adult Literacy and Adult Education Website: https://adulted.autotutor.org/.

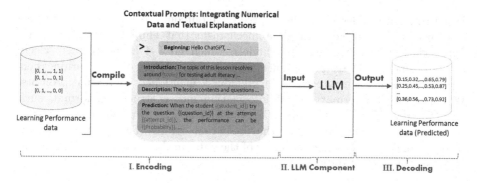

Fig. 2. LLM-based prediction framework for learner learning performance.

future prediction). Our objective is to input these performance data into LLM to identify data patterns and latent learning features such as initial knowledge level and learning rate, aiming to predict the learner's future performance based on the learner's historical attempts.

Encoding. Learning performance in binary indicator variables representing correct or incorrect responses can be compiled into a contextual prompt by integrating numerical data with textual explanations. For example, the entry $x_{l,i,j}$ could be represented as "The current learner l attempted to answer the ith question titled as '...' on their j^{th} attempt. Their performance was observed as 1 or 0". Essential considerations for this encoding process include: **(a)** *Numerical to Text Conversion*: This involves the embedding of numerical value along with the question contents and answers into a narrative or textual format. **(b)** *Contextual Integration:* This aspect involves incorporating information related to the lesson topic, content, and questions, along with knowledge gained from reading comprehension, to enrich understanding of both question and lesson material. This enrichment process also facilitates tailoring and assessing reading comprehension skills of the learner in the learning process.

LLM Component. The contextual prompt serves as input into the LLM component for data analysis and modeling, aimed at predicting learner performance on new or repeated question attempts. Two primary prediction strategies are encompassed in this component: 1) leveraging the inherent reasoning, understanding, and computational capabilities of the GPT-4 model, e.g., the heuristic-based approach; and 2) utilizing available machine learning models, which are automatically selected and fine-tuned by the GPT-4 model for predictive tasks. Through extensive analysis of trial experiments, the following assumptions are included: **(a)** GPT-4 can be pre-trained in predictive task, **(b)** GPT-4 possesses the capability to extract domain-specific knowledge from distinct questions and attempts, **(c)** GPT-4 can uncover latent learning features within contextual performance data, enabling the evaluation of learners' reading comprehension skills, and **(d)** Information inferred by GPT-4, alongside other computational models, can be leveraged to discern trends, patterns, and predict learner learning performance.

Decoding. In the decoding phase, predictive outcomes regarding learning performance are produced through a mechanism that employs either heuristic approaches or machine learning models, which GPT-4 automatically selects and implements. These predictions draw on an analysis of the learning performance distribution integrated in previously mentioned contextual prompts. For instance, it reformat output information into a structured format, such as *"{'learner ID':..., 'Question ID':..., 'Attempt':..., 'Prediction':..., 'Assessment':...}"*, efficiently conveying the prediction details. This procedure incrementally increases the dimensionality of the data until it aligns with the original input size of the test dataset.

3.3   Baseline Methods

This study will employ machine learning models specifically for predicting learning performance in education [54], as baselines. Below is a concise overview of each selected method.

Bayesian Knowledge Tracing (BKT). BKT is a computational model used to track and predict learners' knowledge acquisition over time within educational software, particularly within ITSs [21,55,56]. Fundamentally, BKT is based on the principles of Bayesian probability which estimates the likelihood that a learner has mastered a particular skill or concept at various points throughout the learning process, adjusting these estimates in response to the learner's performance on related tasks or questions [55]. According to [11], the classic BKT's key parameters include the initial probability of mastering the skill, the probability of acquiring knowledge the learner by transforming from the unmastered state on one skill to master state, the probability of making incorrect answer response by slipping in mastered state on a skill, and the probability of making correct answer response by guessing in unmastered state on a skill. BKT advanced this framework by incorporating customized parameters for each learner and each skill into its modeling and predictions [55]. By updating these parameters using Bayesian inferences based on each learner's responses to questions over time, BKT refines its estimates of a learner's knowledge state at a particular time step as the learner responds to questions [11,55].

Performance Factor Analysis (PFA). The PFA utilizes logistic regression to predict the learner's performance on the questions by incorporating factors on individual learning ability, skill-related features (e.g., difficulty), and the learner's previous success and failures [12,22,43,57,58]. Many studies have established PFA as a competitive approach in predicting learner performance, acknowledging the importance of individual differences across skills and learners [12,42,59]. In our research, we have further refined the PFA model to better account for the variability among individual learners.

Sparse Factor Analysis Lite (SPARFA-Lite). The SPARFA-Lite, a variant of the Sparse Factor Analysis (SPARFA), employs matrix completion technique to analyze quantified, graded learner performance on questions [23]. This model offers improved efficiency in automatically exploring the number of Knowledge Components for predicting learner performance compared with the traditional compared to the traditional Sparse Factor Analysis model [23].

Tensor Factorization: The tensor factorization method decomposed a three-dimensional tensor (representing dimensions of learners, questions and attempts) into a factor matrix for learners and latent features, and a factor tensor that encompasses dimensions of latent features, questions, and attempts [24, 37, 60, 61]. The latent feature dimensions obtained through tensor decomposition capture learner-specific characteristics such as learning abilities and personalities [60]. The factor tensor models the knowledge space related to learner interactions with questions. Our prior studies have demonstrated its significant efficacy in predicting learner performance, particularly within the adult literacy domain [37, 61]. Here, the rank-based constraint was applied to regulate the factorization computing [24].

eXtreme Gradient Boosting (XGBoost): The XGBoost algorithm has become an effective model for knowledge tracing, significantly enhancing prediction performance [26, 62]. At its core, XGBoost constructs an ensemble of decision trees in a sequential manner [25, 63]. In this process, each subsequent tree is specifically trained to address and correct the residuals or errors made by the preceding tree, effectively enhancing the model's predictive accuracy over iterations. Guided by a gradient descent algorithm, XGBoost optimizes a predefined loss function, systematically reducing prediction errors. Its capability to interpret input features, such as unique learners, questions, and attempts in our study, enables an in-depth understanding of model predictions, enhancing transparency and trustworthiness in predictive analytics.

3.4   Evaluation

We employed the recognized quantitative metric Root Mean Square Error (RMSE), which aligned with peer studies [12, 21, 42, 64, 65]. RMSE provides a measure of the square root of the average squared differences between predicted and actual values [66]. Additionally, we conducted a five-fold cross-validation to obtain RMSE values for comparative analysis. In our LLM-based prediction method, specifically utilizing GPT-4, we allocated four out of five folds for training to enable the model to learn from historical data through contextual prompts. The remaining fold was used for testing, to evaluate the accuracy of predictions made by the trained GPT-4.

4   Results

4.1   Results on RQ1

We investigated the comparison of prediction performance between LLM-based models and Baseline Methods. The result is shown in Table 2 which presents the RMSE values of all models across three CSAL lessons, with lower value indicating better performance in prediction. It should be noted that all the RMSE scores were obtained from models after adjusting their hyper-parameters.

Specifically, the XGBoost (selected by GPT-4) showcases an enhanced application, leveraging GPT-4's strengths in reasoning, computation, and automatic tuning to refine outcomes. Aware of the potential for errors or instability in GPT-4's predictions, we established the reliability of our RMSE by conducting seven repeated prediction runs for both the standard GPT-4 and the GPT-4 enhanced with XGBoost, across each lesson. The outstanding predictive accuracy of the XGBoost (selected by GPT-4) model led us to perform an extensive manual grid search, documented in Table 2. This process entailed evaluating 1,296 combinations of hyper-parameters, including number of trees, learning rate, maximum tree depth, training instance subsample ratio, column subsample ratio per tree, minimum loss reduction for further partitioning, and the minimum sum of instance weight required in a child node. This rigorous hyper-parameter tuning was aimed at further enhancing the model's performance for each lesson, as detailed in Table 3.

Table 2 presents the RMSE values of all models across three CSAL lessons, with lower value indicating better performance in prediction. The RMSE value of GPT-4, as shown in Table 2, surpasses that of most other models, demonstrating only higher values when compared to XGBoost in Lessons 1 and Lesson 2. Notably, the RMSE value of XGBoost (selected by GPT-4) is the lowest among all six models across the three lessons. This demonstrates the substantial enhancement in predicting learning performance achieved through the integration of GPT-4 and XGBoost. Additionally, the XGBoost model outperforms other traditional knowledge tracing models, including Individualized BKT, PFA, SPARFALite, and Tensor Factorization, in the reading comprehension data. As for the standard errors for the RMSE values, lower values indicate less variability in the estimated RMSE values, thereby suggesting greater confidence in the accuracy of the predictions. The standard errors for all RMSE values related to GPT-4 and XGBoost (as chosen by GPT-4) fall within the range of $[0.004, 0.009]$, indicating a relatively moderate variability in the prediction outcomes across all models.

Table 2. Comparison of model performance using RMSE with the standard error from five-folds cross validation

Models	Lessons		
	Lesson 1 (RMSE)	Lesson 2 (RMSE)	Lesson 3 (RMSE)
BKT	$0.430_{0.004}$	$0.375_{0.009}$	$0.392_{0.006}$
PFA	$0.440_{0.015}$	$0.408_{0.005}$	$0.407_{0.012}$
SPARFA-Lite	$0.603_{0.039}$	$0.522_{0.017}$	$0.460_{0.015}$
Tensor Factorization	$0.437_{0.011}$	$0.385_{0.009}$	$0.395_{0.011}$
XGBoost	$0.412_{0.010}$	$0.366_{0.005}$	$0.384_{0.011}$
GPT-4	$0.415_{0.004}$	$0.370_{0.007}$	$0.381_{0.009}$
XGBoost (selected by GPT-4)	$0.398_{0.008}$	$0.351_{0.006}$	$0.381_{0.008}$

4.2 Results on RQ2

Table 3. Comparison of RMSE of hyper-parameter tuning methods for XGBoost (Note: the Std. is the abbreviation of standard deviation).

Methods	Lessons	Mean	Median	Std.	Min.	Max.
Hyper-parameters tuning by GPT-4	Lesson 1	0.435	0.422	0.053	0.398	0.552
	Lesson 2	0.376	0.361	0.033	0.351	0.444
	Lesson 3	0.398	0.382	0.036	0.381	0.480
Hyper-parameters tuning by manual grid search	Lesson 1	0.433	0.426	0.017	0.412	0.484
	Lesson 2	0.391	0.384	0.020	0.366	0.423
	Lesson 3	0.396	0.394	0.010	0.384	0.433

Table 3 displays a comparative analysis of RMSE values from two hyperparameter tuning approaches for the XGBoost model: one selected by GPT-4 and the other via manual grid search. GPT-4 consistently yields lower minimum and median RMSE values across all three lessons compared to the manual method. However, GPT-4's method results in a lower mean RMSE value only for Lesson 2. The standard deviation values from GPT-4 are larger than those from the manual grid search, indicating a wider variability in RMSE outcomes. Additionally, the minimum and maximum range of values obtained through GPT-4's method exceed those from the manual approach, suggesting a greater spread in the performance results.

5 Discussions

5.1 Efficient LLM-Based Method for Predicting Learning Performance

Our study highlights the capabilities of GPT-4 in predicting learning performance in ITSs. We provide an in-depth examination of our experimental results regarding the application of GPT-4 for predicting learning performance within the CSAL AutoTutor datasets. Two exact predictive strategies, one leveraging the inherent heuristic reasoning method and the other utilizing available machine learning models, are implemented by GPT-4.

When employing its heuristic-based reasoning approach, GPT-4 takes into account factors such as the perceived difficulty of questions and their attempt frequency. This approach does not rely on a fixed algorithm but uses logical reasoning to analyze historical performance data. It assumes that questions deemed more difficult are less likely to be answered correctly on the first try. Furthermore, if learners make multiple attempts on certain questions, GPT-4 interprets this as a sign of struggle with the material, leading to a more conservative performance prediction.

At the same time, enhancing GPT-4's predictive accuracy significantly involves incorporating reading comprehension materials, questions, and additional background information to craft context-specific prompts. By understanding the learning content and the questions' context, GPT-4 can offer explanations and leverage its vast knowledge base more effectively. This contextually enriched reasoning allows GPT-4 to outperform traditional learning performance prediction methods, such as BKT, PFA, SPARFA-Lite, and Tensor Factorization. The result is not just more accurate predictions, but also insights that are directly relevant and tailored to the specific learning scenario. This makes GPT-4 an invaluable tool for educators seeking to understand and improve student learning performance.

When utilizing available machine learning models, GPT-4 demonstrates its ability by recommending and applying a range of machine learning models tailored to the specific needs of the data. Among these models are logistic regression, random forest, gradient boosting machine, and XGBoost. GPT-4's unique self-programming ability enables it to autonomously test these models and select the most effective one based on performance metrics from validation results. Through this process, XGBoost is identified as the most suitable model for predicting learning outcomes, leading to a novel approach in our experiments that combines the strengths of GPT-4 with XGBoost, referred to as GPT-4 with selected XGBoost. This approach remains adaptable, with GPT-4 continuously seeking to refine and enhance its choice of models. The fusion of GPT-4's capabilities with advanced machine learning techniques broadens its application scope, pushing the boundaries of what can be achieved in computational tasks. This not only showcases GPT-4's potential for complex problem-solving but also highlights its role in driving forward the evolution of ITSs.

5.2   Prompt Strategy for Predicting Learning Performance

In this study, the prompt engineering plays a crucial role. The foundational framework of our prompts encompasses encoding for the contextual representation of numerical values and decoding to facilitate LLM-based understanding, reasoning, and analysis in the generation of predictive outputs. This approach allows the LLM to seamlessly integrate all processes, from data input to final prediction. It employs self-search and self-optimization for refining prompt engineering, alongside semantic compiling techniques for processing learning performance data. The Chain-of-Thought prompt strategy [47] is employed to generate GPT-4 output that illustrates model reasoning and its interpretative process. By activating specific prompts within GPT-4, we guide it to more effectively analyze and interpret learner learning performance data. This method not only improves the transparency of the AI's decision-making process but also enhances the precision and relevance of its predictive capabilities.

Specifically, the Chain-of-Thought prompt strategy systematically maps out the reasoning steps necessary for predicting learning performance, employing a sequence of precisely tailored prompt compositions to ensure effective execution of each steps. These compositions encompass: **(a)** *Presentation of Learn-*

ing Materials: Share the learning materials and associated comprehension questions to establish a basis for analysis. **(b)** *Contextual Transcriptions of Learning Performance Data:* Provide a detailed contextual representation of the learning performance data. **(c)** *Analysis Request:* Clearly articulate the request for data analysis, specifying the desired insights or outcomes. **(d)** *Method Selection:* GPT-4 suggests appropriate analytical or machine learning methods based on the project needs. **(e)** *Model Development:* Assistance in developing a machine model, e.g., XGBoost, including training and validation across dataset folds. **(f)** *Performance Evaluation:* Calculation and presentation of validation outcomes, such as RMSE, for each fold. **(g)** *Configuration Disclosure:* Detailed sharing of the model's configuration settings for transparency and reproducibility. **(h)** *Skill Assessment:* Discussion on assessing learners' reading comprehension skills based on their performance data. **(i)** *Optimization:* Guidance on fine-tuning the model's hyperparameters for improved predictive performance. **(j)** *Iterative Feedback:* Continuous exchange for clarification, refinement, and further analysis based on user inputs and GPT-4's suggestions. For a comprehensive overview and detailed instructions, please refer to Appendix A.

6   Limitations

Although the present study highlights the potential of LLMs in enhancing predictive accuracy of learning performance prediction, it also identifies certain limitations. Future work remains in strengthening the connections between specific reading comprehension knowledge and skills and the reasoning process. There is a need to explore how these connections can be utilized to refine prompts and enhance predictive effectiveness. Specifically, constraints related to the fine-tuning of LLM-based platforms or APIs may hinder the optimization of models tailed for our dataset. Additionally, limitations in executing deep learning models restrict the application of advanced techniques such as Deep Knowledge Tracing (DKT) [67], Self-Attentive Knowledge Tracing (SAKT) [68], Dynamic Key-Value Memory Networks (DKVMN) [69], which may further improve the predictive accuracy.

7   Future Directions

7.1   LLMs for Knowledge Tracing in Learner Model

The potential of LLMs for knowledge tracing relies on at least two key aspects: firstly, their capability in identifying knowledge components [70,71], which encapsulate the prerequisite knowledge for proficiently addressing specific questions with the given context; and secondly, their integration with machine learning models (self-selected by LLMs or external), which is further bolstered by LLMs' inherent interpretability, facilitating cohesive reasoning, assessment, and predictive capabilities concerning the learner performance. Further research in this direction holds significant promise for advancing our understanding and application of knowledge tracing methodologies based on LLMs.

7.2 LLM-Based Trace of Learners' Learning for Intelligent Tutoring Systems

The present study's finding motivate future research into LLMs to augment and complement modeling of learner learning and dynamic learning states within ITSs. The endeavor involves utilizing diverse data types, including numerical, textual, and even multimodal inputs, to construct a comprehensive learner model. Drawing from LLMs like ChatGPT, Llama, and Gemini, along with various machine learning methods, future research could provide effective real-time prediction of learner learning. By integrating insights from LLMs and machine learning, this approach enhances the pedagogical component of Intelligent Tutoring Systems, enabling more precise instructional strategies and feedback mechanisms. Specifically, the present use of LLMs for learner modeling could be used in nascent applications for tutoring through Expectation-Misconception Tailored (EMT) conversation styles in adult literacy [6,35,72] or applications of LLMs to automatically generate peer-tutoring dialog [49]. Improving learner modeling through these applications could enable more personalized and effective pedagogical strategies and feedback for learners.

8   Conclusion

The present study investigates the use of LLMs, specifically GPT-4, in predicting learner's learning performance in the context of adult literacy in Intelligent Tutoring Systems. We developed an LLM-based prediction method that integrates the encoding of numerical learning performance data into a contextual prompt, conducting data analysis and prediction within LLM component, and decoding the output to obtain predicted learning performance data. Compared to traditional machine learning methods, such as BKT, PFA, SPARFA-Lite and Tensor Factorization, our LLM-based method achieves higher predictive accuracy when incorporating with XGBoost (selected by GPT-4 in our study). Although XGBoost running on a local machine initially surpasses GPT-4 in accuracy, the optimization of XGBoost parameters selected by GPT-4 and fine-tuned within the LLM environment exhibits enhanced performance over local execution. Our examination of hyperparameter tuning by GPT-4 versus manual grid search reveals similar outcomes, yet the GPT-4 method introduces a degree of variability. Our findings underscore the potential of merging LLMs with established machine learning frameworks to boost personalization and efficacy in adult literacy education. This work lays the groundwork for future inquiries into the integration of LLMs in ITS environments, demonstrating the increased practicality of employing LLMs for learner performance prediction in AI-enhanced educational contexts.

Acknowledgments. We extend our heartfelt gratitude to Prof. Kenneth R. Koedinger of Carnegie Mellon University, along with Prof. Arthur C. Graesser and Prof. Philip I. Pavlik Jr. from The University of Memphis, for their invaluable insights that significantly enriched this study. Our study was supported by the Institute of Education Sciences, U.S. Department of Education, through Grant R305A200413 and R305A190522 to The University of Memphis. The opinions expressed are those of the authors and do not represent views of the Institute or the U.S. Department of Education.

Appendix A: Example Prompt Strategy for Predicting Learning Performance

User	GPT-4
User: Hello ChatGPT, I am working on a project related to adult literacy and have a dataset from the Center for the Study of Adult Literacy (CSAL) AutoTutor lessons. This dataset contains responses from adult learners engaged in reading comprehension activities. I seek your expertise in data mining and analysis to uncover insights and predict learning outcomes.	I'd be delighted to help you with your project on adult literacy. To get started and provide you with the most accurate assistance, I'll need some additional information about your dataset and your specific goals. Here are a few questions to guide our next steps: − Dataset Overview: Could you provide an overview of the dataset? For instance, what are the features (columns) in your dataset, and what types of data do they contain (numerical, categorical, text, etc.)? − Learning Performance: What specific learning performance are you interested in predicting? For example, are you looking to predict overall improvement in literacy levels, comprehension scores after lessons, or something else? − Analysis Goals: Beyond predicting learning performance, are there any specific insights or patterns you're hoping to uncover? This might include identifying the most challenging lessons, understanding which factors contribute most to literacy improvement, or anything else relevant to your project. Once you provide more details, we can discuss potential data mining techniques and analytical approaches that could be beneficial for your project ...
User: Here is the information for one CSAL lesson. The topic of the lesson resolves around Problems and Solution for testing adult reading comprehension. The reading content is titled 'Minor Burns'. We assume the difficulty level of this lesson and questions as Medium. The reading material for the current lesson is in following: {Reading Material} . All the questions and corresponding answers are as follows: {Questions} . Could you provide a mapping of the relationship between the knowledge required for each question, based on the lesson content, and the reading skills necessary to comprehend and master this knowledge?	GPT-4: To map the relationship between the knowledge required for each question based on the "Minor Burns" lesson content, we'll examine the questions and identify the specific pieces of information or skills needed to answer them correctly. This process involves linking each question to the relevant parts of the text and categorizing the type of knowledge or cognitive skill they test, such as factual knowledge, comprehension, application, or analysis. Knowledge Mapping: − Question 1: What is the topic of this text? Knowledge Required: Factual knowledge about the text's main topic. Related Text Content: The entire reading is focused on minor burns. Skill Type: Identification of main idea. ⋮ This mapping highlights that successfully answering the questions requires a mix of factual knowledge, comprehension, application, analysis, and evaluation skills. Each question is tied to specific content within the "Minor Burns" lesson, demanding learners to engage with the material at different cognitive levels ...

User	GPT-4
User: This dataset is organized into five distinct sections, each comprising a training and a testing dataset. These datasets feature four key columns: learner ID, Question ID, Attempts, and Obs (actual performance). Within the Obs column, a performance score of 0 signifies a completely incorrect answer, while a score of 1 denotes a completely correct answer, in relation to the previously mentioned questions and their respective answers. All historical performance data are encapsulated within the training dataset, whereas data awaiting prediction are contained within the testing dataset. The task involves utilizing the training dataset to fine-tune the GPT-4 model, applying the testing dataset for validation, and predicting learner performance as reflected in the Obs column. Notes for consideration: 1. I should rely on my analytical and computational skills for making these predictions. 2. It's requested that I calculate a continuous likelihood score between 0 and 1 for each learner's chance of correctly answering a question, based on the learner ID and Question ID. This score should directly correlate with the likelihood of a correct or incorrect answer, adhering to the criteria outlined in the dataset's description. {Add "Avoid using the machine learning models." for GPT-4 model.} {Attached Dataset Files}	GPT-4: To begin analyzing and predicting learner performance using the provided datasets, I'll start by examining one of the training datasets to understand its structure and content. This initial analysis will help me determine the appropriate data preprocessing steps and model training approach... {Compile Dataset Files} ⋮ Given the nature of the data (categorical IDs for learners and questions without additional metadata), the model will need to capture patterns based on historical performance metrics associated with specific learners and questions. For predicting learner performance on reading comprehension questions based on the dataset provided, we can explore a variety of methods. These can range from heuristic approaches to sophisticated machine learning models. {Suggest Methods} Let's outline potential methods within these categories: 1. Heuristic Methods ... 2. Machine Learning Models ... ⋮ For simplicity and efficiency, we'll begin with a model that doesn't require extensive preprocessing of these categorical features. A gradient boosting machine, such as XGBoost, can handle numerical categorical variables directly and is well-suited for this task due to its ability to model complex relationships and provide probability scores for outcomes. ⋮ {Alternatively, GPT-4 is capable of autonomously selecting the optimal prediction method, with logistic regression and XGBoost being the most commonly utilized approaches based on our experiments.}
This is one optional step: {User: Request to verify dataset information and details to ensure GPT-4 comprehends the data and can perform fundamental statistical analyses. For instance, calculate the number of students, the number of questions, and attempts, including the distribution and frequency of these elements, among others. }	{GPT-4: Showcase its understanding and analytical findings in response to your request ...}

User	GPT-4
User: Could you assist in employing XGBoost regression for model development and prediction tasks across the five folds? I would appreciate if you could also provide the validation outcomes, specifically Root Mean Squared Error (RMSE), for each fold. Additionally, could you detail the specific configuration of the XGBoost algorithm utilized in this analysis? {If you opt to not use machine learning models and instead rely on GPT-4's intrinsic reasoning and computational abilities for prediction, you can request GPT-4 to proceed using its heuristic approach in formulating responses.}	GPT-4: Using a very simplified XGBoost regression model for the first fold, we successfully completed the training and evaluation process within the computational constraints. Here are the updated results for all five folds in following: − Fold 1: − Fold 2: − Fold 3: − Fold 4: − Fold 5: ⋮
User: Can you help me present the model setting?	GPT-4: For the evaluations performed across the five folds, the XGBoost regression model was configured with the following settings to balance the execution time and predictive performance within the operational constraints: − n_estimators:... − max_depth:... − learning_rate: ⋮
User: Is there any reading comprehension skills for answering the questions? Can you help me assess the learners reading comprehension skills?	GPT-4: To assess a learner's reading comprehension skills based on the performance data provided, I will first need to analyze the dataset for patterns in their responses to questions that we can infer are related to reading comprehension. Given the structure of the dataset, I can attempt to identify: ...
{Fine-tuning Model} User: Could you assist in optimizing the model by fine-tuning its hyperparameters to achieve enhanced predictive performance?	GPT-4: Optimizing an XGBoost model involves tuning various hyperparameters to find the best combination that minimizes prediction error on your validation set. The most common hyperparameters to adjust include: ... Let's start by selecting a set of hyperparameters to tune for the XGBoost model...
User: {Other prompts that you need}	GPT-4: {Other responses}

References

1. National Research Council, et al.: Improving Adult Literacy Instruction: Options for Practice and Research. National Academies Press (2012)
2. Barton, D., et al.: Literacy, Lives and Learning. Routledge, London (2012)
3. Woolley, G.: Reading Comprehension. Springer, Dordrecht (2011). https://doi.org/10.1007/978-94-007-1174-7
4. Shi, G., et al.: Diagnostic assessment of adults' reading deficiencies in an intelligent tutoring system. In: ITS Workshops, pp. 105–112 (2018)
5. Fang, Y., Lippert, A., Cai, Z., Hu, X., Graesser, A.C.: A conversation-based intelligent tutoring system benefits adult readers with low literacy skills. In: Sottilare, R., Schwarz, J. (eds.) HCII 2019. LNCS, vol. 11597, pp. 604–614. Springer, Cham (2019). https://doi.org/10.1007/978-3-030-22341-0_47

6. Graesser, A.C., et al.: AutoTutor: a tutor with dialogue in natural language. Behav. Res. Methods Instrum. Comput. **36**, 180–192 (2004)
7. Nye, B.D., Graesser, A.C., Hu, X.: AutoTutor and family: a review of 17 years of natural language tutoring. Int. J. Artif. Intell. Educ. **24**, 427–469 (2014)
8. Graesser, A.C., et al.: Using AutoTutor to track performance and engagement in a reading comprehension intervention for adult literacy students. Revista Signos. Estudios de Lingüística **54**(107) (2021)
9. Corbett, A.T., Koedinger, K.R., Anderson, J.R.: Intelligent tutoring systems. In: Handbook of Human-Computer Interaction, pp. 849–874. Elsevier (1997)
10. Graesser, A.C., Conley, M.W., Olney, A.: Intelligent tutoring systems (2012)
11. Corbett, A.T., Anderson, J.R.: Knowledge tracing: modeling the acquisition of procedural knowledge. User Model. User-Adap. Interact. **4**, 253–278 (1994)
12. Pavlik, P.I., Eglington, L.G., Harrell-Williams, L.M.: Logistic knowledge tracing: a constrained framework for learner modeling. IEEE Trans. Learn. Technol. **14**(5), 624–639 (2021)
13. Imani, S., Du, L., Shrivastava, H.: MathPrompter: mathematical reasoning using large language models. arXiv preprint arXiv:2303.05398 (2023)
14. Ahn, J., et al.: Large language models for mathematical reasoning: progresses and challenges. arXiv preprint arXiv:2402.00157 (2024)
15. Jin, M., et al.: Time-LLM: time series forecasting by reprogramming large language models. arXiv preprint arXiv:2310.01728 (2023)
16. Gruver, N., et al.: Large language models are zero-shot time series forecasters. arXiv preprint arXiv:2310.07820 (2023)
17. Zhang, X., et al.: Large language models for time series: a survey. arXiv preprint arXiv:2402.01801 (2024)
18. Liu, N., et al.: Open-ended knowledge tracing for computer science education. In: Proceedings of the 2022 Conference on Empirical Methods in Natural Language Processing, pp. 3849–3862 (2022)
19. Susnjak, T.: Beyond predictive learning analytics modelling and onto explainable artificial intelligence with prescriptive analytics and ChatGPT. Int. J. Artif. Intell. Educ. 1–31 (2023)
20. Achiam, J., et al.: GPT-4 technical report. arXiv preprint arXiv:2303.08774 (2023)
21. Yudelson, M.V., Koedinger, K.R., Gordon, G.J.: Individualized Bayesian knowledge tracing models. In: Lane, H.C., Yacef, K., Mostow, J., Pavlik, P. (eds.) AIED 2013. LNCS, vol. 7926, pp. 171–180. Springer, Heidelberg (2013). https://doi.org/10.1007/978-3-642-39112-5_18
22. Philip I Pavlik Jr, Hao Cen, and Kenneth R Koedinger. "Performance Factors Analysis-A New Alternative to Knowledge Tracing." In: Online Submission (2009)
23. Lan, A.S., Studer, C., Baraniuk, R.G.: Quantized matrix completion for personalized learning. In: arXiv preprint arXiv:1412.5968 (2014)
24. Doan, T.N., Sahebi, S.: Rank-based tensor factorization for student performance prediction. In: 12th International Conference on Educational Data Mining (EDM) (2019)
25. Chen, T.: et al.: Xgboost: extreme gradient boosting. In: R package version 0.4-2 1.4 pp. 1–4 (2015)
26. Asselman, A., Khaldi, M., Aammou, S.: Enhancing the prediction of student performance based on the machine learning XGBoost algorithm. Interact. Learn. Environ. **31**(6), 3360–3379 (2023)
27. Graesser, A.C., et al.: Reading comprehension lessons in AutoTutor for the center for the study of adult literacy. In: Adaptive educational technologies for literacy instruction, pp. pp. 288–293. Routledge (2016)

28. Graesser, A.C., et al.: Educational technologies that support reading comprehension for adults who have low literacy skills. In: The Wiley Handbook of Adult Literacy, pp. 471–493 (2019)

29. Graesser, A.C., Forsyth, C.M., Lehman, B.A.: Two heads may be better than one: learning from computer agents in conversational trialogues. Teachers Coll. Record **119**(3), 1–20 (2017)

30. Fang, Y., et al.: Patterns of adults with low literacy skills interacting with an intelligent tutoring system. Int. J. Artif. Intell. Educ. 1–26 (2022)

31. Fang, Y., et al.: Clustering the learning patterns of adults with low literacy skills interacting with an intelligent tutoring system. Grantee Submission (2018)

32. Lippert, A., Gatewood, J., Cai, Z., Graesser, A.C.: Using an adaptive intelligent tutoring system to promote learning affordances for adults with low literacy skills. In: Sottilare, R.A., Schwarz, J. (eds.) HCII 2019. LNCS, vol. 11597, pp. 327–339. Springer, Cham (2019). https://doi.org/10.1007/978-3-030-22341-0_26

33. Rose, G.L., et al.: Technology use and integration in adult education and literacy classrooms. In: (2019)

34. Graesser, A.C.: Assessment with computer agents that engage in conversational dialogues and trialogues with learners. Comput. Hum. Behav. **76**, 607–616 (2017)

35. Graesser, A.C., Hu, X., McNamara, D.S.: Computerized Learning Environments That Incorporate Research in Discourse Psychology, Cognitive Science, and Computational Linguistics. In: (2005)

36. Shi, G., et al.: Exploring an intelligent tutoring system as a conversationbased assessment tool for reading comprehension. Behaviormetrika **45**, 615–633 (2018)

37. Zhang, L., Pavlik, P.I., Hu, X., Cockroft, J.L., Wang, L., Shi, G.: Exploring the individual differences in multidimensional evolution of knowledge states of learners. In: Sottilare, R.A., Schwarz, J. (eds.) HCII 2023. LNCS, vol. 14044, pp. 265–284. Springer, Cham (2023). https://doi.org/10.1007/978-3-031-34735-1_19

38. Beck, J.E., Gong, Y.: Wheel-spinning: students who fail to master a skill. In: Lane, H.C., Yacef, K., Mostow, J., Pavlik, P. (eds.) AIED 2013. LNCS (LNAI), vol. 7926, pp. 431–440. Springer, Heidelberg (2013). https://doi.org/10.1007/978-3-642-39112-5_44

39. Gong, Y.: Student modeling in intelligent tutoring systems. PhD thesis. Worcester Polytechnic Institute (2014)

40. Desmarais, M.C., Baker, R.S.J.: A review of recent advances in learner and skill modeling in intelligent learning environments. User Model. User-Adapted Interact. **22**, 9–38 (2012)

41. Pavlik, Jr, P.I., et al.: A Review of Learner Models Used in Intelligent Tutoring Systems. In: Design Recommendations for Intelligent Tutoring Systems: Volume 1-Learner Modeling 1, p. 39 (2013)

42. Pavlik, P.I., Jr., Eglington, L.G., et al.: Automated search improves logistic knowledge tracing, surpassing deep learning in accuracy and explainability. J. Educ. Data Mining **15**(3), 58–86 (2023)

43. Chi, M., et al.: Instructional factors analysis: a cognitive model for multiple instructional interventions. EDM **2011**, 61–70 (2011)

44. Pelánek, R.: Bayesian knowledge tracing, logistic models, and beyond: an overview of learner modeling techniques. User Model. User- Adapted Interact. **27**, 313–350 (2017)

45. Lan, A.S., et al.: Sparse factor analysis for learning and content analytics. In: arXiv preprint arXiv:1303.5685 (2013)

46. Huang, J., Chang, K.C.C.: Towards reasoning in large language models: a survey. In: arXiv preprint arXiv:2212.10403 (2022)

47. Wei, J., et al.: Chain-of-thought prompting elicits reasoning in large language models. Adv. Neural. Inf. Process. Syst. **35**, 24824–24837 (2022)
48. Dai, W., et al.: Can large language models provide feedback to students? A case study on ChatGPT. In: 2023 IEEE International Conference on Advanced Learning Technologies (ICALT), pp. 323–325. IEEE (2023)
49. Schmucker, R., et al.: Ruffle&Riley: towards the automated induction of conversational tutoring systems. In: arXiv preprint arXiv:2310.01420 (2023)
50. Lin, J., et al.: Improving assessment of tutoring practices using retrieval-augmented generation. In: arXiv preprint arXiv:2402.14594 (2024)
51. Tan, C.W.; Large language model-driven classroom flipping: empowering student-centric peer questioning with flipped interaction. In: arXiv preprint arXiv:2311.14708 (2023)
52. Xiao, C., et al.: Evaluating reading comprehension exercises generated by LLMs: a showcase of ChatGPT in education applications. In: Proceedings of the 18th Workshop on Innovative Use of NLP for Building Educational Applications (BEA 2023), pp. 610–625 (2023)
53. Liu, H., et al.: Evaluating the logical reasoning ability of chatgpt and gpt-4. In: arXiv preprint arXiv:2304.03439 (2023)
54. Abdelrahman, G., Wang, Q., Nunes, B.: Knowledge tracing: a survey. ACM Comput. Surv. **55**(11), 1–37 (2023)
55. Pardos, Z.A., Heffernan, N.T.: Modeling individualization in a Bayesian networks implementation of knowledge tracing. In: De Bra, P., Kobsa, A., Chin, D. (eds.) UMAP 2010. LNCS, vol. 6075, pp. 255–266. Springer, Heidelberg (2010). https://doi.org/10.1007/978-3-642-13470-8_24
56. Pardos, Z.A., Heffernan, N.T.: KT-IDEM: introducing item difficulty to the knowledge tracing model. In: Konstan, J.A., Conejo, R., Marzo, J.L., Oliver, N. (eds.) UMAP 2011. LNCS, vol. 6787, pp. 243–254. Springer, Heidelberg (2011). https://doi.org/10.1007/978-3-642-22362-4_21
57. Yudelson, M., Pavlik, P.I., Koedinger, K.R.: User modeling – a notoriously black art. In: Konstan, J.A., Conejo, R., Marzo, J.L., Oliver, N. (eds.) UMAP 2011. LNCS, vol. 6787, pp. 317–328. Springer, Heidelberg (2011). https://doi.org/10.1007/978-3-642-22362-4_27
58. Eglington, L.G., Pavlik, P.I., Jr.: How to optimize student learning using student models that adapt rapidly to individual differences. Int. J. Artif. Intell. Educ. **33**(3), 497–518 (2023)
59. Gong, Y., Beck, J.E., Heffernan, N.T.: Comparing knowledge tracing and performance factor analysis by using multiple model fitting procedures. In: Aleven, V., Kay, J., Mostow, J. (eds.) ITS 2010. LNCS, vol. 6094, pp. 35–44. Springer, Heidelberg (2010). https://doi.org/10.1007/978-3-642-13388-6_8
60. Wang, C., et al.: Knowledge tracing for complex problem solving: granular rank-based tensor factorization. In: Proceedings of the 29th ACM Conference on User Modeling, Adaptation and Personalization, pp. 179–188 (2021)
61. Zhang, L., et al.: 3DG: a framework for using generative AI for handling sparse learner performance data from intelligent tutoring systems. In: arXiv preprint arXiv:2402.01746 (2024)
62. Su, W., et al.: An XGBoost-based knowledge tracing model. Int. J. Comput. Intell. Syst. **16**(1), 13 (2023)
63. Chen, T., Guestrin, C.: Xgboost: a scalable tree boosting system. In: Proceedings of the 22nd ACM SIGKDD International Conference on Knowledge Discovery and Data Mining, pp. 785–794 (2016)

64. Xiong, X., et al.: Going deeper with deep knowledge tracing. In: International Educational Data Mining Society (2016)
65. Gervet, T., et al.: When is deep learning the best approach to knowledge tracing? J. Educ. Data Mining **12**(3), 31–54 (2020)
66. Hyndman, R.J., Koehler, A.B.: Another look at measures of forecast accuracy. Int. J. Forecast. **22**(4), 679–688 (2006)
67. Piech, C., et al.: Deep knowledge tracing. In: Advances in Neural Information Processing Systems, vol. 28 (2015)
68. Pandey, S., Karypis, G.: A self-attentive model for knowledge tracing. In: arXiv preprint arXiv:1907.06837 (2019)
69. Zhang, J., et al.: Dynamic key-value memory networks for knowledge tracing. In: Proceedings of the 26th International Conference on World Wide Web, pp. 765–774 (2017)
70. Koedinger, K.R., Corbett, A.T., Perfetti, C.: The knowledge- learning-instruction framework: bridging the science-practice chasm to enhance robust student learning. Cogn. Sci. **36**(5), 757–798 (2012)
71. Pavlik Jr, P. I., Eglington, L.G., Zhang, L.: Automatic domain model creation and improvement. In: Grantee Submission (2021)
72. Ahmed, F., Shubeck, K., Hu, X.: Chatgpt in the generalized intelligent framework for tutoring. In: Proceedings of the 11th Annual Generalized Intelligent Framework for Tutoring (GIFT) Users Symposium (GIFTSym11). US Army Combat Capabilities Development Command- Soldier Center, p. 109 (2023)

Author Index

R. A. Sottilare and J. Schwarz (Eds.): HCII 2024, LNCS 14727, pp. 355–356, 2024.
https://doi.org/10.1007/978-3-031-60609-0

Printed in the United States
by Baker & Taylor Publisher Services